中 国 手 工 纸 文 库

Library of Chinese Handmade Paper

中 国 手 工 纸 文 库

Library of Chinese Handmade Paper

中国手工纸文库

Library of Chinese Handmade Paper

汤书昆

总主编

《中国手工纸文库》编撰委员会

总主编

汤书昆

编　委

（按拼音顺序排列）

陈　彪　　陈敬宇　　达尔文·尼夏

方媛媛　　郭延龙　　黄飞松　　蓝　强

李宪奇　　刘　靖　　彭长贵　　沈佳斐

汤书昆　　杨建昆　　张燕翔　　郑久良

朱　赟　　朱正海　　朱中华

Library of Chinese Handmade Paper
Editorial Board

Editor-in-Chief	Tang Shukun
Members	Chen Biao, Chen Jingyu, Darwin Nixia, Fang Yuanyuan, Guo Yanlong, Huang Feisong, Lan Qiang, Li Xianqi, Liu Jing, Peng Changgui, Shen Jiafei, Tang Shukun, Yang Jiankun, Zhang Yanxiang, Zheng Jiuliang, Zhu Yun, Zhu Zhenghai, Zhu Zhonghua (in alphabetical order)

安徽卷·下卷

Anhui III

汤书昆　黄飞松

主　编

中国科学技术大学出版社

University of Science and Technology of China Press

图书在版编目（CIP）数据

中国手工纸文库.安徽卷.下卷/汤书昆，黄飞松主编.—合肥：中国科学技术大学出版社，2021.5
国家出版基金项目
"十三五"国家重点出版物出版规划项目
ISBN 978-7-312-04638-4

Ⅰ.中… Ⅱ.①汤… ②黄… Ⅲ.手工纸—介绍—安徽 Ⅳ.TS766

中国版本图书馆CIP数据核字（2018）第300901号

中国手工纸文库

安徽卷·下卷

项目负责	伍传平 项赟飚
责任编辑	韩继伟 姚 硕 高哲峰
艺术指导	吕敬人
书籍设计	敬人书籍设计 吕 旻+黄晓飞
出版发行	中国科学技术大学出版社 地址 安徽省合肥市金寨路96号 邮编 230026
印　　刷	北京雅昌艺术印刷有限公司
经　　销	全国新华书店
开　　本	880 mm×1230 mm　1/16
印　　张	23.25
字　　数	720千
版　　次	2021年5月第1版
印　　次	2021年5月第1次印刷
定　　价	1800.00元

《中国手工纸文库·安徽卷》编撰委员会

主 编

汤书昆　黄飞松

副主编

朱　赟　朱正海

翻译主持

方媛媛

统稿主持

汤书昆

技术分析统筹

朱　赟

编　委

（按拼音顺序排列）

陈　彪　陈　琪　陈　龑　陈敬宇　程　曦　方媛媛
郭延龙　黄飞松　刘　靖　刘　伟　罗文伯　沈佳斐
汤书昆　王圣融　郑久良　朱　赟　朱正海

Library of Chinese Handmade Paper: *Anhui*
Editorial Board

Editors-in-Chief	Tang Shukun, Huang Feisong
Deputy Editors-in-Chief	Zhu Yun, Zhu Zhenghai
Chief Translator	Fang Yuanyuan
Director of Modification	Tang Shukun
Director of Technical Analysis	Zhu Yun
Members	Chen Biao, Chen Qi, Chen Yan, Chen Jingyu, Cheng Xi, Fang Yuanyuan, Guo Yanlong, Huang Feisong, Liu Jing, Liu Wei, Luo Wenbo, Shen Jiafei, Tang Shukun, Wang Shengrong, Zheng Jiuliang, Zhu Yun, Zhu Zhenghai (in alphabetical order)

总　序

　　造纸技艺是人类文明的重要成就。正是在这一伟大发明的推动下，我们的社会才得以在一个相当长的历史阶段获得比人类使用口语的表达与交流更便于传承的介质。纸为这个世界创造了五彩缤纷的文化记录，使一代代的后来者能够通过纸介质上绘制的图画与符号、书写的文字与数字，了解历史，学习历代文明积累的知识，从而担负起由传承而创新的文化使命。

　　中国是手工造纸的发源地。不仅人类文明中最早的造纸技艺发源自中国，而且中华大地上遍布着手工造纸的作坊。中国是全世界手工纸制作技艺提炼精纯与丰富的文明体。可以说，在使用手工技艺完成植物纤维制浆成纸的历史中，中国一直是人类造纸技艺与文化的主要精神家园。下图是中国早期造纸技艺刚刚萌芽阶段实物样本的一件遗存——西汉放马滩古纸。

西汉放马滩古纸残片
纸上绘制的是地图
1986年出土于甘肃省天水市
现藏于甘肃省博物馆

Map drawn on paper from
Fangmatan Shoals
in the Western Han Dynasty
Unearthed in Tianshui City,
Gansu Province in 1986
Kept by Gansu Provincial Museum

Preface

Papermaking technique illuminates human culture by endowing the human race with a more traceable medium than oral tradition. Thanks to cultural heritage preserved in the form of images, symbols, words and figures on paper, human beings have accumulated knowledge of history and culture, and then undertaken the mission of culture transmission and innovation.

Handmade paper originated in China, one of the largest cultural communities enjoying advanced handmade papermaking techniques in abundance. China witnessed the earliest papermaking efforts in human history and embraced papermaking mills all over the country. In the history of handmade paper involving vegetable fiber pulping skills, China has always been the dominant centre. The picture illustrates ancient paper from Fangmatan Shoals in the Western Han Dynasty, which is one of the paper samples in the early period of papermaking techniques unearthed in China.

一

本项目的缘起

从2002年开始，我有较多的机缘前往东邻日本，在文化与学术交流考察的同时，多次在东京的书店街——神田神保町的旧书店里，发现日本学术界整理出版的传统手工制作和纸（日本纸的简称）的研究典籍，先后购得近20种，内容包括日本全国的手工造纸调查研究，县（相当于中国的省）一级的调查分析，更小地域和造纸家族的案例实证研究，以及日、中、韩等东亚国家手工造纸的比较研究等。如：每日新闻社主持编撰的《手漉和纸大鉴》五大本，日本东京每日新闻社昭和四十九年（1974年）五月出版，共印1 000套；久米康生著的《手漉和纸精髓》，日本东京讲谈社昭和五十年（1975年）九月出版，共印1 500本；菅野新一编的《白石纸》，日本东京美术出版社昭和四十年（1965年）十一月出版等。这些出版物多出自几十年前的日本昭和年间（1926~1988年），不仅图文并茂，而且几乎都附有系列的实物纸样，有些还有较为规范的手工纸性能、应用效果对比等技术分析数据。我阅后耳目一新，觉得这种出版物形态既有非常直观的阅读效果，又散发出很强的艺术气息。

1. Origin of the Study

Since 2002, I have been invited to Japan several times for cultural and academic communication. I have taken those opportunities to hunt for books on traditional Japanese handmade paper studies, mainly from old bookstores in Kanda Jinbo-cho, Tokyo. The books I bought cover about 20 different categories, typified by surveys on handmade paper at the national, provincial, or even lower levels, case studies of the papermaking families, as well as comparative studies of East Asian countries like Japan, Korea and China. The books include five volumes of *Tesukiwashi Taikan* (*A Collection of Traditional Handmade Japanese Papers*) compiled and published by Mainichi Shimbun in Tokyo in May 1974, which released 1 000 sets, *The Essence of Japanese Paper* by Kume Yasuo, which published 1 500 copies in September 1975 by Kodansha in Tokyo, Japan, *Shiraishi Paper* by Kanno Shinichi, published by Fine Arts Publishing House in Tokyo in November 1965. The books which were mostly published between 1926 and 1988 among the Showa reigning years, are delicately illustrated with pictures and series of paper samples, some even with data analysis on performance comparison. I was extremely impressed by the intuitive and aesthetic nature of the books.

我几乎立刻想起在中国看到的手工造纸技艺及相关的研究成果，在我们这个世界手工造纸的发源国，似乎尚未看到这种表达丰富且叙述格局如此完整出色的研究成果。对中国辽阔地域上的手工造纸技艺与文化遗存现状，研究界尚较少给予关注。除了若干名纸业态，如安徽省的泾县宣纸、四川省的夹江竹纸、浙江省的富阳竹纸与温州皮纸、云南省的香格里拉东巴纸和河北省的迁安桑皮纸等之外，大多数中国手工造纸的当代研究与传播基本上处于寂寂无闻的状态。

此后，我不断与国内一些从事非物质文化遗产及传统工艺研究的同仁交流，他们一致认为在当代中国工业化、城镇化大规模推进的背景下，如果不能在我们这一代人手中进行手工造纸技艺与文化的整体性记录、整理与传播，传统手工造纸这一中国文明的结晶很可能会在未来的时空中失去系统记忆，那真是一种令人难安的结局。但是，这种愿景宏大的文化工程又该如何着手？我们一时觉得难觅头绪。

《手漉和纸精髓》
附实物纸样的内文页
A page from *The Essence of Japanese Paper* with a sample

《白石纸》
随书的宣传夹页
A folder page from *Shiraishi Paper*

The books reminded me of handmade papermaking techniques and related researches in China, and I felt a great sadness that as the country of origin for handmade paper, China has failed to present such distinguished studies excelling both in presentation and research design, owing to the indifference to both papermaking technique and our cultural heritage. Most handmade papermaking mills remain unknown to academia and the media, but there are some famous paper brands, including Xuan paper in Jingxian County of Anhui Province, bamboo paper in Jiajiang County of Sichuan Province, bamboo paper in Fuyang District and bast paper in Wenzhou City of Zhejiang Province, Dongba paper in Shangri-la County of Yunnan Province, and mulberry paper in Qian'an City of Hebei Province.

Constant discussion with fellow colleagues in the field of intangible cultural heritage and traditional craft studies lead to a consensus that if we fail to record, clarify, and transmit handmade papermaking techniques in this age featured by a prevailing trend of industrialization and urbanization in China, regret at the loss will be irreparable. However, a workable research plan on such a grand cultural project eluded us.

2004年，中国科学技术大学人文与社会科学学院获准建设国家"985工程"的"科技史与科技文明哲学社会科学创新基地"，经基地学术委员会讨论，"中国手工纸研究与性能分析"作为一项建设性工作由基地立项支持，并成立了手工纸分析测试实验室和手工纸研究所。这一特别的机缘促成了我们对中国手工纸研究的正式启动。

2007年，中华人民共和国新闻出版总署的"十一五"国家重点图书出版规划项目开始申报。中国科学技术大学出版社时任社长郝诗仙此前知晓我们正在从事中国手工纸研究工作，于是建议正式形成出版中国手工纸研究系列成果的计划。在这一年中，我们经过国际国内的预调研及内部研讨设计，完成了《中国手工纸文库》的撰写框架设计，以及对中国手工造纸现存业态进行全国范围调查记录的田野工作计划，并将其作为国家"十一五"规划重点图书上报，获立项批准。于是，仿佛在不经意间，一项日后令我们常有难履使命之忧的工程便正式展开了。

2008年1月，《中国手工纸文库》项目组经过精心的准备，派出第一个田野调查组（一行7人）前往云南省的滇西北地区进行田野调查，这是计划中全中国手工造纸田野考察的第一站。按照项目设计，将会有很多批次的调查组走向全中国手工造纸现场，采集能获

In 2004, the Philosophy and Social Sciences Innovation Platform of History of Science and S&T Civilization of USTC was approved and supported by the National 985 Project. The academic committee members of the Platform all agreed to support a new project, "Studies and Performance Analysis of Chinese Handmade Paper". Thus, the Handmade Paper Analyzing and Testing Laboratory, and the Handmade Paper Institute were set up. Hence, the journey of Chinese handmade paper studies officially set off.

In 2007, the General Administration of Press and Publication of the People's Republic of China initiated the program of key books that will be funded by the National 11th Five-Year Plan. The former President of USTC Press, Mr. Hao Shixian, advocated that our handmade paper studies could take the opportunity to work on research designs. We immediately constructed a framework for a series of books, *Library of Chinese Handmade Paper*, and drew up the fieldwork plans aiming to study the current status of handmade paper all over China, through arduous pre-research and discussion. Our project was successfully approved and listed in the 11th Five-Year Plan for National Key Books, and then our promising yet difficult journey began.

The seven members of the *Library of Chinese Handmade Paper* Project embarked on our initial, well-prepared fieldwork journey to the northwest area of Yunnan

取的中国手工造纸的完整技艺与文化信息及实物标本。

2009年，国家出版基金首次评审重点支持的出版项目时，将《中国手工纸文库》列入首批国家重要出版物的资助计划，于是我们的中国手工纸研究设计方案与工作规划发育成为国家层面传统技艺与文化研究所关注及期待的对象。

此后，田野调查、技术分析与撰稿工作坚持不懈地推进，中国科学技术大学出版社新一届领导班子全面调动和组织社内骨干编辑，使《中国手工纸文库》的出版工程得以顺利进行。2017年，《中国手工纸文库》被列为"十三五"国家重点出版物出版规划项目。

二
对项目架构设计的说明

作为纸质媒介出版物的《中国手工纸文库》，将汇集文字记

调查组成员在香格里拉县
白地村调查
2008年1月

Researchers visiting Baidi Village of Shangri-la County
January 2008

Province in January 2008. After that, based on our research design, many investigation groups would visit various handmade papermaking mills all over China, aiming to record and collect every possible papermaking technique, cultural information and sample.

In 2009, the National Publishing Fund announced the funded book list gaining its key support. Luckily, *Library of Chinese Handmade Paper* was included. Therefore, the Chinese handmade paper research plan we proposed was promoted to the national level, invariably attracting attention and expectation from the field of traditional crafts and culture studies.

Since then, field investigation, technical analysis and writing of the book have been unremittingly promoted, and the new leadership team of USTC Press has fully mobilized and organized the key editors of the press to guarantee the successful publishing of *Library of Chinese Handmade Paper*. In 2017, the book was listed in the 13th Five-Year Plan for the Publication of National Key Publications.

2. Description of Project Structure

Library of Chinese Handmade Paper compiles with many forms of ideography language: detailed descriptions and records, photographs, illustrations of paper fiber structure and transmittance images, data analysis, distribution of the papermaking sites, guide map

录与描述、摄影图片记录、样纸纤维形态及透光成像采集、实验分析数据表达、造纸地分布与到达图导引、实物纸样随文印证等多种表意语言形式，希望通过这种高度复合的叙述形态，多角度地描述中国手工造纸的技艺与文化活态。在中国手工造纸这一经典非物质文化遗产样式上，《中国手工纸文库》的这种表达方式尚属稀见。如果所有设想最终能够实现，其表达技艺与文化活态的语言方式或许会为中国非物质文化遗产研究界和保护界开辟一条新的途径。

项目无疑是围绕纸质媒介出版物《中国手工纸文库》这一中心目标展开的，但承担这一工作的项目团队已经意识到，由于采用复合度很强且极丰富的记录与刻画形态，当项目工程顺利完成后，必然会形成非常有价值的中国手工纸研究与保护的其他重要后续工作空间，以及相应的资源平台。我们预期，中国（计划覆盖34个省、市、自治区与特别行政区）当代整体的手工造纸业态按照上述记录与表述方式完成后，会留下与《中国手工纸文库》伴生的中国手工纸图像库、中国手工纸技术分析数据库、中国手工纸实物纸样库，以及中国手工纸的影像资源汇集等。基于这些伴生的集成资源的丰富性，并且这些资源集成均为首次，其后续的价值延展空间也不容小视。中国手工造纸传承与发展的创新拓展或许会给有志于继续关注中国手工造纸技艺与文化的同仁提供

to the papermaking sites, and paper samples, etc. Through such complicated and diverse presentation forms, we intend to display the technique and culture of handmade paper in China thoroughly and vividly. In the field of intangible cultural heritage, our way of presenting Chinese handmade paper was rather rare. If we could eventually achieve our goal, this new form of presentation may open up a brand-new perspective to research and preservation of Chinese intangible cultural heritage.

Undoubtedly, the *Library of Chinese Handmade Paper* Project developed with a focus on paper-based media. However, the team members realized that due to complicated and diverse ways of recording and displaying, there will be valuable follow-up work for further research and preservation of Chinese handmade paper and other related resource platforms after the completion of the project. We expect that when contemporary handmade papermaking industry in China, consisting of 34 provinces, cities, autonomous regions and special administrative regions as planned, is recorded and displayed in the above mentioned way, a Chinese handmade paper image library, a Chinese handmade paper technical data library, a Chinese handmade paper sample library, and a Chinese handmade paper video information collection will come into being, aside from the *Library of Chinese Handmade Paper*. Because of the richness of these byproducts, we should not overlook these possible follow-up

更多元的机遇。

　　毫无疑问,《中国手工纸文库》工作团队整体上都非常认同这一工作的历史价值与现实意义。这种认同给了我们持续的动力与激情,但在实际的推进中,确实有若干挑战使大家深感困惑。

三
我们的困惑和愿景

困惑一:

　　中国当代手工造纸的范围与边界在国家层面完全不清晰,因此无法在项目的田野工作完成前了解到中国到底有多少当代手工造纸地点,有多少种手工纸产品;同时也基本无法获知大多数省级区域手工造纸分布地点的情况与存活、存续状况。从调查组2008~2016年集中进行的中国南方地区(云南、贵州、广西、四川、广东、海南、浙江、安徽等)的田野与文献工作来看,能够提供上述信息支持的现状令人失望。这导致了项目组的田野工作规划处于"摸着石头过河"的境地,也带来了《中国手工纸文库》整体设计及分卷方案等工作的不确定性。

developments. Moving forward, the innovation and development of Chinese handmade paper may offer more opportunities to researchers who are interested in the techniques and culture of Chinese handmade papermaking.

Unquestionably, the whole team acknowledges the value and significance of the project, which has continuously supplied the team with motivation and passion. However, the presence of some problems have challenged us in implementing the project.

3. Our Confusions and Expectations

Problem One:

From the nationwide point of view, the scope of Chinese contemporary handmade papermaking sites is so obscure that it was impossible to know the extent of manufacturing sites and product types of present handmade paper before the fieldwork plan of the project was drawn up. At the same time, it is difficult to get information on the locations of handmade papermaking sites and their survival and subsisting situation at the provincial level. Based on the field work and literature of South China, including Yunnan, Guizhou, Guangxi, Sichuan, Guangdong, Hainan, Zhejiang and Anhui etc., carried out between 2008 and 2016, the ability to provide the information mentioned above is rather difficult. Accordingly, it placed the planning of the project's fieldwork into an obscure unplanned route,

困惑二：

中国正高速工业化与城镇化，手工造纸作为一种传统的手工技艺，面临着经济效益、环境保护、集成运营、技术进步、消费转移等重要产业与社会变迁的压力。调查组在已展开了九年的田野调查工作中发现，除了泾县、夹江、富阳等为数不多的手工造纸业态聚集地，多数乡土性手工造纸业态都处于生存的"孤岛"困境中。令人深感无奈的现状包括：大批造纸点在调查组到达时已经停止生产多年，有些在调查组到达时刚刚停止生产，有些在调查组补充回访时停止生产，仅一位老人或一对老纸工夫妇在造纸而无传承人……中国手工造纸的业态正陷于剧烈的演化阶段。这使得项目组的田野调查与实物采样工作处于非常紧迫且频繁的调整之中。

困惑三：

作为国家级重点出版物规划项目，《中国手工纸文库》在撰写开卷总序的时候，按照规范的说明要求，应该清楚地叙述分卷的标准与每一卷的覆盖范围，同时提供中国手工造纸业态及地点分布现

贵州省仁怀市五马镇取缔手工造纸作坊的横幅
2009年4月

Banner of a handmade papermaking mill in Wuma Town of Renhuai City in Guizhou Province, saying "Handmade papermaking mills should be closed as encouraged by the local government"
April 2009

which also led to uncertainty in the planning of *Library of Chinese Handmade Paper* and that of each volume.

Problem Two:
China is currently under the process of rapid industrialization and urbanization. As a traditional manual technique, the industry of handmade papermaking is being confronted with pressures such as economic benefits, environmental protection, integrated operation, technological progress, consumption transfer, and many other important changes in industry and society. During nine years of field work, the project team found out that most handmade papermaking mills are on the verge of extinction, except a few gathering places of handmade paper production like Jingxian, Jiajiang, Fuyang, etc. Some handmade papermaking mills stopped production long before the team arrived or had just recently ceased production; others stopped production when the team paid a second visit to the mills. In some mills, only one old papermaker or an elderly couple were working, without any inheritor to learn their techniques... The whole picture of this industry is in great transition, which left our field work and sample collection scrambling with hasty and frequent changes.

Problem Three:
As a national key publication project, the preface of *Library of Chinese Handmade Paper* should clarify the standard and the scope of each volume according to the research plan. At the same time, general information such as the map with locations of Chinese handmade

状图等整体性信息。但由于前述的不确定性，开宗明义的工作只能等待田野调查全部完成或进行到尾声时再来弥补。当然，这样的流程一定程度上会给阅读者带来系统认知的先期缺失，以及项目组工作推进中的迷茫。尽管如此，作为拓荒性的中国手工造纸整体研究与田野调查就在这样的现状下全力推进着！

当然，我们的团队对《中国手工纸文库》的未来仍然满怀信心与憧憬，期待着通过项目组与国际国内支持群体的协同合作，尽最大努力实现尽可能完善的田野调查与分析研究，从而在我们这一代人手中为中国经典的非物质文化遗产样本——中国手工造纸技艺留下当代的全面记录与文化叙述，在中国非物质文化遗产基因库里绘制一份较为完整的当代手工纸文化记忆图谱。

汤书昆

2017年12月

papermaking industry should be provided. However, due to the uncertainty mentioned above, those tasks cannot be fulfilled, until all the field surveys have been completed or almost completed. Certainly, such a process will give rise to the obvious loss of readers' systematic comprehension and the team members' confusion during the following phases. Nevertheless, the pioneer research and field work of Chinese handmade paper have set out on the first step.

There is no doubt that, with confidence and anticipation, our team will make great efforts to perfect the field research and analysis as much as possible, counting on cooperation within the team, as well as help from domestic and international communities. It is our goal to keep a comprehensive record, a cultural narration of Chinese handmade paper craft as one sample of most classic intangible cultural heritage, to draw a comparatively complete map of contemporary handmade paper in the Chinese intangible cultural heritage gene library.

Tang Shukun

December 2017

编撰说明

1

关于类目的划分标准，《中国手工纸文库·安徽卷》（以下简称《安徽卷》）在充分考虑安徽地域当代手工造纸高度聚集于泾县一地，而且手工纸的历史传承品种相对丰富的特点后，决定不按地域分布划分类目，而是按照宣纸、书画纸、皮纸、竹纸、加工纸、工具划分第一级目类，形成"章"的类目单元，如第二章"宣纸"、第三章"书画纸"。章之下的二级类目以造纸企业或家庭纸坊为单元，形成"节"的类目，如第二章第一节"中国宣纸股份有限公司"、第四章第三节"潜山县星杰桑皮纸厂"。

2

《安徽卷》成书内容丰富，篇幅较大，从适宜读者阅读和装帧牢固角度考虑，将其分为上、中、下三卷。上卷内容为第一章"安徽省手工造纸概述"、第二章"宣纸"；中卷内容为第三章"书画纸"、第四章"皮纸"、第五章"竹纸"；下卷内容为第六章"加工纸"、第七章"工具"以及"附录"。

3

《安徽卷》第一章为概述，其格式与先期出版的《中国手工纸文库·云南卷》（以下简称《云南卷》）、《中国手工纸文库·贵州卷》（以下简称《贵州卷》）等类似。其余各章各节的标准撰写格式则因有手工纸业态高度密集的县级区域存在，所以与《云南卷》《贵州卷》所用的单一标准撰写格式不同，分为三类撰写标准格式。

第一类与《云南卷》《贵州卷》相近，适应一个县域内手工造纸厂坊不密集、品种相对单纯的业态分布。通常分为七个部分，即"××××纸的基础信息及分布""××××纸生产的人文地理环境""××××纸的历史与传承""××××纸的生产工艺与技术分析""××××纸的用途与销售情况"

Introduction to the Writing Norms

1. Referring to the categorization standards, *Library of Chinese Handmade Paper: Anhui* will not be categorized based on location, but the paper types, i.e. Xuan Paper, Calligraphy and Painting Paper, Bast Paper, Bamboo Paper, Processed Paper and Tools, due to the fact that papermaking sites in the region cluster around Jingxian County, and the diverse paper types historically inherited in the area. Each category covers a whole chapter, e.g. Chapter II "Xuan Paper", Chapter III "Calligraphy and Painting Paper". Each chapter consists of sections based on different papermaking factories or family-based papermaking mills. For instance, first section of the second chapter is "China Xuan Paper Co., Ltd.", and the third section of Chapter IV is "Xingjie Mulberry Bark Paper Factory in Qianshan County".

2. Due to its rich content and great length, *Library of Chinese Handmade Paper: Anhui* is further divided into three sub-volumes (I, II, III) for convenience of the readers and bookbinding. *Anhui* I consists of Chapter I "Introduction to Handmade Paper in Anhui Province", Chapter II "Xuan Paper"; *Anhui* II contains Chapter III "Calligraphy and Painting Paper", Chapter IV "Bast Paper" and Chapter V "Bamboo Paper"; *Anhui* III is composed of two chapters, i.e. Chapter VI "Processed Paper", Chapter VII "Tools", and "Appendices".

3. First chapter of *Library of Chinese Handmade Paper: Anhui* is introduction, which follows the volume format of *Yunnan* and *Guizhou*, which have already been released. Sections of other chapters follow three different writing norms, because of the concentrated distribution of county-level handmade papermaking practice, and this is different from two volumes that have been published.

First type of volume writing norm is similar to that of *Yunnan* and *Guizhou*: each section consists of seven sub-sections introducing various aspects of each kind of handmade paper, namely, Basic Information and Distribution, The Cultural and Geographic

"××××纸的品牌文化与习俗故事""××××纸的保护现状与发展思考"。如遇某一部分田野调查和文献资料均未能采集到信息,则将按照实事求是原则略去标准撰写格式的相应部分。

第二类主要针对泾县宣纸与书画纸企业以及少数加工纸企业的特征,手工造纸厂坊在一个小地区聚集度特别高,或者纸品非常丰富,不适合采用第一类撰写格式时采用。通常的格式及大致名称为:"××××纸(纸厂)的基础信息与生产环境""××××纸(纸厂)的历史与传承情况""××××纸(纸厂)的代表纸品及其用途与技术分析""××××纸(纸厂)生产的原料、工艺与设备""××××纸(纸厂)的市场经营状况""××××纸(纸厂)的品牌文化与习俗故事""××××纸(纸厂)的业态传承现状与发展思考"。

第三类主要针对当代世界最大的手工造纸企业——中国宣纸股份有限公司,由于其从业人数多达1 300余人,工艺、产品、制度与文化的丰富性独具一格,因此专门设计了撰写类目形式,分为:"中国宣纸股份有限公司的基础信息与生产环境""中国宣纸股份有限公司的历史与传承情况""中国宣纸股份有限公司的关键岗位和产量变更情况""'红星'宣纸制作技艺的基本形态""原料、辅料、人员配置、工具和用途""'红星'宣纸的分类与品种""'红星'宣纸的价格、销售、包装信息""社会名流品鉴'红星'宣纸的重要掌故""中国宣纸股份有限公司保护宣纸业态的措施"。

4

《安徽卷》专门安排一节讲述的手工纸的入选标准是:(1)项目组进行田野调查时仍在生产;(2)项目组田野调查时虽已不再生产,但保留着较完整的生产环境与设备,造纸技师仍能演示或讲述完整技艺和相关知识。

考虑到竹纸在安徽省历史上曾经是大宗民生产品,而其当代业态萎缩特别明显,处于几近消亡状态,因此对调查组所能够找到的很少的竹纸产地中的泾县竹纸放宽了"保留着较完整的生产环境与设备"这一项标准。

5

《安徽卷》调查涉及的造纸点均参照国家地图标准绘制两幅示意图:一幅为造纸点在安徽省和所属县的地理位置图,另一幅为由该县县城前往造纸点的路线图,但在具体出图时,部分节会将两图合一呈现。在标示地名时,均统一标示出

Environment, History and Inheritance, Papermaking Technique and Technical Analysis, Uses and Sales, Brand Culture and Stories, Preservation and Development. Omission is also acceptable if our fieldwork efforts and literature review fail to collect certain information. This writing norm applies to the handmade papermaking practice in the area where factories and papermaking mills are not dense, and the paper produced is of single variety.

The second writing norm is applied to Xuan paper, and calligraphy and painting paper factories in Jingxian County, and a few processed paper factories, which all cluster in a small area, and produce diverse paper types. In such chapter, sections are: Basic Information and Production Environment; History and Inheritance; Representative Paper and Its Uses and Technical Analysis; Raw Materials, Papermaking Techniques and Tools; Marketing Status; Brand Culture and Stories; Current Status of Business Inheritance and Thoughts on Development.

The third writing norm is applied to China Xuan Paper Co., Ltd., which boasts the largest handmade papermaking factory around the world. It harbors over 1,300 employees and unique papermaking techniques, products, and colorful management system and culture. In this chapter, sections are listed differently: Basic Information and Production Environment of China Xuan Paper Co., Ltd.; History and Inheritance of China Xuan Paper Co., Ltd.; Key Positions and Production Profile of China Xuan Paper Co., Ltd.; "Red Star" Xuan Papermaking Techniques; Types and Varieties of "Red Star" Xuan Paper; Celebrities and "Red Star" Xuan Paper; Preservation of Xuan Paper by China Xuan Paper Co., Ltd.

4. The handmade paper included in each section of this volume conforms to the following standards: firstly, it was still under production when the research group did their fieldwork. Secondly, the papermaking equipment and major sites were well preserved, and the handmade papermakers were still able to demonstrate the papermaking techniques and relevant knowledge, in case of ceased production.

县城、乡镇两级，乡镇下一级则直接标示造纸点所在村，而不再做行政村、自然村、村民组之区别。示意图上的行政区划名称及编制规则均依据中国地图出版社、国家基础地理信息中心的相关地图。

6

《安徽卷》原则上对每一个所调查的造纸厂坊的代表纸品，均在珍稀收藏版书中相应章节后附调查组实地采集的实物纸样。采样量足的造纸点代表纸品附全页纸样；由于各种限制因素，采样量不足的则附2/3、1/2、1/4或更小规格的纸样；个别因近年停产等导致未能获得纸样或采样严重不足的，则不附实物纸样。

7

《安徽卷》原则上对所有在章节中具体描述原料与工艺的代表纸品进行技术分析，包括实物纸样可以在书中呈现的类型，以及个别只有极少量纸样遗存，可以满足测试要求而无法在"珍稀收藏版"中附上实物纸样的类型。

全卷对所采集纸样进行的测试参考了中国宣纸的技术测试分析标准（GB/T 18739—2008），并根据安徽地域手工纸的多样性特色做了必要的调适。实测、计算了所有满足测试分析标示足量需求的已采样的手工纸中的宣纸类、书画纸类、皮纸类的厚度、定量、紧度、抗张力、抗张强度、撕裂度、湿强度、白（色）度、耐老化度下降、尘埃度、吸水性(数种熟宣未测该指标)、伸缩性、纤维长度和纤维宽度共14个指标；加工纸类的厚度、定量、紧度、抗张力、抗张强度、撕裂度、色度、吸水性共8个指标；竹纸类的厚度、定量、紧度、抗张力、抗张强度、色度、纤维长度和纤维宽度共8个指标。由于所采集的安徽省各类手工纸样的生产标准化程度不同，因而若干纸种纸品所测数据与机制纸、宣纸的标准存在一定差距。

8

测试指标说明及使用的测试设备如下：

（1）厚度 ▶ 所测纸的厚度指标是指纸在两块测量板间受一定压力时直接测

Because bamboo paper used to be mass produced in Anhui Province, while the practice shrank greatly or even is lingering on extinction in current days, the research team decided to omit the requirement of comparatively complete preservation of production environment and equipment.

5. For each handmade papermaking site, we draw two standard illustrations, i.e. distribution map and roadmap from the county center to the papermaking sites (in some sections, two figures are combined). We do not distinguish the administrative village, natural village or villagers' group, and we provide county name, town name and village name of each site based on standards released by Sinomaps Press and National Geomatics Center of China.

6. For each type of paper included in Special Edition, we attach a piece of paper sample (a full page, 2/3, 1/2 or 1/4 of a page, or even smaller if we do not have sufficient sample available) to the corresponding section. For some sections, no sample is attached for the shortage of sample paper (e.g. the papermakers had ceased production).

7. All the paper samples elaborated on in this volume, in terms of raw materials and papermaking techniques, were tested, including those attached to the special edition, or not attached to this volume due to scarce sample which only enough for technical analysis.

The test was based on the technical analysis standards of Chinese Xuan paper (GB/T 18739—2008), with modifications adopted according to the specific features of the handmade paper in Anhui Province. All paper with sufficient sample, such as Xuan paper, calligraphy and painting paper, bast paper, was tested in terms of 14 indicators, including thickness, mass per unit area, tightness, resistance force, tensile strength, tear resistance, wet strength, whiteness, ageing resistance, dirt count, absorption of water (several processed Xuan paper was not tested on the indicator), elasticity, fiber length and fiber width. Processed paper was tested in terms of 8 indicators, including thickness, mass per unit area, tightness, resistance force, tensile strength, tear resistance, whiteness,

量得到的厚度。根据纸的厚薄不同，可采取多层指标测量、单层指标测量，以单层指标测量的结果表示纸的厚度，以mm为单位。

所用仪器▶长春市月明小型试验机有限责任公司JX-HI型纸张厚度仪、杭州品享科技有限公司PN-PT6厚度测定仪。

(2) 定量▶所测纸的定量指标是指单位面积纸的质量，通过测定试样的面积及质量，计算定量，以g/m²为单位。

所用仪器▶上海方瑞仪器有限公司3003电子天平。

(3) 紧度▶所测纸的紧度指标是指单位体积纸的质量，由同一试样的定量和厚度计算而得，以g/cm³为单位。

(4) 抗张力▶所测纸的抗张力指标是指在标准试验方法规定的条件下，纸断裂前所能承受的最大张力，以N为单位。

所用仪器▶杭州高新自动化仪器仪表公司DN-KZ电脑抗张力试验机、杭州品享科技有限公司PN-HT300卧式电脑拉力仪。

(5) 抗张强度▶所测纸的抗张强度指标一般用在抗张强度试验仪上所测出的抗张力除以样品宽度来表示，也称为纸的绝对抗张强度，以kN/m为单位。

《安徽卷》采用的是恒速加荷法，其原理是使用抗张强度试验仪在恒速加荷的条件下，把规定尺寸的纸样拉伸至撕裂，测其抗张力，计算出抗张强度。公式如下：

$$S=F/W$$

式中，S为试样的抗张强度（kN/m），F为试样的绝对抗张力（N），W为试样的宽度（mm）。

(6) 撕裂度▶所测纸张撕裂强度的一种量度，即在测定撕裂度的仪器上，拉开预先切开一小切口的纸达到一定长度时所需要的力，以mN为单位。

所用仪器▶长春市月明小型试验机有限责任公司ZSE-1000型纸张撕裂度测定仪、杭州品享科技有限公司PN-TT1000电脑纸张撕裂度测定仪。

(7) 湿强度▶所测纸张在水中浸润规定时间后，在润湿状态下测得的机械强度，以mN为单位。

and absorption of water. Bamboo paper was tested in terms of 8 indicators, including thickness, mass per unit area, tightness, resistance force, tensile strength, whiteness, fiber length and fiber width. Due to the various production standards involved in papermaking in Anhui Province, the data might vary from those standards of machine-made paper and Xuan paper.

8. Test indicators and devices:
(1) Thickness: the values obtained by using two measuring boards pressing the paper. In the measuring process, single layer or multiple layers of paper were employed depending on the thickness of the paper, and its measurement unit is mm. The thickness measuring instruments employed are produced by Yueming Small Testing Instrument Co., Ltd., Changchun City (specification: JX-HI) and Pinxiang Science and Technology Co., Ltd., Hangzhou City (specification: PN-PT6).
(2) Mass per unit area: the sample mass divided by area, with the measurement unit g/m². The measuring instrument employed is 3003 electronic balance produced by Shanghai Fangrui Instrument Co., Ltd.
(3) Tightness: mass of paper per volume unit, obtained by measuring the mass per unit area and thickness, with the measurement unit g/cm³.
(4) Tensile strength: the resistance of sample paper to a force tending to tear it apart, measured as the maximum tension the material can withstand without tearing. The resistance force testing instrument (specification: DN-KZ) is produced by Gaoxin Technology Company, Hangzhou City and PN-HT300 horizontal computer tensiometer by Pinxiang Science and Technology Co., Ltd., Hangzhou City.
(5) Unit tensile strength: the resistance of one unit sample paper to a force, with the measurement unit kN/m. In *Library of Chinese Handmade Paper*: *Anhui*, constant loading method was employed to measure the tensile strength. The sample's maximum resistance force against the constant loading was tested, then we divided the maximum force by the sample width. The formula is:

$$S=F/W$$

S stands for tensile strength (kN/m) for each unit, F is resistance force (N) and W represents sample width (mm).
(6) Tear resistance: a measure of how well a piece of paper can

所用仪器▶长春市月明小型试验机有限责任公司ZSE-1000型纸张撕裂度测定仪、杭州品享科技有限公司PN-TT1000电脑纸张撕裂度测定仪。

(8) 白(色)度▶白度测试针对白色纸，色度测试针对其他颜色的纸。白度是指被测物体的表面在可见光区域内与完全白（标准白）的物体漫反射辐射能的大小的比值，用百分数来表示，即白色的程度。所测纸的白度指标是指在D65光源、漫射/垂射照明观测条件下，以纸对主波长475 nm蓝光的漫反射因数表示白度的测定结果。

所用仪器▶杭州纸邦仪器有限公司ZB-A色度测定仪、杭州品享科技有限公司PN-48A白度颜色测定仪。

(9) 耐老化度下降▶指所测纸张进行高温试验的温度环境变化后的参数及性能。本测试采用105 ℃高温恒温放置72小时后进行测试，以百分数（%）表示。

所用仪器▶上海一实仪器设备厂3GW-100型高温老化试验箱、杭州品享科技有限公司YNK/GW100-C50耐老化度测试箱。

(10) 尘埃度▶所测纸张单位面积上尘埃涉及的黑点、黄茎和双浆团个数。测试时按照标准要求计算出每一张试样正反面每组尘埃的个数，将4张试样合并计算，然后换算成每平方米的尘埃个数，计算结果取整数，以个/m²为单位。

所用仪器▶杭州品享科技有限公司PN-PDT尘埃度测定仪。

(11) 吸水性▶所测纸张在水中能吸收水分的性质。测试时使用一条垂直悬挂的纸张试样，其下端浸入水中，测定一定时间后的纸张吸液高度，以mm为单位。

所用仪器▶四川长江造纸仪器有限责任公司J-CBY100型纸与纸板吸收性测定仪、杭州品享科技有限公司PN-KLM纸张吸水率测定仪。

(12) 伸缩性▶指所测纸张由于张力、潮湿，尺寸变大、变小的倾向性。分为浸湿伸缩性和风干伸缩性，以百分数（%）表示。

所用仪器▶50 cm × 50 cm × 20 cm长方体容器。

withstand the effects of tearing. It measures the strength the test specimen resists the growth of any cuts when under tension. The measurement unit is mN. Paper tear resistance testing instrument (specification: ZSE-1000) is produced by Yueming Small Testing Instrument Co., Ltd., Changchun City and computer paper tear resistance testing instrument (specification: PN-TT1000) produced by Pinxiang Science and Technology Co., Ltd., Hangzhou City.

(7) Wet strength: a measure of how well the paper can resist a force of rupture when the paper is soaked in the water for a set time. The measurement unit is mN. Paper tear resistance testing instrument (specification: ZSE-1000) is produced by Yueming Small Testing Instrument Co., Ltd., Changchun City and computer paper tear resistance testing instrument (specification: PN-TT1000) produced by Pinxiang Science and Technology Co., Ltd., Hangzhou City.

(8) Whiteness: degree of whiteness, represented by percentage, which is the ratio obtained by comparing the radiation diffusion value of the test object in visible region to that of the completely white (standard white) object. Whiteness test in our study employed D65 light source, with dominant wavelength 475nm of blue light, under the circumstances of diffuse reflection or vertical reflection. The whiteness testing instrument (specification: ZB-A) is produced by Zhibang Instrument Co., Ltd., Hangzhou City and whiteness tester (specification: PN-48A) produced by Pinxiang Science and Technology Co., Ltd., Hangzhou City respectively.

(9) Ageing Resistance: the performance and parameters of paper sample when put in high temperature. In our test, temperature is set 105 degrees centigrade, and the paper is put in the environment for 72 hours. It is measured in percentage(%). The high temperature ageing test box (specification: 3GW-100) is produced by Yishi Testing Instrument Factory and ageing test box (specification: YNK/GW100-C50) produced by Pinxiang Science and Technology Co., Ltd., Hangzhou City.

(10) Dirt count: fine particles (black dots, yellow stems, fiber knots) in the test paper. It is measured by counting fine particles in every side of four pieces of paper sample, adding up and then calculate the number (integer only) of particles every square meter. It is measured by the number of particles/m². Dust tester (specification: PN-PDT) is produced by Pinxiang Science and Technology Co., Ltd., Hangzhou City.

（13）纤维长度/宽度▸所测纸的纤维长度/宽度是指从所测纸里取样，测其纸浆中纤维的自身长度/宽度，分别以mm和μm为单位。测试时，取少量纸样，用水湿润，用Herzberg试剂染色，制成显微镜试片，置于显微分析仪下采用10倍及20倍物镜进行观测，并显示相应纤维形态图各一幅。

所用仪器▸珠海华伦造纸科技有限公司生产的XWY-VI型纤维测量仪和XWY-VII型纤维测量仪。

9

《安徽卷》对每一种调查采集的纸样均采用透光摄影的方式制作成图像，以显示透光环境下的纸样纤维纹理影像，作为实物纸样的另一种表达方式。其制作过程为：先使用透光台显示纯白影像，作为拍摄手工纸纹理透光影像的背景底，然后将纸样铺平在透光台上进行拍摄。拍摄相机为佳能5DIII。

10

《安徽卷》引述的历史与当代文献均以当页脚注形式标注。所引文献原则上要求为一手文献来源，并按统一标准注释，如"[宋]罗愿.《新安志》整理与研究[M].萧建新，杨国宜，校.合肥:黄山书社,2008:371." "民国三年（1913年）泾县小岭曹氏编撰.曹氏宗谱[Z].自印本." "魏兆淇.宣纸制造工业之调查:中央工业试验所工业调查报告之一[J].工业中心,1936(10):8."等。

11

《安徽卷》所引述的田野调查信息原则上要求标示出调查信息的一手来源，如："据访谈中刘同烟的介绍，星杰桑皮纸厂年产5 000多刀纸，年销售额约100万元" "按照访谈时沈维正的说法，以他为核心的这个团队专注造纸新技术的研发和传统技艺的保护"等。

(11) Absorption of water: it measures how sample paper absorbs water by dipping the paper sample vertically in water and testing the level of water. It is measured in mm. Paper and Paper Board Water Absorption Tester (specification: J-CBY100) is produced by Changjiang Papermaking Instrument Co., Ltd., Sichuan Province, and Water absorption tester (specification: PN-KLM) produced by Pinxiang Science and Technology Co., Ltd., Hangzhou City.

(12) Elasticity: continuum mechanics of paper that deform under stress or wet. It is measured in percentage(%), consists of two types, i.e. wet elasticity and dry elasticity. Testing with a rectangle container (50cm×50cm×20cm).

(13) Fiber length and width: analyzed by dying the moist paper sample with Herzberg reagent, and the fiber pictures were taken through ten times and twenty times objective lens of the microscope, with the measurement unit mm and μm. We used the fiber testing instruments (specifications: XWY-VI and XWY-VII) produced by Hualun Papermaking Technology Co., Ltd., Zhuhai City.

9. Each paper sample included in this volume was photographed against a luminous background, which vividly demonstrated the fiber veins of the sample. This is a different way to present the status of our paper sample. Each piece of paper sample was spread flat-out on the light table giving white light, and photographs were taken with Canon 5DIII camera.

10. All the quoted literature are original first-hand resources and the footnotes are used for documentation with a uniform standard. For instance, "[Song Dynasty] Luo Yuan. *Xin'an Records* [M]. Proofread by Xiao Jianxin and Yang Guoyi. Hefei: Huangshan Publishing House, 2008:371." and "*Genealogy of The Caos* [Z]. compiled by the Caos in Xiaoling Village of Jingxian County, 1913. Self-printed." and "Wei Zhaoqi. *Investigation of Xuan paper industry: One of the national industrial investigation report series* [J]. Industrial Center, 1936 (10):8." etc.

11. Sources of field investigation information were attached in this volume. For instance, "According to Liu Tongyan, annual output of Xingjie Mulberry Bark Factory exceeded 5,000 *dao* each year, with annual sales about one million RMB." "According to Shen Weizheng,

12

《安徽卷》所使用的摄影图片主体部分为调查组成员在实地调查时所拍摄的图片，也有项目组成员在既往田野工作中积累的图片，另有少量属撰稿过程中所采用的非项目组成员的摄影作品。由于项目组成员在完成全卷过程中形成的图片的著作权属集体著作权，且在调查过程中多位成员轮流拍摄或并行拍摄为工作常态，因而全卷对图片均不标示项目组成员作者。项目组成员既往积累的图片，以及非项目组成员拍摄的图片在图题文字或后记中特别说明，并承认其个人图片著作权。

13

考虑到《安徽卷》中文简体版的国际交流需要，编著者对全卷重要或提要性内容同步给出英文表述，以便英文读者结合照片和实物纸样领略全卷的基本语义。对于文中一些晦涩的古代文献，英文翻译采用意译的方式进行解读。英文内容包括：总序、编撰说明、目录、概述、图目、表目、术语、后记，以及所有章节的标题，全部图题、表题与实物纸样名。

"安徽省手工造纸概述"为全卷正文第一章，为保持与后续各章节体例一致，除保留章节英文标题及图表标题英文名外，全章的英文译文作为附录出现。

14

《安徽卷》的名词术语附录兼有术语表、中英文对照表和索引三重功能。其中收集了全卷中与手工纸有关的地理名、纸品名、原料与相关植物名、工艺技术和工具设备、历史文化等5类术语。各个类别的名词术语按术语的汉语拼音先后顺序排列。每条中文名词术语后都以英文直译，可以作中英文对照表使用，也可以当作名词索引使用。

---

he played a key role in the team which focused on papermaking techniques R&D, and preserving the traditional skills".

12. The majority of photographs included in the volume were taken by the researchers when they were doing fieldworks of the research. Others were taken by our researchers in even earlier fieldwork errands, or by the photographers who were not involved in our research. We do not give the names of the photographers in the book, because almost all our researchers are involved in the task and they agreed to share the intellectual property of the photos. Yet, as we have claimed in the epilogue or the caption, we officially admit the copyright of all the photographers, including those who are not our researchers.

13. For the purpose of international academic exchange, English version of some important chapters is provided, so that the English readers can have a basic understanding of the volume based on the English parts together with photos and samples. For the ancient literature which is hard to understand, free translation is employed to present the basic idea. English part includes Preface, Introduction to the Writing Norms, Contents, Introduction, Figures, Tables, Terminology, Epilogue, and section titles, figure and table captions and paper sample names.

Among them, "Introduction to Handmade Paper in Anhui Province" is the first chapter of the volume and its translation is appended in the appendix part, apart from the section titles and table and figure titles.

14. Terminology is appended in *Library of Chinese Handmade Paper: Anhui*, which covers five categories of places, paper names, raw materials and plants, techniques and tools, history and culture, etc., relevant to our handmade paper research. All the terms are listed following the alphabetical order of the first Chinese character. The Chinese and English parts in the Terminology can be used as check list and index.

目 录
Contents

总 序
Preface
I

编撰说明
Introduction to the Writing Norms
XI

第六章　　加工纸
Chapter VI　Processed Paper
0 0 1

0 0 2　第一节　安徽省掇英轩书画用品有限公司
　　　　Section 1　Duoyingxuan Calligraphy and Painting Supplies Co., Ltd. in Anhui Province

0 6 2　第二节　泾县艺英轩宣纸工艺品厂
　　　　Section 2　Yiyingxuan Xuan Paper Craft Factory in Jingxian County

0 8 6　第三节　泾县艺宣阁宣纸工艺品有限公司
　　　　Section 3　Yixuange Xuan Paper Craft Co., Ltd. in Jingxian County

1 0 8　第四节　泾县宣艺斋宣纸工艺厂
　　　　Section 4　Xuanyizhai Xuan Paper Craft Factory in Jingxian County

1 3 2　第五节　泾县贡玉堂宣纸工艺厂
　　　　Section 5　Gongyutang Xuan Paper Craft Factory in Jingxian County

154	第六节 泾县博古堂宣纸工艺厂
	Section 6　Bogutang Xuan Paper Craft Factory in Jingxian County

168	第七节 泾县汇宣堂宣纸工艺厂
	Section 7　Huixuantang Xuan Paper Craft Factory in Jingxian County

186	第八节 泾县风和堂宣纸加工厂
	Section 8　Fenghetang Xuan Paper Factory in Jingxian County

第七章　工具
Chapter VII　Tools

217

218	第一节 泾县明堂纸帘工艺厂
	Section 1　Mingtang Papermaking Screen Craft Factory in Jingxian County

232	第二节 泾县全勇纸帘工艺厂
	Section 2　Quanyong Papermaking Screen Craft Factory in Jingxian County

244	第三节 泾县后山大剪刀作坊
	Section 3　Houshan Shears Workshop in Jingxian County

附　录
Appendices

259

259	Introduction to Handmade Paper in Anhui Province
287	图目　Figures
314	表目　Tables
316	术语　Terminology

后　记
Epilogue

327

第六章
加工纸

Chapter VI
Processed Paper

第一节
安徽省掇英轩书画用品有限公司

Section 1
Duoyingxuan Calligraphy and Painting Supplies Co., Ltd. in Anhui Province

Subject
Processed Paper of Duoyingxuan Calligraphy and Painting Supplies Co., Ltd. in Anhui Province in Huanglu Town

调查对象
黄麓镇
安徽省掇英轩书画用品有限公司
加工纸

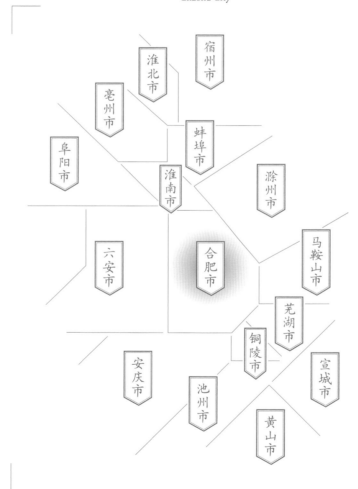

一 安徽省掇英轩书画用品有限公司的基础信息与生产环境

1
Basic Information and Production Environment of Duoyingxuan Calligraphy and Painting Supplies Co., Ltd. in Anhui Province

安徽省掇英轩书画用品有限公司（以下简称"掇英轩"）坐落于距合肥市区54 km的巢湖市（原为地级市，2011年7月巢湖地级市撤销，原市府所在地居巢区改为县级巢湖市并划归合肥市管辖）黄麓镇，地理坐标为东经117°35′9″，北纬31°37′51″。公司前身为1993年创办的合肥市掇英轩文房用品研究所，1995年研究所的生产基地从合肥市迁往巢湖市。2002年公司注册成立巢湖市掇英轩文房用品厂，2008年入选国家级非物质文化遗产"纸笺加工技艺"保护单位；2014年易名为安徽省掇英轩书画用品有限公司，2015年3月入选安徽省级非物质文化遗产传习基地。"掇英轩"是中国当代利用宣纸与书画纸原纸再加工的著名传统纸笺制作品牌。

2016年3月18～19日，在历经10余轮前期访谈调查后，调查组对"掇英轩"进行了正式记录式的田野调查；2017年8月5～6日，调查组再入"掇英轩"生产厂区进行了专项补充访谈与信息核实。获得的基础信息如下："掇英轩"的员工近3年保持在26名左右，主要产品为粉蜡笺、绢本宣、泥金笺、木版水印笺、金银印花笺等加工纸笺。据"掇英轩"总经理刘靖介绍，2014～2016年，掇英轩书画用品有限公司年销售额300余万元，利润率在15%左右，年动态利润在50万元上下。

黄麓镇隶属于合肥市辖的巢湖县级市，南滨中国五大淡水湖之一的巢湖，西南接地方文化旅游景区中庙街道（原中庙镇），西北接以古街区闻名的长临河镇，东北和北面与留下神秘水底古城故事的烔炀镇相邻。黄麓镇1914年建镇时名桐荫镇，因镇区北缘有一座小黄山，镇区位于小黄山南麓而得今名。代表性文化景点有：安徽省第一所乡村师范——历史名校黄麓师范学校、安徽省文物保护单位——张治中将军故居、江淮地区著名寺院——相隐寺、中国传统工艺知名工坊——掇英轩。

1/2 "掇英轩"非物质文化遗产荣誉牌匾
Honorary plaques of the intangible cultural heritage of "Duoyingxuan"

3 中国科学技术大学人文与社会科学学院与"掇英轩"共建"传统纸笺产学研基地"揭牌仪式
Opening ceremony of the "Traditional Paper Production, Education and Research Base", jointly built by School of Humanities and Social Sciences, USTC and "Duoyingxuan"

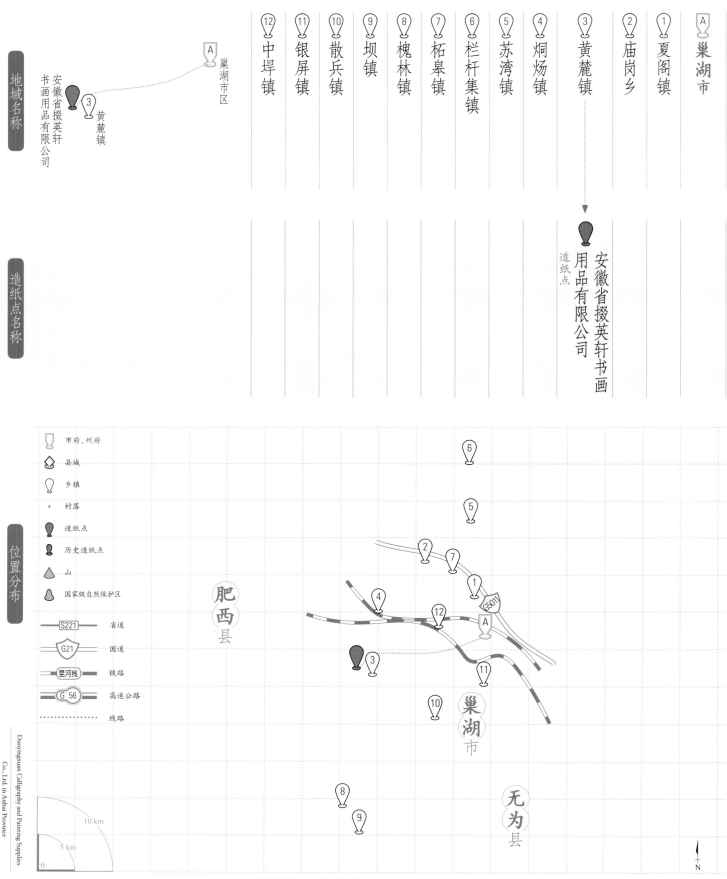

1 江淮名寺——相隐寺
Famous temple in Jianghuai area—Xiangyin Temple

2 张治中故居
Former house of general Zhang Zhizhong

3 黄麓镇洪家疃村的乡贤文化展馆
County Sage Cultural Exhibition Hall in Hongjiatuan Village of Huanglu Town

4 1985年刘锡宏在日本大玄堂试纸旧照（"掇英轩"供图）
Old photo of Liu Xihong testing paper in Daxuan Hall in Japan in 1985 (offered by "Duoyingxuan")

二 安徽省掇英轩书画用品有限公司的历史与传承情况

2 History and Inheritance of Duoyingxuan Calligraphy and Painting Supplies Co., Ltd. in Anhui Province

安徽省掇英轩书画用品有限公司注册于2014年，前身是1993年创办的合肥市掇英轩文房用品研究所，更早以前则可追溯到1989年创建的合肥汉韵堂工艺厂。"掇英轩"第一代经营者为刘锡宏和杨宁英夫妻，第二代主要经营负责人为两人的二儿子刘靖。

据访谈时安徽省掇英轩书画用品有限公司负责人刘靖介绍，调查组了解到的企业沿革发展信息如下：

刘锡宏，1941年生，合肥师范学院工艺美术专业毕业。1979年,在改革开放的新环境下,开始主持安徽省工艺品进出口公司的文房四宝产品出口业务。

1985年,刘锡宏作为中国文房四宝小组交流团的成员去日本访问时,日本一位书法家及收藏家向代表团展示了中国古代制作的粉蜡笺,诚挚地表示这是中国最好的纸,并对其在近现代中国的技艺失传表示非常痛惜。

刘锡宏受到这件事的刺激与启发,回国后先后找到当时的合肥工艺美术厂、合肥十竹斋、古稀斋等从事纸笺制作的工厂,开始制作粉蜡笺的试验。刘锡宏组织研制的"粉蜡笺"数年后即有了产品,而且看起来质感挺好,刚开始销往日本时受到了日本消费者的喜爱。但没有料到好景不长,售出后仅3年的时间,这些"粉蜡笺"就纷纷出现严重的纸张老化,脆、裂、易折断等问题,就此宣告了"粉蜡笺"复制的失败。

"粉蜡笺"研制的失败,没有挫伤刘锡宏对传统纸笺工艺文化的喜爱与传承的执着。1989年,刘锡宏动员妻子杨宁英与他人联合组建了合肥汉韵堂工艺厂,专事传统纸笺加工与销售。据2017年8月访谈时刘靖介绍,"汉韵堂"由他们家与原合肥十竹斋员工杨桂英、合肥工艺美术厂原厂长缪世发、泾县百岭轩纸厂的老厂长沈伯泉四家合组而成,租赁的厂房地址在安徽省立儿童医院后面(今合肥市望江东路39号附近)。技术、销售、财务、生产四家分别负责,至1992年下半年因合作不顺利,"汉韵堂"解体。

"汉韵堂"解体后,杨宁英带领3名从"汉韵堂"分离出来的老家及泾县来的员工,又从老家招来四五人,在合肥市宿松路(现科苑新村的位置)租赁了几间民房,继续纸笺的加工、生产,起名"掇英轩"。

1993年5月,"掇英轩"搬到租赁的安徽省轻工干校的几间教舍内。6月14日,合肥市掇英轩文房用品研究所挂牌成立,挂靠在当时的安徽省文房四宝研究所名下,以从事纸笺的研究与生产为主。

杨宁英,1945年生,曾在安徽省工艺品进出口公司一个存放文房四宝产品的中转仓库从事管理、检验工作10年时间,熟悉安徽地域文房四宝各品牌及其性能、特点,逐渐积累了较多的传统纸笺产品知识。1993年合肥市掇英轩文房用品研究所成立后,任法人、所长,负责"掇英轩"的全面管理。

至1994年底,在充分考虑了在省城租房、雇工的成本以及多数技术工人来自黄麓镇杨岗村的现实情况后，杨宁英和刘锡宏决定将"掇英轩"搬到巢湖市黄麓镇杨岗村自己老家的老屋里。

2017年8月底调查组驻村调查时，刘靖描述了老屋的建筑格局：老屋当时有三排，三进布局，中间有通道，每排间有个天井。因为祖辈弟兄三个，一家一进，爷爷排行老大，住北边一进。到父亲这一代的时候，男丁就只有父亲（刘锡宏）一个人，三爷爷家已没有人了，二爷爷家有两个女儿，2017年调查时，一个在合肥，一个在芜湖。爷爷经商，在南京开过杂货店，叫刘德才。

访谈时刘靖回忆："1994年'掇英轩'搬回杨岗村老屋的时候，做的主要是色宣、洒金宣、

⊙3

木版水印信笺及丝网水印笺等几类，90%以上销往日本市场。一同从合肥回来的技术工人有6～7个，只有3个留下来在老屋纸坊做，分别是杨慧、方玉红、刘霞。我（刘靖）1995年10月回杨岗村，年把时间（当地方言指一年左右）后，曾在合肥干过的一对姐弟（杨小波、杨小进）也来了。当时就母亲一个人管理生产，父亲仍在合肥上班，经常节假日赶回老家做技术指导，当时

⊙4

乘车单程要2个多小时，很辛苦。"

2000年，由于国家企业实体挂靠政策的调整与限制，"掇英轩"与安徽省文房四宝研究所脱钩。2002年，合肥市掇英轩文房用品研究所注销，巢湖市掇英轩文房用品厂成立。2008年，注册的商标"掇英轩"获国家工商总局商标局批准。

⊙5　⊙6

刘靖，1972年生，现任安徽省掇英轩书画用品有限公司总经理，高级工艺美术师，中国艺术研究院民间艺术创作研究员，中国科学技术大学手工纸研究所兼职研究员，国家级非物质文化遗产代表性传承人、中国文房四宝纸笺艺术大师。

⊙7

⊙3 刘家老屋旧照（"掇英轩"供图）
Old photo of the aged house of the Liu's family (offered by "Duoyingxuan")
⊙4 老屋里检纸员工旧照（"掇英轩"供图）
Old photo of staff checking paper (offered by "Duoyingxuan")
⊙5 2000年挂靠脱钩申请（"掇英轩"供图）
Application for disconnection in 2000 (offered by "Duoyingxuan")
⊙6 商标注册证
Registered trademark
⊙7 中国艺术研究院聘书
Letter of appointment of the Chinese Academy of Arts

1994年，刘靖在合肥联合大学工业产品造型专业毕业后，先后在合肥、芜湖等地从事广告和装潢设计工作。1995年10月，刘靖回到"掇英轩"，跟着父母学习纸笺加工技艺；1996年，没有安定下来的心又让刘靖回到合肥，在合肥市太湖路小学担任了1年的小学数学、英语代课教师。在代课期间，刘靖经常利用节假日，回到老家帮忙。1995～1996年，"掇英轩"也曾试制过粉蜡

⊙1 在老屋小院中接受访谈的刘靖 Interviewing Liu Jing in the yard of old house
⊙2 第一批粉蜡笺小样旧照（"掇英轩"供图）Old photo of the first batch of Fenla paper produced (offered by "Duoyingxuan")
⊙3 手绘描金粉蜡笺初进"荣宝斋"时旧照（"掇英轩"供图）Old photo of hand-painted golden Fenla paper first introduced to "Rongbaozhai" (offered by "Duoyingxuan")

笺，但都没有取得理想效果。

1997年7月，刘靖再次回到"掇英轩"，受父亲的嘱托，专心尝试攻克粉蜡笺加工技术这一难题。在总结父辈此前研发经验教训的基础上，刘靖潜心揣摩并向专家请教，经过4个月的反复试验，1997年11月研制出一批粉蜡笺样纸。据刘靖介绍，这批样品受到了时任安徽省文房四宝研究所所长田恒铭的赞赏，认为可与明清时的粉蜡笺媲美。

样品试制成功之后，刘靖带领"掇英轩"开始尝试批量生产粉蜡笺，此后，刘靖请多位书法名家试纸，不断改进粉蜡笺纸性与工艺。在成功复制了"素粉蜡笺"的基础上，"掇英轩"陆续恢复生产出了梅花玉版笺、云龙粉蜡笺、五龙盘珠粉蜡笺、暗八仙粉蜡笺等数十种花色的手绘描金粉蜡笺。通过安徽省进出口公司"掇英轩"逐渐接到一些粉蜡笺的外贸订单，主要对象是日本的客户。不过批量生产时，由于之前在试验时没有将温度、湿度的变化考虑进去，数次造成大量返工重做。

又经过了近2年时间，粉蜡笺制作技艺及手绘描金技艺不断完善，熟练员工的培养也渐有成效，至1999年"掇英轩"手绘描金粉蜡笺制作技艺已基本成熟定型，开始批量稳定生产。在2017年8月访谈中刘靖回忆道：1999年"掇英轩"首次将包括描金粉蜡笺在内的纸笺产品带到了北京全国文房四宝春季博览会上，受到了各地文房四宝经销商的关注。年底，他带着手绘描金粉蜡笺前往北京"荣宝斋"，当时"荣宝斋"业务科负责人袁良观纸后立即决定："荣宝斋"将销售大厅中间最显眼的两节柜台腾空，放上手绘描金粉蜡笺；西厅中间几根柱子上挂着的字画也随即被拆下，挂上了四张手绘描金粉蜡笺。

2002年，"荣宝斋"这样评价"掇英轩"生产的描金粉蜡笺："做工精细，用料考究，其制

作技艺在继承传统的基础上又有所创新，为目前同类产品的佳者，可代表中国传统手工制纸之最高技艺。"2003年，时任"荣宝斋"顾问的米景扬题词，评价"掇英轩"的描金粉蜡笺为"纸中重宝"。

⊙4

在"粉蜡笺"系列获成功后，刘靖带领的团队又相继成功恢复与发展了金银印花笺、泥金笺、朱砂笺、绢本宣、羊脑笺、流沙笺、刻画笺等纸笺名品。"掇英轩"生产的系列纸笺产品，受到了"荣宝斋"的充分认可，成为"荣宝斋"纸笺销售的主导产品。

"掇英轩"的纸笺产品通过每年的全国文房四宝博览会和"荣宝斋"的销售平台，逐渐为更多的业界人士及消费者所了解，开始在全国有了影响力。日本客户来信要求经销的也多了起来。

"掇英轩"体系的传承脉络中，刘锡宏和杨宁英为第一代；第二代以国家级"非遗"传承人刘靖为技艺传习核心，传承人包括：方玉红、杨慧、刘霞等；第三代包括：袁琴芳、肖红梅、方

春希、刘娇娇、周海霞等；第四代包括：杨芳、杨娟、黄超、张玮、张素素等。由于该企业成立的时间不长，"掇英轩"一代代的传承顺序不是以年龄来划分的，而是以进入"掇英轩"的先后顺序以及师徒关系来划分的。从1992年至今，

"掇英轩"先后培养纸笺加工技艺的技术骨干40余人，形成了一个师徒传习有序的群体。

就创始人家庭的传承来源而言，调查中了解的情况是：刘锡宏和杨宁英的技艺学习有部分来自常秀峰（系刘锡宏在原合肥师范学院的绘画兼

⊙4 "荣宝斋"对"掇英轩"纸笺的反馈材料（"掇英轩"供图）
Feedback material from "Rongbaozhai" to "Duoyingxuan" paper (offered by "Duoyingxuan")

⊙5 米景扬题"纸中重宝"原件照
Original photo of Mi Jingyang's inscription "Treasure of the Paper"

⊙6 日本客户要求经销的信（"掇英轩"供图）
A letter from a Japanese customer demanding for paper (offered by "Duoyingxuan")

⊙7 杨岗村老屋的"掇英轩"旧址
Former site of "Duoyingxuan" in Yanggang Village

工艺美术教师）、汪叙伦（原合肥师范学院从徽州聘请的木雕老师）和沈伯泉（原泾县百岭轩纸厂厂长）。

⊙1

⊙2

⊙3

⊙4

刘靖同辈家人的技艺传习情况为：哥哥刘干，1970年出生，访谈时在合肥市住房公积金管理中心工作，受家庭影响，也曾学过一点纸笺加工技艺。

妹妹刘晴，1975年出生，与妹夫徐金松两人经营合肥留香阁美术用品商行。据刘靖介绍，妹妹没出嫁前参与过纸笺加工技艺的研究与生产，她的手绘技术很好，手快，一小时多点一张四尺对开的龙纹就画完了。

刘锡宏、杨宁英夫妇培养的徒弟方玉红与刘靖在纸笺加工中有了好感，方玉红于1999年与刘靖结婚。访谈中刘靖表示："方玉红对纸笺加工技艺掌握得很全面，技艺娴熟，我自己刚回厂时，有些技术还是跟方玉红学的，今天的成功有她的一半。"

刘靖的女儿刘子嘉，15周岁，调查时在合肥市第五十中学读初三。据刘靖介绍，刘子嘉从小在厂里跟着父母和工人后面玩，有时还有模有样地跟在后面干，帮着擦板、贴纸条等，有时还"指挥"员工。

⊙5

三 安徽省掇英轩书画用品有限公司的代表纸品及其用途与技术分析

3 Representative Paper and Its uses and Technical Analysis of Duoyingxuan Calligraphy and Painting Supplies Co., Ltd.

（一）代表纸品及其用途

调查组实地调查的纸品信息如下："掇英轩"加工纸笺种类相当丰富，各类纸笺又有较多细目，细目间还相互穿插，较难准确分类。如粉蜡笺里有真金手绘粉蜡笺、仿金手绘粉蜡笺、丝网印金银花粉蜡笺、洒金粉蜡笺等；而洒金粉蜡笺有洒真金、仿金之分，又按洒金片的大小分为雨雪粉蜡笺、金粟粉蜡笺、雾雨粉蜡笺及各色金银混合的彩金粉蜡笺等。

6 样品室陈列的各色粉蜡笺
Various Fenla paper displayed in the sample room

7 琳琅满目的"掇英轩"纸品陈列
Various "Duoyingxuan" paper types

"掇英轩"现有粉蜡笺、泥金笺、木版水印笺、金银印花笺、绢本宣、流沙笺、刻画笺、羊脑笺、洒金笺、砑花笺、拱花笺、色宣、丝网水印笺等大类加工笺纸品类。调查组对粉蜡笺、绢本宣、泥金笺、木版水印笺、流沙笺、金银印花笺六类进行了详细的信息采集。

1. 粉蜡笺

根据古代史料的记述，粉蜡笺诞生于唐代，

性能分析

是在魏晋南北朝时期的粉笺与唐代蜡笺的基础上发展而成的。宋代书法名家米芾《书史》记载："唐中书令褚遂良《枯树赋》是粉蜡纸搨。"粉蜡笺是以皮纸或宣纸为原料，经过拖染、施粉、施蜡、研光、托裱等复杂工艺制作而成的。粉蜡笺融合了吸水的"粉"和防水的"蜡"两种材料，是以原材料特征起名的。

粉蜡笺纸质平滑温润，气质高雅华贵，具有较强的抗老化性能。访谈中刘靖介绍：在粉蜡笺上描金勾银，融入描绘工艺，则制成手绘描金粉蜡笺；若洒以金箔、银箔，则制成洒金、银粉蜡笺；若和拱花、研花工艺结合，可制作成拱花粉蜡笺、研花粉蜡笺；若以木版或丝网印金银图案，则制成金银花粉蜡笺等。另外，粉蜡笺还有单面和双面之分。双面粉蜡笺是为了获得更好的防潮、防蛀性能，双面施粉、施蜡、研光，一般是正面绘图案，背面饰以金箔。

粉蜡笺由于制作精细、造价高昂，自诞生之日起，历代多为皇家御用纸，用于宫廷内府书写匾、联及屏风、壁贴等。根据传世的实物评价，通常认为以清代康熙、乾隆年间制作的粉蜡笺最为精良，史称"库蜡笺"。"掇英轩"生产的粉蜡笺的常规规格有44 cm×44 cm、66 cm×66 cm、34 cm×136 cm、68 cm×136 cm、94 cm×175 cm、45 cm×175 cm、52 cm×230 cm、26.5 cm×232 cm等，最大规格的可以做到97 cm×234 cm。

⊙1 真金手绘仿明代云龙纹双面粉蜡笺
Golden Double-faced Fenla paper hand-painted with cloud and dragon decorations and imitating the style of Ming Dynasty
⊙2 真金雨雪粉蜡笺
Golden Fenla paper with rain and snow decorations
⊙3 刘靖向调查组成员讲解洒金粉蜡笺工艺
Liu Jing is explaining the procedure of producing golden Fenla paper to researchers
⊙4 手绘描金『龙腾如意』纹粉蜡笺
Golden Fenla paper hand-painted with "Flying Dragon is Good Fortune" decorations

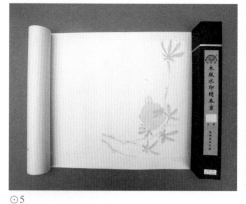
⊙5

⊙6

⊙7

2. 绢本宣

"掇英轩"绢本宣是在绢的底层以黏合剂托裱一层宣纸而成的。2016年访谈时据刘靖介绍,一开始是应日本客户的要求这样制作的,1998年左右,他将这种一面绢一面纸的产品命名为绢本宣。

绢作为书画载体古即有之,尤以唐、宋时期为盛。以绢书画,多以胶矾刷在绢上,制成熟绢后再绷平使用。绢本宣改变了在绢上书画的使用习惯,使绢像纸一样直接铺开就可使用。

绢本宣既有绢的特性,又有纸的特性,增强了纸的拉力,同时又增强了在绢面书画的吸墨性。"掇英轩"的绢本宣制有生、半熟、熟等不同的书画效果品种。

调查时,"掇英轩"绢本宣有素色绢本宣、木版水印绢本宣、绢本泥金笺、绢本手绘描金笺、绢本洒金笺等系列产品。

3. 泥金笺

泥金笺在唐代即已出现。简单说,泥金笺就是将整幅纸面全部涂上金粉制成的,主要在书法、工笔重彩、小写意及装裱等中使用。

泥金笺制作分真金、仿金。真金制作色呈赤黄,经久不变;古代仿金材料多为铜粉,制作的泥金笺如遇潮湿闷热,颜色会很快被氧化变黑。

"掇英轩"除保留传统泥金笺外,2000年,刘靖用现代合成金黄珠光粉、金色珠光粉分别替代铜粉,制成了2种色相的仿金泥金笺,克服了其易氧化变黑的缺点。同时,刘靖发现珠光粉有系列色彩,就从中选择银白珠光粉制成了泥银笺,又选择古铜、酒红2色珠光粉,制成了古铜泥金笺、酒红泥金笺;金黄、金色、银白、古铜、酒红珠光粉制作的纸笺,又称为金黄、金色、银白、古铜、酒红珠光笺。

调查时,"掇英轩"泥金笺有宣纸泥金笺、皮纸泥金笺、绢本泥金笺三大类。

4. 木版水印笺

木版水印古时被称为饾版术,即根据书画原稿,分色刻版,然后以水调色,通过对刻版的套印、叠印等,再现原作的风貌。"掇英轩"生产的木版水印笺多用作书法作品的

⊙5 木版水印绢本宣(石榴图案) Woodblock printing Silk Xuan paper (with figure of pomegranate)

⊙6 绢本宣手卷(八十七神仙图案) Silk Xuan paper hand scroll (with figure of eighty-seven immortal)

⊙7 五色珠光笺(从左往右为银白、金黄、金色、古铜、酒红珠光笺) Paper with five-color of pearly luster (from left to right: silvery, golden yellow, golden, bronze, claret pearly luster paper)

载体，其图案作为烘托书法作品的装饰性底图，色彩印制淡雅。

调查时，"掇英轩"木版水印笺有诗笺、信笺、条幅、对联及绢本木版水印笺等多个品种。

的原料有所不同。调查时，"掇英轩"制作的流沙笺有单色、复色，色彩丰富，纹饰多样。规格以四尺、四尺对开、四尺四开、四尺三开为主。

⊙1

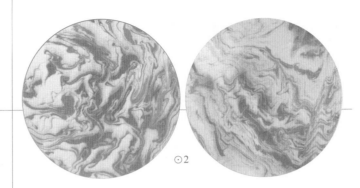

⊙2

5. 流沙笺

流沙笺在唐朝即已出现，因其纸面色彩斑斓，似流动的沙而得名，常在书画、书籍装帧、签条及装饰等中使用。流沙笺制作的原理是利用水与色料的密度不同，使色料浮于水的表面，轻轻搅动水面，令色料在水面自由流淌幻化，然后用纸覆于水面拖曳而成。宋人苏易简《文房四谱》记载："亦有煮皂荚子膏并巴豆油，浮于水面，能点墨或丹青于其上，以姜搵之则散，以狸须拂头垢引之则聚，然后绘之为人物，压之为云霞及鸷鸟翎羽之状，繁缛可爱，以纸布其上而受彩焉。必须虚窗幽室，明盘净水，澄神虚而制之，以增其妙也，近有江表僧于内廷造而进之，御毫一洒，光彩焕发。"[1]

流沙笺又名墨流笺。据刘靖介绍，现在的水拓画就是利用流沙笺原理制成的，主要是使用

⊙3

6. 金银印花笺

明代屠隆的《考槃余事》、高濂的《遵生八笺》、项元汴的《蕉窗九录》等文献中记载了一种"造金银印花笺法"：将云母与通草、苍术、生姜等，经过煮料、布包揉洗、过滤、灰堆干燥等工艺过程制成细腻的银粉，再调入白芨胶制成银色浆料；如果制作金色浆料，还需在银色浆料中加入姜黄汁。然后雕版，将制得的浆料涂于花版

[1] （宋）苏易简. 文房四谱[M]. 朱学博，校点. 上海：上海书店出版社，2015: 57.

上，再以纸覆之，即可制成金银印花笺。

据刘靖介绍，以上述方法制作的金、银粉具有耐氧化、不变黑的特点。现代化工生产的金银色珠光粉的制作原理及主要原料与此方法相同。

明代"造金银印花笺法"失传已久，2000年，"掇英轩"协助中国科学技术大学科技史与科技考古系张秉伦、樊嘉禄课题组，成功复原了造金银印花笺制作技艺。

据刘靖介绍，因造金银印花笺法制作费工、费时，造价较高，市场接受能力差。"掇英轩"在掌握造金银印花笺法后，以现代丝网版替代手工雕版，以现代金、银色珠光粉替代自制金、银粉，运用丝网印刷技术生产出了物美价廉的"金银印花笺"。

"掇英轩"金银印花笺图案多选用传统吉祥图案，仿金、银图案耐氧化、不泛黑。纸性多为生宣效果，亦有少量加工成半熟或全熟的。

"掇英轩"金银印花笺使用的原纸多为宣纸和泾县书画纸，也有少量楮皮原纸制作的产品，用宣纸制作的"掇英轩"称为"精品金银印花笺"。调查时，"掇英轩"生产的金银印花笺有15种颜色，金银印花图案四尺对开的有20种，四尺的有12种，尺八屏的有12种，六尺的约有5种，六尺对开的有5种。

（二）
代表纸品泥金笺技术分析

"掇英轩"加工纸产品相当丰富，而且原纸的相似性较高，调查组经过研讨，选择原纸为泾县产四尺棉料单宣的泥金笺一种进行原纸特性的技术测试。

测试小组对采样自安徽省掇英轩书画用品有限公司的泥金笺原纸所做的性能分析，主要包括厚度、定量、紧度、抗张力、抗张强度、撕裂度、色度、吸水性等。按相应要求，每一指标都需重复测量若干次后求平均值。其中，定量抽取5个样本进行测试，厚度抽取10个样本进行测试，抗张力抽取20个样本进行测试，撕裂度抽取10个样本进行测试，色度抽取10个样本进行测试，吸水性抽取10个样本进行测试。对"掇英

⊙4 各色尺八屏金银印花笺
Various eight-foot gold and silver paper with flower figures

⊙5 精品金银印花笺包装
Package of high quality gold and silver paper with flower figures

"轩"泥金笺进行测试分析所得到的相关性能参数见表6.1。表中列出了各参数的最大值、最小值及测量若干次所得到的平均值或者计算结果。

表6.1 "掇英轩"泥金笺相关性能参数
Table 6.1 Performance parameters of "Duoyingxuan" Nijin paper

指标		单位	最大值	最小值	平均值	结果
定量		g/m²	—	—	—	37.3
厚度		mm	0.084	0.063	0.069	0.069
紧度		g/cm³	—	—	—	0.541
抗张力	纵向	N	12.7	11.1	11.8	11.8
	横向	N	13.5	9.4	10.8	10.8
抗张强度		kN/m	—	—	—	0.753
撕裂度	纵向	mN	80	50	65	65
	横向	mN	110	100	104	104
撕裂指数		mN·m²/g	—	—	—	2.1
色度		%	74.5	72.3	73.0	73.0
吸水性		mm	—	—	—	很难吸水

由表6.1中的数据可知，泥金笺最厚约是最薄的1.33倍，经计算，其相对标准偏差为0.007，纸张厚薄较为一致。所测"掇英轩"泥金笺的平均定量为37.3 g/m²。通过计算可知，"掇英轩"泥金笺紧度为0.541 g/cm³，抗张强度为0.753 kN/m，抗张强度较大。所测泥金笺撕裂指数为2.1 mN·m²/g，撕裂度较大。

所测"掇英轩"泥金笺平均色度为73.0%，色度最大值约是最小值的1.03倍，相对标准偏差为1.025，色度差异相对较小。由于制作泥金笺刷金时添加了黏合材料，加上表面附有珠光闪亮的云母粉，且原纸经过加工已经成为熟纸等因素，泥金笺的吸水性弱，表现出很强的拒水性。

四 "掇英轩"加工纸生产的原料、工艺流程与工具设备

4 Raw Materials, Papermaking Techniques and Tools of "Duoyingxuan" Processed Paper

(一) "掇英轩"手绘描金粉蜡笺

1. "掇英轩"手绘描金粉蜡笺生产的原料

主要包括宣纸原纸、粉、蜡、颜料、动物胶、矾、金箔、银箔、金粉、银粉、水等。

（1）宣纸。宣纸是从泾县定制的均匀度好、表面光洁、无包点杂质、无褶皱的棉料单宣。订制规格尺寸通常有145 cm×74 cm、242 cm×57 cm、188 cm×101 cm、242 cm×101 cm等。

（2）粉。一种矿物粉，调查时，因"掇英轩"保密要求而不方便透露。

（3）蜡。使用的是蜂蜡还是石蜡，因处于保密状态，故"掇英轩"没有具体透露。

（4）颜料。多使用矿物色、植物色。

（5）动物胶。使用的是明胶和骨胶。因处于保密状态，故调查时"掇英轩"选用的明胶、骨胶产地、价格等均未透露。

（6）矾。明矾，多使用合肥市庐江县矾山镇的块状明矾。

（7）金箔。"掇英轩"的真金箔、仿金箔均从南京金线金箔总厂购买。真金箔最常用的是"98金"，规格为9.33 cm×9.33 cm。截至2016年3月入厂调查时，"98金"金箔的市场价格约为10元/张。

（8）金粉。"掇英轩"使用的金粉分真金粉和仿金粉两种。真金粉是由自己买来的真金箔研制的；仿金粉在2000年前多使用铜粉，其后使用的是金黄色珠光粉，珠光粉在合肥购得，2016年入厂调查时价格为100元/kg。

⊙1 胡粉、朱砂、胭脂、石绿、石青 Chinese white, cinnabar, rouge, mineral green, azurite
⊙2 明胶原料 Gelatin material
⊙3 骨胶原料 Bone glue material
⊙4 明矾 Alum
⊙5 购自南京的成品金箔 Gold foil bought from Nanjing
⊙6 加胶调配好的仿金粉 Gold-like powder mixed with glue

(9) 水。调查时生产车间里使用的是自来水。据刘靖介绍,在村子里不通自来水前,多使用雨水,在车间旁建有储蓄雨水的蓄水池;有时遇到雨水少,就选择到周边干净的池塘挑水来打矾沉淀后使用。水质清、pH接近7的中性水才能使用。

2. "掇英轩"手绘描金粉蜡笺生产的工艺流程

据调查组入厂调查时观察及刘靖介绍,归纳"掇英轩"粉蜡笺生产的工艺流程为:

壹	贰	叁	肆	伍	陆	柒
拣纸	备桔	施胶	配料	施粉、晾	挑晾	施蜡、研光

拾叁	拾贰	拾壹	拾	玖	捌
包装	裁切	手绘描金	下挣	上挣	托裱

壹 拣纸 1

将买回的宣纸一刀刀、一张张地挑拣,剔除包点杂质,将不符合加工要求的宣纸挑出。

贰 备桔 2

用糨糊将宣纸的一头一张张粘在木棍上,挂在晾杆上备用。

⊙1 拣纸 Picking the paper
⊙2 备桔 Sticking the paper

叁
施 胶
3

将骨胶和明矾按3∶1的比例和水调配好倒入拖纸盆中,取下晾杆上的宣纸,一张张从胶矾水中拖浸,然后挂在晾杆上晾干。

肆
配 料
4

将制作粉蜡笺的关键颜料与粉等原料调配好待用,配方保密。

⊙3

伍
施 粉
5

将晾干的纸取下,用底纹笔蘸取粉色配料均匀涂布在纸面上。

⊙4

陆　挑晾

6

将刷好粉色颜料的纸挑晾在晾杆上。

⊙5

柒　施蜡、砑光

7

晾干的纸经过施蜡、砑光，纸面会更加紧密、光洁。

施蜡与砑光是"掇英轩"很核心的工艺秘密，没有拍到工序图。

捌　托裱

8

将砑光后的纸正面朝下扣在桌面上，取排笔蘸配制好的糨糊水涂布其上，再取将另一张生宣托裱在上面，然后挑晾。

⊙6

玖　上挣

9

将经托裱晾干的纸背面朝上放在挣板上，取喷壶均匀喷水，使纸湿润后用排笔将纸刷平，然后将纸翻身，正面朝上，四周平整后，再取小毛笔蘸糨糊在纸的四周涂约0.5 cm宽的糨糊，同时相交的板上也要涂上0.5 cm左右宽的糨糊。然后取约1.5 cm宽的纸条沿纸的四周糨糊处粘上，再将纸板拿到一边靠墙阴干。纸张湿时，表面积会增大，干时纸面会收缩，湿时将四周用干纸条固定，利用干湿原理，使纸面绷平。

⊙7

该工序所需完整的时间视气候条件影响而不等，夏季大约4小时，秋冬季至少需10个小时才可将纸绷平，如遇梅雨季节一般避免此道工序。

拾　下挣

10

用竹起沿四周将绷平的粉蜡笺从挣板上取下。

拾壹 手绘描金

11　⊙8～⊙10

一般将设计好的图案墨稿平铺在拷贝台上,然后将制作好的素色粉蜡笺正面朝上平铺在上面,打开拷贝台的灯光,透过灯光,图案的影子便显现在粉蜡笺上;然后取细小紫毫笔,一笔笔蘸取配制好的真金水或仿金水依图案影子描绘。描绘完成后,一张手绘描金粉蜡笺便制作完成了。

⊙8

⊙9

真金水与仿金水的调制均是先用水将金粉浸湿再加胶,胶与水的比例由工人在试笔后根据经验确定,以试笔顺畅、线条干后不掉色为合适。据刘靖介绍,以四尺手绘描金粉蜡笺为例,图案简单的,熟练员工1～2小时就能画一张,图案复杂的画的时间就长,最复杂的图案约1个月才能画完一张。

⊙10

⊙8 刘靖展示真金手绘粉蜡笺
Liu Jing is showing Fenla paper hand-painted with gold

⊙9 手绘描金
Hand-painted with golden powder

⊙10 描金车间的女工们
Female workers in the workshop

拾贰 裁切
12

根据客户要求的尺寸手工裁切。常规尺寸有136 cm×68 cm、136 cm×34 cm、175 cm×94 cm、175 cm×45 cm、230 cm×52 cm、44 cm×44 cm、66 cm×66 cm等。"掇英轩"目前做的最大粉蜡笺是234 cm×97 cm。

拾叁 包装
13

在每张粉蜡笺的正面左下角以蜡研上葫芦形"掇英轩"标志暗记，并在反面左下角盖上相应手绘图案的货号，如果是真金手绘的则在粉蜡笺反面左上角盖上篆书"真金手绘掇英轩制"章。章盖好后，将每张（或对）手绘描金粉蜡笺附上使用说明一起卷成筒状，装入盒中，并在盒中放入产品简介。真金手绘的还放入一双白手套，合上盖子，才包装完成。调查时，"掇英轩"真金手绘粉蜡笺的包装盒为特制的楠木盒，仿金的为桐木盒包装。

⊙11

 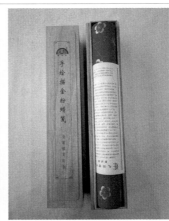

⊙12　⊙13

3. "掇英轩"粉蜡笺的使用

调查中了解到"掇英轩"粉蜡笺的使用有讲究。

（1）粉蜡笺使用前，应将手洗净。手上若有油脂、香脂，接触了粉蜡笺，易造成纸难以上墨，墨迹会花，出现拒墨现象。

（2）宿墨（隔夜的墨），不能在粉蜡笺上书写，因为宿墨中的胶轻，粉蜡笺比较光滑，胶轻的墨留不住，易花。

（3）粉蜡笺是熟宣，适合用浓墨书写，用研磨的油烟墨书写效果最好。

（4）粉蜡笺书写时笔迹处不宜积墨太多，否则，墨迹干后积墨多的地方墨容易龟裂。

（5）粉蜡笺纸面上若沾了油污，为解决油污拒墨问题，"掇英轩"试验在研好的或瓶装墨汁中加入几滴洗洁精、肥皂水之类，搅匀静置约30分钟后再书写，效果更好。

⊙14

4. "掇英轩"手绘描金粉蜡笺生产的工具设备

壹 毛笔 1

"掇英轩"手绘描金常用细小的紫毫（兔尖）笔勾线，用羊毫笔晕染、填画等。

⊙15

贰 底纹笔 2

刷粉色使用。"掇英轩"所用底纹笔是从江苏江都做毛笔的厂家定制的，宽33 cm，约100元/把。

⊙16

叁 竹起 3

刀形，用竹子制作，长约40 cm，宽约1.5 cm，为"掇英轩"自制工具。主要用于将裱贴在木板上的纸挑开，便于揭下。

⊙17

肆 棕把 4

用棕编织而成，从泾县购买。用于将干纸平整地贴在刷满糨糊的湿纸上。

⊙18

伍 排笔 5

用于粉蜡笺的托裱，由一支支管状羊毛笔并列穿制而成，"掇英轩"多用26管羊毛排笔。

⊙19

陆 拖纸盆 6

塑料制，盛稀释后的矾、胶水器物，四边光滑，便于拖染纸。现场实测尺寸为长110 cm，宽70 cm，高15 cm。在一长边镶一根光洁铝合金方条。据刘靖介绍，以前还用过铁皮盆。

⊙20

柒 挣板 7

用于将纸挣平，内用木条做框，两面蒙上平整洁净的烤漆板。"掇英轩"常用挣板尺寸有170 cm×85 cm×3.5 cm、244 cm×60 cm×3.5 cm、244 cm×122 cm×3.5 cm三种规格。

⊙21

⊙15 手绘描金常用的毛笔（「掇英轩」供图）Brushes used for painting with golden powder (offered by "Duoyingxuan")
⊙16 底纹笔 Brush
⊙17 竹起 Bamboo tool for peeling the paper
⊙18 棕把 Brush made of palm
⊙19 多管羊毛排笔 Wool brush with a row of pipes
⊙20 拖纸盆 Papermaking basin
⊙21 码放着的挣板 Boards for smoothing the paper

(二)"掇英轩"绢本宣

1. "掇英轩"绢本宣生产的原料

主要包括绢、宣纸、糨糊、水。

（1）绢。蚕丝制，从浙江湖州双林镇定制。"掇英轩"常用的绢有7种颜色：白、浅灰、淡黄、仿古、绿仿古、佛黄、紫色。

（2）宣纸。多使用棉料单宣和棉料棉连，从安徽泾县定制。

（3）糨糊。托裱绢、纸所用，较稠。以小麦粉为原料，洗去面筋，打制成糨糊再添加保密配方中的物质后，制成"掇英轩"绢本宣所用的特制糨糊。

（4）水。厂区接入的公用自来水。

⊙1

⊙2

⊙3

2. "掇英轩"绢本宣生产的工艺流程

据调查中刘靖介绍，以及调查组现场的观察，归纳"掇英轩"绢本宣生产的工艺流程为：

壹	贰	叁	肆	伍	陆	柒	捌	玖
绷绢	上浆	托裱	挂晾	上挣	下挣	检验	裁切	包装

⊙1 码放着的各色绢
⊙2 从泾县定制的宣纸原纸
⊙3 托裱绢用的糨糊

壹 绷绢

1

将绢放在光洁的台面上打湿,用手将其绷平绷直。绷绢时要注意将绢的纹理绷平绷直。

贰 上浆

2

用棕把将糨糊均匀刷涂于绷平的绢上。

叁 托裱

3

两人配合,一主一次,两人双手各牵宣纸一头,此时在主方双手牵宣纸的同时,将棕把也同时拿在右手,然后将宣纸正面朝下与上浆平铺的绢悬空对齐,主方将所牵的宣纸一头落在绢上,旋即用右手的棕刷将宣纸的一头压平压实在绢上,然后再转换位置从右往左以棕刷将宣纸托裱于绢上。与此同时,次方要随棕刷的排压,缓慢地将宣纸往下落,使其与绢黏合,直至主方将宣纸完全压实在绢上。宣纸与绢覆合后,再用棕刷从左往右、从右往左各压一遍,将宣纸压实、压平在绢上,注意不能将纸与绢压皱或出现没压实的情况。

肆 挂晾

4

将已托裱好的潮湿的绢本宣一头夹在木棍上,双手抓住木棍将绢本宣揭离,挂在晾杆上晾干。

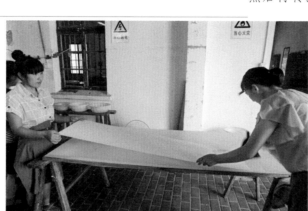

伍 上挣

5

将晾干的绢本宣从木棍上取下,此时表面不平整,将其正面朝下平铺在挣板上,反面喷水后用羊毛排笔将绢本宣刷平,再将其翻身,正面朝上,用糨糊、纸条将四周平实地粘在挣板上。然后将挣板搬到墙边阴凉处自然晾干,使绢本宣绷平在挣板上。

陆 下挣

6

用竹起将绢本宣从挣板上揭离。

柒 检验

7

将揭下挣平的绢本宣一张张检验,剔除次品。

捌 裁切

8

将正品绢本宣一张张理齐,按照客户需求裁切成相应尺寸。

④ 在挣板上铺绢 Spreading silken cloth on the board for smoothing the paper

⑤ 托裱 Mounting the paper

玖 包 装

绢本宣多卷起包装，按规格大小有10张一盒、5张一盒、2张一盒不等，装入自制的盒中。

如客户需要熟的绢本宣，可在做好的绢本宣的表面再刷上熟纸胶矾水配料，再经晾晒、上挣、下挣等制得。

3."掇英轩"绢本宣生产的工具设备

与手绘描金粉蜡笺制作中的工具相同，不再一一赘述。

⊙6

（三）"掇英轩"皮纸仿金泥金笺

1."掇英轩"皮纸仿金泥金笺生产的原料

主要包括皮纸、仿金粉、胶、水。

（1）皮纸（或绢本宣）。选均匀、光洁、无杂质、无褶皱的楮皮纸，从泾县定购。

（2）仿金粉。2000年前使用的仿金粉多为铜粉，之后多使用金黄珠光粉或金色珠光粉制作泥金笺。

（3）胶。胶的使用在泥金笺制作中至关重要，属于"掇英轩"机密而未透露。

（4）水。厂区接入的公用自来水。

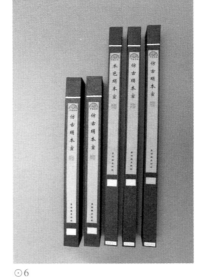

⊙7

2."掇英轩"皮纸仿金泥金笺生产的工艺流程

据调查与访谈信息，以"掇英轩"四尺仿金泥金笺为例，归纳"掇英轩"皮纸仿金泥金笺生产的工艺流程为：

壹 备料楮 → 贰 施胶 → 叁 配料 → 肆 涂布 → 伍 挑晾 → 陆 再次涂布 → 柒 再次挑晾 → 捌 上挣 → 玖 下挣 → 拾 第三次挑晾 → 拾壹 检验 → 拾贰 裁切 → 拾叁 包装

⊙6 绢本宣包装
⊙7 仿金材料金黄珠光粉

壹 备楮
1
将皮纸一张张粘在木棍上,备用。

贰 施胶
2 ⊙8
将骨胶和明矾按3∶1的比例和水调配好倒入拖纸盆中,取下晾杆上的绢本宣或备好楮的皮纸,一张张从胶矾水中拖浸,然后提起挂在晾杆上晾干。

⊙8

叁 配料
3
将金黄珠光粉(或金色珠光粉)加水、胶等,按一定比例调配好待用。配料是制作泥金笺成败的关键。

肆 涂布
4 ⊙9
将制得的熟皮纸从木棍上揭下,平铺在涂布台上,取底纹笔蘸泥金配料均匀涂布于皮纸上。

⊙9

伍 挑晾
5 ⊙10
先将一根木棍夹在均匀涂布泥金粉的皮纸一头,然后左手慢慢平提皮纸,右手握另一木棍从刚掀起的皮纸底部中间位置将皮纸挑离台面,走到晾杆前;随即左手将夹着皮纸的木棍搭在晾杆上,右手将木棍慢慢撤离纸背,让纸自然悬挂晾干。

⊙10

⊙8 拖染 Dying the paper
⊙9 涂布 Applying color on the paper
⊙10 挑晾 Drying the paper

陆　再次涂布
6

为增强泥金纸的亮度和色彩饱和度等，将晾干的泥金纸再次涂布泥金配料。

柒　再次挑晾
7 ⊙11

将二次涂布的泥金纸再次挑起晾干。

⊙11

捌　上挣
8 ⊙12

将晾干的泥金纸正面朝下放在挣板上，喷水刷平，然后将泥金纸翻面，在泥金纸四周贴上纸条，再将挣板靠墙使本来不平的泥金纸自然阴干绷平。

⊙12

玖　下挣
9

泥金纸绷平后，取竹起先沿一角将一短边挑起，然后再顺边将其余三边揭起，一张平整的泥金纸便会脱离挣板。

拾　第三次挑晾
10

晾干时长视气候的情况不等。温度高、空气干燥时需3～4小时，温度低、湿度大时需10多个小时；如湿度太大，数天都不会干，比如逢上梅雨季节会造成纸张上霉。如温度高，空气过于干燥，有的纸会在挣板上绷裂；下挣时竹起挑揭纸面时也极易造成炸裂。保持车间的湿度、温度是该道工序的关键。

拾壹　检验
11

将制得的泥金笺一张张检验，去除破损、通洞、皱、包点多或泥金刷得不匀等次品。

拾贰　裁切
12

将正品泥金笺一张张理齐，根据客户需要将泥金笺裁切成相应尺寸。35 cm×138 cm、70 cm×138 cm、94 cm×175 cm、53 cm×234 cm是"掇英轩"泥金笺的常规尺寸。

拾叁　包装
13 ⊙13

35 cm×138 cm、70 cm×138 cm泥金笺多为每10张1卷，94 cm×175 cm、53 cm×234 cm泥金笺多为每5张一卷，将其卷起，装入特定泥金笺盒中。

⊙13

3."掇英轩"皮纸仿金泥金笺生产的工具设备

与手绘描金粉蜡笺制作中的工具相同，不再一一赘述。

（四）"掇英轩"套色木版水印信笺

1. "掇英轩"套色木版水印信笺生产的原料

主要包括宣纸、梨木板、国画色。

（1）宣纸。从泾县定制或协议购入。"掇英轩"木版水印信笺使用的多为棉料单宣，也有根据客户需求使用净皮、特种净皮及特皮扎花宣等。

印制时，多将宣纸裁成按所需信笺尺寸的2倍大小且上下左右各放大0.5 cm的头对头的双联以便使用，印制时一般每100张为一沓夹起。

（2）梨木板。2017年8月访谈时从刘靖处获得的信息为："掇英轩"通常从淮北、宿州等地购得梨木，2015～2016年大约5 000 元/m³。将梨木剖开，经过泡、锯、刨等工序后制作成一面更为光洁的板备用。选择梨木是因为梨木材质比较细密，相对收缩率小，不容易开裂。

（3）国画色。木版水印的印色"掇英轩"多选用市场上销售的中高档的国画色，以水调色配制所需印色。

2. "掇英轩"套色木版水印信笺生产的工艺流程

据调查时刘靖介绍，以及调查组成员现场的观察，归纳"掇英轩"套色木板水印信笺生产的工艺流程为：

壹 设计信笺 → 贰 绘稿 → 叁 分色雕版 → 肆 配色版 → 伍 校版 → 陆 镶版水印 → 柒 裁切 → 捌 包装

⊙14
"马利"牌国画色
Marie's pigment for Chinese painting

壹 设计信笺 1

"掇英轩"根据不同的客户和市场的需求，或客户提供图案，或仿制、复制书画作品、历代笺谱，或自行设计图案等，设计好信笺的图案及其在信笺上的位置，以及信笺的规格、用纸等。

贰 绘稿 2

根据设计好的信笺图案分色，按分色的图案一一绘制出墨稿。

叁 分色雕版 3　⊙1～⊙3

图案有多少套色，就雕多少块分版。取梨木板，选择或锯成与墨稿大小匹配的一块块分版，将分版光洁一面再次打磨，磨光后将墨稿翻印到分版上，然后用木刻刀刻除墨线之外的部分，刻的深度一般在4 mm左右，不宜过深，也不宜过浅；待一块分版刻好后，再刻第二块分版，直至将分版雕完；之后再将一块块分版分别粘在长方形薄板上，制成印版备用。

雕好的一块梨木板可印2万次左右，之后需用木刻刀修版，不能再修的破损版就需重新刻版。

肆 配色 4

按设计的色稿配色，用国画色和水一一调配至所需要的水色。

伍 校版 5

将一块块印版排好顺序，以水色套印，试印一次，确认雕刻是否到位、套色是否准确，若有不足，再修版或重刻；在试印时，印色如有不足，则需重新调色。各分版与水色试印、调整，直至达到设计要求为止，此为校版。

陆 饾版水印 6　⊙4～⊙6

将裁好的双联宣纸正面朝下用木条与铁夹固定在木版水印架右侧，然后将水印架放在木版水印桌上。取第一块印版放在水印架左侧，先校对好印版图案在纸上的位置，固定印版，然后左手取毛刷蘸水色涂刷于版上，右手牵起一张纸递于左手，随即左手将纸牵平，覆在版上，右手在相应纸背平抚，将水色印入纸中，随后将纸揭起自然垂落在水印桌的孔洞中，再开始第二张、第三张、第四张……的循环印制。待第一套色完成后，将印版取下，取第二块印版，更换相应水色，然后校准位置与颜色，重复上述工艺。待第二套色完成后再换第三块印版，换色，再重复上述工艺。依此反复直至将分色印版套印结束，一张套色木版水印信笺便套印完成了。这种以水调色分版套印的方法称饾版水印。

木版水印对空气的湿度有一定的要求，在制作木版水印笺时要保持室内空气较为湿润，这样才能使印出的图案有书画的润、晕、透的水墨效果。否则，在空气干燥的环境下印木版水印，印出的图案往往会出现干、涩以及断线等现象。在印木版水印笺时为保持室内湿度，"掇英轩"购买了加湿器，给室内空气加湿。

柒 裁切

待两头都已印好，将双联纸从水印架上取下，自然阴干后，取裁纸刀将双联从中间裁断，然后一张张检验所印信笺，去除次品信笺，再根据市场需求一袋或一盒数量配套完成后，再取裁纸刀将信笺裁至所需尺寸。"掇英轩"套色木版水印信笺的主要尺寸有18 cm×28 cm、24 cm×33 cm、21 cm×29.7 cm、14 cm×24 cm、19.6 cm×26.9 cm等。

⊙4

⊙5

⊙6

捌 包装

裁切好后，按一袋或一盒在一叠信笺中上部位封一环状封条后装入相应的袋或盒中。"掇英轩"套色木版水印信笺多为30张一袋，也有20张一袋，有塑料袋包装和纸袋包装，另一种木版水印信笺锦盒装的多为50张或100张装入一盒。

⊙7

⊙8

⊙9

⊙ 4 套印：对版 Overprint: register
⊙ 5 套印：上色 Overprint: coloring
⊙ 6 套印：覆印 Overprint: overlay print
⊙ 7 裁切 Cutting the paper
⊙ 8/9 锦盒包装 Package of brocade box

3. "掇英轩"套色木版水印信笺生产的工具设备

主要包括木刻刀、加湿器、毛刷、木版水印桌、木版水印架等。

壹 木刻刀 1

雕刻木版使用,市场购买。

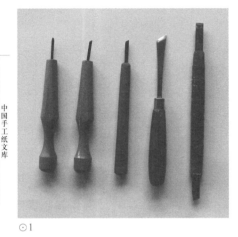

⊙1

贰 加湿器 2

木版水印笺的制作对环境湿度有要求,干燥的环境会影响印制效果,使颜色发干发涩。2016年和2017年入厂调查时,"掇英轩"均用普通市售加湿器对制作车间进行加湿。

叁 毛刷 3

"掇英轩"原使用棕老虎刷色,后发现一种刷鞋的毛刷正反面及侧面都有鬃毛,宽窄大小不同,可满足不同面积的雕版刷色使用,且鬃毛含色量适中,弹性好,手柄也使用方便,已使用多年。

⊙2

肆 木版水印桌 4

"掇英轩"木版水印桌是刘锡宏自己设计后请木匠打制的,桌面有孔(28 cm×26 cm)。调查时实测木版水印桌尺寸为长85 cm,宽50 cm,高72 cm。

伍 木版水印架 5

由刘锡宏自己设计并制作,木制,与1根木夹条和2个铁夹配合,用于固定所印双联纸张,与木版水印桌配套使用,简单方便而且实用。

⊙3

⊙4

（五）"掇英轩"流沙笺

1. "掇英轩"流沙笺生产的原料

主要包括原纸、糨糊、颜料、水。

（1）原纸。"掇英轩"制作流沙笺时，使用的原纸多为吸水性良好的宣纸、皮纸，也有用书画纸（龙须草浆、木浆原料）制作的流沙笺。

（2）糨糊。制作流沙笺的关键配料是糨糊。糨糊是"掇英轩"用小粉浆（小麦粉通过揉面、水洗等去除面筋制得）自己制作的。据刘靖介绍，在制作糨糊时不能加入明矾。糨糊与水的配比浓度不同，可制作出流动感各异的流沙笺。

另据刘靖介绍，除糨糊外，还有多种黏性材料都可以制作流沙笺，不同的黏性材料与水混合，再辅以理梳等工具，能制作出不同肌理效果的流沙笺。

（3）颜料。多选用"马利"牌膏状国画色，以水调色制成所需浓度的颜色备用。

（4）水。厂区接入的公用自来水。

⊙5

⊙6

2. "掇英轩"流沙笺生产的工艺流程

苏易简《文房四谱》记载："亦有作败面糊，和以五色，以纸曳过令沾濡，流离可爱，谓之流沙笺。"[1]

"掇英轩"流沙笺生产的工艺流程为：

壹	贰	叁	肆	伍	陆	柒	捌	玖
备糨	配料	滴色	和色	覆纸	晾晒	下糨	裁切	包装

⊙5 揉面（"掇英轩"供图）
Kneading dough (offered by "Duoyingxuan")

⊙6 变幻多彩的流沙笺系列（"掇英轩"供图）
Series of Liusha paper with different figures (offered by "Duoyingxuan")

[1]（宋）苏易简.文房四谱[M].朱学博，校点.上海：上海书店出版社，2015：57.

工艺流程

壹 备梠 1

将制作流沙笺的纸的一头用糯糊粘在木棍上,然后挂到晒杆上将粘接处晾干备用。备梠时,要注意纸与木棍一定要粘得平整服帖。

贰 配料 2 ⊙1

先在小盆里将制作好的糯糊加水调稀,然后取制作流沙笺的长方形盆盛上约三分之二的水,再将调稀的糯糊一勺勺舀入水中搅匀。糯糊与水的配比通常依技术工人的经验而定。可先制小样试配比,在调好的配料中滴入调配好的颜色,若颜色不能浮于水面、下沉或迅速散开,说明糯糊的浓度不够,就需添加糯糊。

叁 滴色 3 ⊙2

将国画色用水调至合适浓度,取毛笔蘸上颜色,然后将笔尖轻触调配好糯糊的水面,有时也可用毛笔洒溅颜色,此时颜色会漂浮在水面上缓慢扩散。流沙笺制作根据需要或单色、或双色、或多色滴入。每做一张流沙笺要重新滴一次颜料。

肆 和色 4

用干净的毛笔或小木棒轻轻搅动水面,此时浮于水面的颜色会自由流淌幻化。如糯糊浓度大,达到一定比例时,浮在水面的颜色基本没有自由扩散,此时可用毛笔或小木棒牵引、控制颜色的走向及融合等。利用此原理可创作美术作品。

⊙1

⊙2 ⊙3 ⊙4

伍 覆纸 5 ⊙3 ⊙4

流淌幻化的颜色呈现出令人较为满意的图形与色彩时,拿起一张备好梠的纸,平提木棍,将纸从盆的一头开始慢慢地平覆在水面上,水色将覆在水面的纸完全浸染至离木棍几厘米处停止平覆。

陆 晾晒 6 ⊙5

双手平提木棍将纸从盆边横担的铝合金条沿口刮过,提起,晾晒。据刘靖介绍,有时流沙笺制作完成后,悬挂一会,流沙图纹颜色渐渐地就消失不见了,这是糯糊少的缘故。

⊙5

⊙1 调制糯糊 Modulating paste
⊙2 滴色 Dropping color
⊙3/4 覆纸试验 Dyeing paper experiment
⊙5 提纸试验 Lifting paper experiment

柒　下梪
7

从晾杆上将晾干的流沙笺取下，并将流沙笺从木棍上剥离。

捌　裁切
8

将制作好的流沙笺一张张检验并理齐后，按所需规格裁切。

玖　包装
9

包装形式有多种，有卷起10张一盒、5张一盒包装，一般盒装的是由宣纸、皮纸原纸制作的；有100张一刀折叠包装，也有10张一袋包装，多为书画纸原纸制作。

3. "掇英轩"流沙笺生产的工具设备

主要包括盆、毛笔、木棍。

壹　盆
1

小圆盆和长方形盆，均为塑料盆。小圆盆为调制糨糊时使用。在长方形盆一条短边的内沿镶一根平直的铝合金方条，用于刮除多余颜色与水。

贰　毛笔
2

用于滴或洒溅颜色和搅动、牵引颜色。

叁　木棍
3

木制。"掇英轩"备梪木棍有多种规格，尺寸为长90 cm，宽2 cm，高2 cm的最常用，用于制作四尺规格流沙笺备梪。

⊙6

⊙7

⊙6 塑料盆 Plastic basin
⊙7 备梪使用的木棍 Sticks for papermaking

（六）"掇英轩"金银印花笺

1. "掇英轩"金银印花笺生产的原料

主要包括书画纸、宣纸、丝网、颜料、感光胶、透明纸、水。

（1）书画纸、宣纸。"掇英轩"制作的金银印花笺分为两种：一种由书画纸制作，称为金银印花笺；另一种由宣纸制作，称为精品金银印花笺。书画纸、宣纸均购于安徽泾县，在用书画纸制作金银印花笺时通常在浆料中加入少量檀皮，以增强纸张的拉力及润墨效果。宣纸伸缩率大，成本高，印制难度大，次品率高。调查中刘靖表示：用宣纸印制时，对纸张要求高，必须平整，不平的要压平后使用。截至2016年第一次调查时，"掇英轩"制作精品金银印花笺的宣纸是从泾县定制的平板纸，因其平整，有利于印制。

（2）丝网。在合肥、泾县印刷品店购买。据刘靖介绍，不同的丝印浆料，使用的丝网的目数不同。如印金粉时，丝网的目数一般为100或120；印颜色时，丝网的目数一般为250或300。数字越大，网孔越密，数字越小，网孔越大。丝网的目数有时也根据客户对产品印制的精细程度来选择。

（3）颜料。使用的是金黄珠光粉、银白珠光粉。

（4）感光胶。在合肥购买北京太平桥产水性感光胶，价格一般为35元一瓶。

⊙1

（5）透明纸。画墨稿时使用，普通市购硫酸纸即可。

（6）水。厂区接入的公用自来水。

2. "掇英轩"金银印花笺生产的工艺流程

据2017年8月刘靖介绍，以及调查组现场的观察，归纳"掇英轩"金银印花笺生产的工艺流程为：

壹 绘稿 → 贰 绷框 → 叁 制版 → 肆 修版 → 伍 裁纸 → 陆 配浆料 → 柒 印刷 → 捌 挑晾 → 玖 检验 → 拾 裁切 → 拾壹 包装

工艺流程

壹 绘稿 1

先构思图稿，然后取大白纸，用铅笔在其上绘制草图，再用硫酸纸覆盖其上，用紫毫笔蘸墨汁勾墨稿。图稿多由刘靖绘制。

⊙2

贰 绷框 2

根据墨稿的大小用木头做相应大小的木框，然后用钉枪将丝网一边固定在木框上，再用绷网夹将一边的丝网绷平绷紧，随后用钉枪将丝网钉牢，将丝网平整地绷在木框上。

⊙3

叁 制版 3

"掇英轩"自己制版。在制版间里，把感光胶均匀地刮满已完成绷框的丝网上，刷胶的过程要在红外线环境下完成，不能见紫外线。感光胶刮好后放在漆黑的环境下视天气来阴干或吹干。在红外线环境下，将墨稿放在里面排满了紫外线灯管的大玻璃晒版台上，正面朝上，然后把感光胶已干的丝网版盖在墨稿上，上面用装满细沙的袋子压实压平。然后打开紫外灯，大约3分钟关灯。取下丝网版，放入水中浸泡。感光胶经紫外线照射后，没有图案的部分透光，感光胶老化、变硬，有图案的部分不透光，感光胶没有老化，放入水中浸泡，没有感光的部分会变软、泡起，水中抖动网版，没有感光的胶会脱落。从水中提起丝网版，用气泵带动的水枪，

⊙4

利用压力水流细心地将感光胶还没有脱落的部分一点点冲掉，感光胶脱落的部分显现丝网空隙，空、亮，直至整个图案清晰地镂空在版子上，制版成功。

肆 修版 4

冲版时往往会将一些本不需冲掉的地方冲掉，待版子干后，用毛笔蘸感光胶手工去描补、修版，填补不该冲掉的部分。

伍 裁纸 5

按照需要的尺寸将书画纸裁剪好待印。

陆 配浆料 6

用珠光粉、增稠剂等调配金或银印花浆料待用。

柒 印刷 7

将丝网版一头用丝网夹固定在台面上,把裁好的纸放在丝网版下面,将丝网版上的图案调整到纸面的设计位置。然后用勺将配好的浆按需舀到丝网框的一头,再用刮刀将浆料从丝网版的一头刮向另一头。

⊙5

捌 挑晾 8

掀起丝网版,抽出已印好的金银印花笺,由另一名工人用木棍将纸挑起挂在晾杆上晾干。挑晾时注意不要将前后挂的两张纸粘在一起,由于不干,粘在一起会形成两张纸图案重叠、糊、脏等现象,造成次品。

玖 检验 9 ⊙6⊙7

纸晾干后,从晾杆的木棍上取下,收纸,然后一张张检查印制效果,将印得不清晰、重影、断线及有破损、通洞、杂质多、脏等的纸去除。

⊙6

拾 裁切 10 ⊙8

将合格品理齐裁切修边。常规尺寸为四尺、四尺对开、六尺和尺八屏。

拾壹 包装 11 ⊙9

金银印花笺有盒装、袋装和整刀折叠包装3种。盒装、袋装多为10张一盒(袋),整刀为100张一盒。

⊙7

⊙8　⊙9

- ⊙5 丝网印刷 Silk screen printing
- ⊙6 收纸 Collecting the paper
- ⊙7 检验合格的五色泥金笺 Checking the paper
- ⊙8 裁切 Cutting the paper
- ⊙9 袋装金银印花笺 Gold and silver paper with flower figures in bags

3. "掇英轩"金银印花笺生产的工具设备

主要包括刮刀、晒版台、木框、气泵。

壹 刮刀 1

由铝合金手柄与PU材料制成的刮条组成，用于刮浆料。"掇英轩"的刮刀为自制，买来材料，自己组合，尺寸依丝网版的规格大小不等。刮刀制作要做到刮条安装平整，无豁口。

⊙10

贰 晒版台 2

"掇英轩"自制。由晒版台面与台面支撑上下两部分组成。晒版台面是在木质U形框内安装一排紫外线灯管，上盖1 cm厚玻璃制成的。晒版台面越大，说明制的丝网版就越大。调查时实测"掇英轩"黄麓镇车间晒版台面长300 cm，宽120 cm，高28 cm。台面支撑由两个储物柜组成。储物柜长45 cm，宽120 cm，高50 cm，在台面下方一头一个，内储放制版用的硫酸纸墨稿。

叁 木框 3

实木制，用于绷平固定丝网。"掇英轩"依印制的金银印花笺规格的不同，木框有多种规格。调查时实测图中木框尺寸为长190 cm，宽52 cm，高4 cm，为制作四尺对开丝网版使用。

肆 气泵 4

冲丝网版时使用。利用气泵压缩空气，经喷枪前部的空气帽喷射出来时，就在与之相连的喷嘴前部产生一个低压区，此时这个压力差就会把与喷嘴相连的储罐中的水吸出来，并在压缩空气高速喷射力的作用下，将水雾化喷在丝网版上，利用水的冲力，将丝网版上没有感光、没有脱落的感光胶冲掉，显露丝网孔隙。

⊙11

⊙ 刮刀 10 Scraper
⊙ 气泵 11 Air pump

五 安徽省掇英轩书画用品有限公司的市场经营状况

5 Marketing Status of Duoyingxuan Calligraphy and Painting Supplies Co., Ltd. in Anhui Province

2017年8月驻厂调查中了解的市场销售现状是："掇英轩"目前尚未组建专门的销售队伍，主要依靠每年两次的全国文房四宝订货会及电话、传真、淘宝、微信等订货方式。

从销售渠道看，"掇英轩"主要通过各地经销商经销与自营两种方式销售。北京"荣宝斋"自1999年与"掇英轩"建立经销关系起，就一直是"掇英轩"最大的经销商。通过"荣宝斋"和全国文房四宝博览会（在文房四宝业内均称"订货会"）的平台，"掇英轩"的纸笺被销往北京、天津、上海、深圳、太原、长沙等大城市。

自营则有线上、线下两个渠道。线下的自营店建在合肥市的城隍庙古玩城外街，由刘靖的夫人方玉红负责打理。线上的网店从2014年开始经营。刘靖的说法是：考虑到对多年合作的经销商利益的保护，并没有积极开拓线上渠道，线上渠道的销售量近两年来至多占总销量的2%。

从销售对象看，分为内销与外销。"掇英轩"成立初期，90%的产品销往国（境）外，主要是日本。1997年刘靖从父母处接手管理"掇英轩"以后，国内订单逐渐增多。2017年8月调查时的数据显示，国内销售占了总销售量的85%左右，外销的主导产品包括手绘描金粉蜡笺、泥金笺、金银印花笺、绢本宣、木版水印信笺等，除了销往日本，美国、德国、新加坡也陆续有了"掇英轩"的客户。

2017年8月访谈时刘靖补充了新的信息：近几年日本客户数量在萎缩，应该与日本市场对这类价位较高的纸需求下降及纸的替代有关，也与"掇英轩"纸定价的上升较快有关，对日本经销商和消费者此前很明显的性价比优势不再保持了，当然这是中国手工造纸成本与售价双向快速上涨的结果，并非"掇英轩"一家出

⊙1
合肥城隍庙掇英轩文房用品商店
Duoyingxuan paper stationary store in Town God's Temple of Hefei

现这种情况。现在每年卖到日本的纸少时约10万元,多时20万~30万元。香港主要是"文联庄""笔墨庄"两家老客户,我父亲与现店主父亲早就认识,香港老牌店,这些客户是靠产品讲话的。台湾地区没有客户,只有个别人买过。我们的产品销往美国、德国少,不是常态,一年就一两次订单。新加坡过去有常态的,这两年断了。

从产品类型看,分为自有产品销售和定制产品销售两类。定制类产品来自部分书画家或企业。定制类产品以信笺居多,金银印花笺、手绘描金粉蜡笺也占一定比例。定制量占全年产量的10%左右。

2017年8月调查时,"掇英轩"正在巢湖市黄麓镇张疃路与规划路(竺城路)西北角新建占地约9 300 m²的"掇英轩"纸笺加工技艺传承基地。交流中刘靖表示:传承基地的建设、销售团队的组建,以及同时打造线上线下的销售渠道是"掇英轩"下一阶段发展规划的重点目标。

2 北京安徽"四宝堂"对"掇英轩"纸的评价材料
Evaluation of "Duoyingxuan" paper by Beijing Anhui "Sibaotang"

3 手绘中的粉蜡笺
Hand-painted Fenla paper

4 规划效果图
Planning effect picture

5 在建中的"掇英轩"纸笺加工基地新厂区
"Duoyingxuan's" new papermaking factory in building

⊙1 ⊙2

至调查组2015年调查时止,"掇英轩"的产品根据种类不同、制作工艺不同,销售价格从几角、几元到上万元一张不等。以四尺纸笺为例,金银印花笺(书画纸)市场价为5~6元;洒金的粉蜡笺,市场价为220~260元;仿金手绘粉蜡笺根据图案的不同,市场价格从500元至2000元不等;真金手绘粉蜡笺一般在700元至上万元不等。手绘粉蜡笺价格的差异与图案的繁简、使用的金银材料及画工水平高低有关。据刘靖介绍,"掇英轩"2014~2015年年销售额为300多万元,利润率为15%~20%。

六 安徽省掇英轩书画用品有限公司的品牌文化与民俗故事

Brand Culture and Stories of Duoyingxuan Calligraphy and Painting Supplies Co., Ltd. in Anhui Province

⊙ 1 / 2
真金手绘粉蜡笺高端纸品
High quality Fenla paper hand-painted with gold

1. "掇英轩"的得名由来

"掇英轩"现在已经是中国手工纸笺加工技艺的经典品牌,当年为什么取了这个名字呢?2016年访谈时刘靖表示:取名"掇英轩",是父母受上海人民美术出版社出版的《艺苑掇英》的启发,20世纪80年代至90年代前期,《艺苑掇英》是深受我国美术与工艺爱好者喜爱的杂志。"掇"是拾起、捡起的意思,"英"为精华,合在一起即拾取传统文化艺术精华的意思。刘锡宏家中有不少《艺苑掇英》杂志,恰巧杨宁英名字中也有一个英字,"母亲看到上海的杂志名受启发,认为'掇英'两字很好,与他们要将失传的传统纸笺制作技艺捡起来的想法很吻合。父亲本为工艺美术的从业者,自然也高度赞同,是父母共同取名的"。

"掇英轩"的品牌由此诞生,现在"掇英轩"是企业品牌,也是注册商标。

刘靖在随后的闲聊中又补充了一则小趣闻:没想到"掇英"虽然意思很好,但"掇"这个字不好认,经常会遇见读错音的客户,也不好意思当面纠正人家。

当展示范曾为"掇英轩"写的题匾时,刘靖补充说道:"我没有和范曾交往过,题匾是当年'荣宝斋'的米景扬帮忙要的。当年范曾和米景扬住隔壁,他俩也是师兄弟,好像两个人在学徒时还在故宫一起临摹过画。"

画纸厂的毛边纸,他以为是四川夹江人在卖毛边纸,就坐下来聊天。一聊知道我们是巢湖人,之后知道我们店里还有一些加工纸产品,看起来质量好像还不错,就问是谁做的,我妈说是我们自己在巢湖老家纸坊做的。樊嘉禄很有兴趣,之后谈到合作,这样就认识和交流起来了。

随后,就有缘与张秉伦先生相识并交往,张秉伦与樊嘉禄也来纸坊考察,结果发现我们家与张秉伦先生家居然有亲戚关系。再往后,由张、樊两位提议并提供相关资料和技术指导,2000年,"掇英轩"协助中国科学技术大学科技史与

⊙ 3

2. 与中国科学技术大学的因缘际会

本调查组自然对"掇英轩"与学校的合作历史很感兴趣,当问起这一问题时,刘靖的说法是:虽然自己在中国科学技术大学附属中学读过书,但与后来的合作并无联系。起源是先认识的手工造纸技艺研究者樊嘉禄教授,他当时正在跟中国科学技术大学科技史与科技考古系的张秉伦教授读研究生,研究纸张与印刷工艺史,后来才认识张秉伦教授。结识樊嘉禄是偶然巧合,1999年的某一天,他到合肥老城隍庙买东西,那天方玉红和我妈(杨宁英)在掇英轩文房用品店里,樊嘉禄看到店门口挂着的纸品销售信息有夹江书

⊙ 4

⊙ 5

⊙3 范曾题"掇英轩"牌匾原件照
⊙4 与樊嘉禄合影旧照(右樊嘉禄)("掇英轩"供图)
⊙5 与张秉伦等专家合影旧照(左张秉伦、中华觉明,右潘吉星)("掇英轩"供图)

科技考古系课题组成功复原了技艺已经失传很久的历史名纸金银印花笺，投放市场后很受欢迎。

"掇英轩"成立20余年来，先后复原了失传多年的若干加工纸品，包括粉蜡笺、金银印花笺、泥金笺、羊脑笺在内的多种古代名贵手工纸。2014年，"掇英轩"与中国科学技术大学手工纸研究所共同组建了"中国历史名纸复原研究中心"，致力于当代有实用价值的古法纸笺制作技艺复原的探究，希望用积极的行动诠释"掇英"二字的价值。

2012年开始，中国科学技术大学人文与社会科学学院和手工纸研究所开始将"掇英轩"作为实践学习的合作基地，经常会有师生及来宾前往黄麓镇的纸笺加工工坊访学交流；2016～2017年，中国科学技术大学手工纸研究所承办国家文化部、教育部"全国非物质文化遗产传承人群研修班"，已有80位中国手工造纸重要传承人到"掇英轩"现场观摩学习。

3. 创业年代艰苦而有趣的故事

2017年8月底驻厂访谈时，兴致很高的刘靖回顾了当年的创业故事：

（1）老屋与蛇的故事。1995年回到杨岗村老屋纸坊时，没自来水没电，回来3个月之后，才用上电。只好点煤油灯，画画的时候用手电筒照着，画的时候不断移动。那时家里乱七八糟的，老鼠也多，睡在床上都能逮到老鼠，夜里睡觉还压死过老鼠。蛇也多，各种蛇都有，赤板蛇、土公蛇，每年都打死不少蛇。

（2）杨宁英的"好梦"。在粉蜡笺研制过程中，铜粉手绘仿金的调配，是我母亲做梦时想到的。一开始什么都不会，不知道铜粉怎么调，用胶粘不起来，后来父亲有经验，试着用酒精，好调一些，但仍不理想；后来使用胶粘，但是用什么胶不知道。有一天夜里母亲做梦，梦里拿一种胶跟铜粉混合，发现很好用，醒来后一试，真的好用。后来

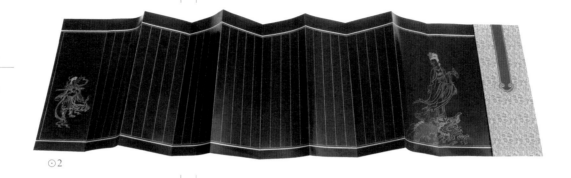

⊙1

⊙2

⊙3

○4 修葺后的老屋后门
The back door of a restored old house

一直就用这个胶来调铜粉仿金,用了很多年,现在不用了。

(3)终生难忘的"加班"。1995年粉蜡笺小样做好的时候,客户反映不错,上海外贸订了一批,安徽省建筑公司订了一批,但不幸出了质量问题。我记不清时间了,仿金手绘粉蜡笺,画了梅花,主要是以梅花玉版笺为主,有斗方、四尺、对联、尺八屏,折合成四尺的话大约有1 000多张。那时没有现在这种原材料,用的是铜粉,夏天一个星期梅雨天气,都变黑了,那批货全部得重新做。

那是1998年,返工加班赶货3个月,每天加班到晚上11点。七八个工人,从原纸开始重新返工,那批货量很大,每个图案都订了几百张。最后离交货还有3天的时候,又赶上我哥结婚,我带着全部工人几乎三天三夜没睡觉。我们搞点鸡蛋吃提神,只有一个年纪大的工人,实在顶不住在那里睡觉。最强的工人,三天三夜没睡觉,干完活,又去干农活了。我是干完活之后,去参加我哥婚礼的。这一次没有出次品,分工做的,我画圈,有人染,有人点画心,那个时候画得特别快,流水作业。当时找的那批工人都是乡里乡亲的,而且全部是女工。

○5 洒金桶
Tool for sprinkling golden powder

后来为什么粉蜡笺等使用金粉,就是这次返工逼的,必须找替代品。我们找过日本的准金,价格也不低,实际上是质量比较好的铜粉,试了之后,也变色,不行。后来我受到香烟嘴上面一圈金纸的启发,直接到卖这些材料的市场找这个原材料,一试可以。刚开始只是画图案,没有做泥金,泥金原先也是用铜粉做的,容易变黑。后来又把这个金粉刷泥金上面,这个金粉叫珠光粉,颜色丰富,我就选择了一些好看的用到泥金纸上。

有了性价比高的金粉,有了与中国科学技术大学合作恢复的造金银印花笺技术,我们开始用丝网印金银印花笺,第一年把金银印花笺拿到全国文房四宝博览会的时候,反响很大,台湾地区著名的纸老板小名叫东五郎来看我的纸,我们的金银印花笺没有打市场之前,他的"绿阳宣"风行一时。我们的金银印花笺在市场上销售两年以后,他主动退出了市场。他觉得我们的纸传统味比他的好,东五郎还特地用我们的金银印花笺写了一幅字给我。

（4）"荣宝斋"与"掇英轩"的深厚交情。"荣宝斋"是"掇英轩"中高端纸的大客户与积极推广者，对企业的发展起了很大作用。

"荣宝斋"是原淮北留香阁毛笔厂朱凌旺带我去的，他跟"荣宝斋"有业务关系。我记得带了粉蜡笺，还带了木版水印的纸，第一次见的是袁良，他看了东西，说我们的产品不错，现场决定留粉蜡笺，不仅把中间最大的两个柜子留给我们展示，而且西厅原先挂字画的四个柱子，也全部给我们挂粉蜡笺了。我去年去看，还在挂着，只是位置变了。

米景扬、唐遹昌、陈长智、冯大彪四个人对我前期的发展贡献很大。冯大彪当时在中新社，陈长智是中学教师，唐遹昌也是教师。那个时候正在研制透光笺，有一天我找冯大彪请另外一位书法家写一副字，唐遹昌搭车聊天时了解到我是做加工纸的，就让去他家看看纸。他的东西真好：一张是乾隆二十七年（1762年）的刻画笺，刻两条龙，现在我册子上就是仿这张刻画笺复制的；还有一张是我经常说的嘉庆年间的"皖纸大升纸庄"刻纸，我判断是宣纸。他又拿出好多木版水印的信笺，全是名家的。他无偿送给我一些纸样，这对我产品的开发有不少启发。

冯大彪也给了我一些古代纸样，记得最清楚的是梅花玉版笺2张（有点残），一张皮纸的，一张宣纸的。

陈长智的父亲是民国年间的著名画家陈少梅。他父亲遗留下来的纸有两捆。陈长智对我印象很好，就把父亲留下来的两捆纸给我。我不太好意思，好的整的纸不挑，挑了边角料。里面有绢本泥金、各种皮纸、各种洒金，有多种清代和民国的纸，还有老汪六吉纸（现在还剩一个纸边），帘纹很细；还有泥金纸，有仿金的，有纯金的。这带回来的一捆纸里加工纸超过10样，原纸种类更多（有皮纸、高丽纸，还有宣纸玉版），高丽纸个人判断是纯桑皮的，尺寸不大。

⊙1

⊙2

这捆纸对我纸样的恢复有很大帮助。

米景扬与陈长智是郎舅关系，陈长智带我去米景扬家介绍我们相识，我们聊得很高兴，米景扬当场给题了"纸上风云"，后来又题过多次。我画的那个龙，就是米景扬让"荣宝斋"打开地库，拍照片回来复制的。那张纸是米景扬20世纪80年代3 000块钱收的。

第二次到"荣宝斋"的地下库房是为了看由日本株式会社竹尾创立80周年纪念出版的《世界手工纸》一书（1979年11月版）。这套书除图文并茂对手工纸进行介绍外，最大的特色是在书中附有世界各地手工纸的实物纸样，其中包含中国、日本、韩国、印度、泰国、尼泊尔、西班牙、意大利、法国、英国、波兰、奥地利、南斯拉夫、瑞典、美国、澳大利亚等国的手工纸样品。

到荣宝斋库房查看资料非常不容易，需要几个人同时开门才能进入。见到这套书我如获至宝，一页一页仔细翻看，学习了半天时间。2005年我在中国科学院自然科学史研究所展厅里又看到这套书，能收藏一套《世界手工纸》成了我的一个心愿。直到2015年，托朋友从日本买到了这套书，它成了我珍贵的收藏。

⊙ 3

⊙ 4

七 安徽省掇英轩书画用品有限公司的传承现状与发展思考

7 Current Status of Business Inheritance and Thoughts on Development of Duoyingxuan Calligraphy and Painting Supplies Co., Ltd. in Anhui Province

据访谈时刘靖表述及强调，"掇英轩"的经营特色是以订单定生产，以市场需求定发展。从基本模式看，属于市场导向鲜明、力求规避盲目扩张、坚持稳健经营的类型。

从生产基地建设方面看，自1997年第二代传承人刘靖接手后，20年的时间里，一直坚持在老家黄麓镇生产，最初是在杨岗村的自家老屋里。2010年在镇上租用弃用的长源小学校舍作为纸笺的制作基地。直到2015～2016年，才买地约9 300 m²开工建设"掇英轩"新厂区。而这时，"掇英轩"已是国内外闻名的中高端纸笺生产基地、纸笺加工技艺国家级非物质文化遗产传承保护单位。

从技艺传承团队的集聚培养看，长期以来一直以黄麓镇老家的乡亲和亲戚为主体形成纸笺加工的工艺制作队伍，优势是传承队伍稳定，一定

⊙ 3 合影：左三陈长智，左四冯大彪，右四杨宁英，右三米景扬，右二唐遇昌（"掇英轩"供图）
Group photo (the third on the left is Chen Changzhi, the fourth is Feng Dabiao, the fourth on the right is Yang Ningying, the third one is Mi Jingyang, the second is Tang Naichang) (offered by "Duoyingxuan")

⊙ 4 米景扬为刘靖题的"纸上风云"
Mi Jingyang inscribed "Legends on Paper" for Liu Jing

○1

程度上消除了生产厂区位于乡间,招聘引进外来较高素质的人员易流失、核心技艺易外泄的担忧;劣势则是较关键的技艺始终只有很少人掌握,高端产品如手绘描真金粉蜡笺等由于工人的艺术素养不足,制作跟不上市场需求。

○2

"掇英轩"生产的加工纸,在全国加工纸行业中一直属于中高档产品定位。2013年后,因为消费市场和礼品治理政策的影响,至调查组调查时的2016年,"掇英轩"的定位已有一定调整,高档纸的产量计划明显下调,增加了中档及普通大众适用的纸笺类产品生产,如丝网水印的印色条屏、状元卷及各种半熟水印格等,这其中既有市场调节因素,也有"掇英轩"在新发展节点自我调节的思考。

调查中看到,"掇英轩"虽然以稳健经营著称,但面临行业发展的低谷时,也并非仅仅是等待市场对发展的裁定。按访谈中刘靖的说法,"掇英轩"目前已经推进的新布局有:

(1)主动地思考各类新型纸笺的生产研制,如2016年创新性地研发制作二十四节气花草纸。

(2)考虑到高水平工艺与设计人员的引进、培养和稳定的压力,2016年在合肥市区的华润国际大厦新购了一处超过170 m²的写字楼作为展示与设计工作室,计划招聘优秀的创意设计与绘画技师提升高端产品的制作能力和创意能力。

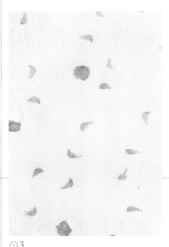

○3

○4

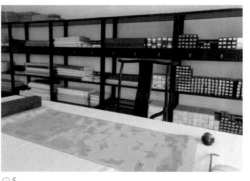

○5

1 租用的原长源小学校舍一角
2 新产品：方格半熟印格纸
3/4 新创制的二十四节气花草纸
5 合肥新办公兼展示空间一角

（3）准备打造企业自己的专职销售团队，启动线上、线下双模式，尝试全渠道开拓市场。

刘靖表示："掇英轩"以恢复历史上的纸笺名品为宗旨和目标，是一家立足传统产品和技艺的手工纸企业，但是，传统手工纸行业要在新的时代背景下适应环境变化，需要不断输入新的知识，给企业带来新的启发和发展动力。

调查组成员最后问刘靖，作为"掇英轩"的领头人，还有哪些复原著名加工纸的心愿与计划时，刘靖表示：

首先当然是名声大到让人无法不动心的由五代南唐皇室主持制作的"澄心堂纸"，但是没看到原物是什么样，无法确定可靠的样本。明清两代仿制的"澄心堂纸"都是粉蜡笺，乾隆年间仿的"澄心堂纸"是粉蜡笺，明代仿的"澄心堂纸"也是粉蜡笺。我和父亲曾合写了一篇文章，得出了对"澄心堂纸"的一个看法：它应该以加工纸为主，应该有很多种纸，不是只有一种纸，粉蜡笺是其中的一个名品。

复制薛涛笺也是很诱人的，但问题在于不知道原物是什么样的。虽然北京、四川、安徽泾县都有人在仿制，但真正的实物看不到，难度会很大。

至于流沙笺、落水纸，它们正在复原中，新工厂盖好后，想用这些参与性强、趣味性高的加工纸工艺开"非遗"体验课。

6　2015年中国台湾『巧手慧心——安徽省传统手工技艺展』演示（『掇英轩』供图）

7　在中央美术学院设计学院的『造纸课』上（『掇英轩』供图）

8　『掇英轩』木版水印笺透光摄影图

9　刘靖谈『掇英轩』恢复历史名纸新思路

安徽省掇英轩书画用品有限公司
加工纸

Processed Paper of Duoyingxuan Calligraphy and Painting Supplies Co., Ltd. in Anhui Province

"掇英轩"植物黄檗染色纸(浓)透光摄影图
A photo of "Duoyingxuan" vegetation Huangbo dyed paper (dense) seen through the light

安徽省掇英轩书画用品有限公司加工纸

Processed Paper of Duoyingxuan Calligraphy and Painting Supplies Co., Ltd. in Anhui Province

「掇英轩」植物黄檗染色纸（淡）透光摄影图
A photo of "Duoyingxuan" vegetation Huangbo dyed paper (light) seen through the light

加工纸

安徽省掇英轩书画用品有限公司

Processed Paper of Duoyingxuan Calligraphy and Painting Supplies Co., Ltd. in Anhui Province

055

『掇英轩』金粟粉蜡笺透光摄影图
A photo of "Duoyingxuan" Jinsu Fenla paper seen through the light

安徽省掇英轩书画用品有限公司 加工纸

Processed Paper of Duoyingxuan Calligraphy and Painting Supplies Co., Ltd. in Anhui Province

「掇英轩」绢本宣透光摄影图
A photo of "Duoyingxuan" silk Xuan paper seen through the light

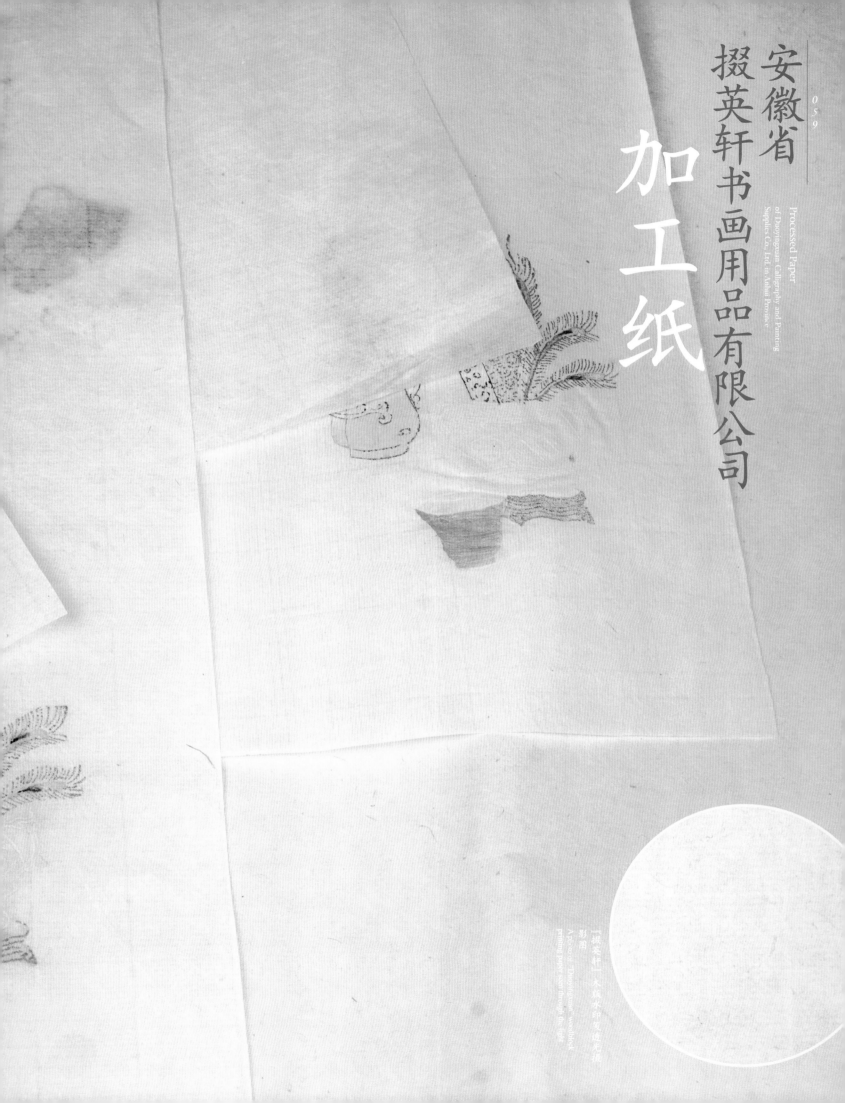

安徽省掇英轩书画用品有限公司
加工纸

Processed Paper
of Duoyingxuan Calligraphy and Painting
Supplies Co., Ltd. in Anhui Province

「掇英轩」木版水印笺透光摄
影图
A picture of Duoyingxuan woodblock
printing paper seen through the light

加工纸

安徽省掇英轩书画用品有限公司

Processed Paper
of Duoyingxuan Calligraphy and Painting
Supplies Co., Ltd. in Anhui Province

061

「掇英轩」泥金纸透光摄影图
A photo of "Duoyingxuan" Nijin paper seen through the light

第二节
泾县艺英轩宣纸工艺品厂

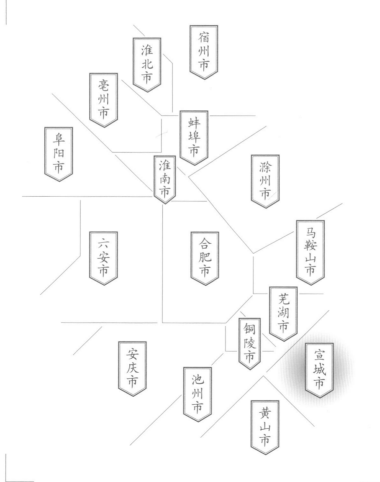

安徽省
Anhui Province

宣城市
Xuancheng City

泾县
Jingxian County

调查对象
泾县艺英轩宣纸工艺品厂
琴溪镇
加工纸

Section 2
Yiyingxuan Xuan Paper Craft Factory in Jingxian County

Subject
Processed Paper of Yiyingxuan Xuan Paper Craft Factory in Jingxian County in Qinxi Town

一

泾县艺英轩宣纸工艺品厂的基础信息与生产环境

1

Basic Information and Production Environment of Yiyingxuan Xuan Paper Craft Factory in Jingxian County

泾县艺英轩宣纸工艺品厂位于泾县琴溪镇赤滩街道粮站内，为租用地方粮站已废弃的旧粮库场地的加工纸厂。地理坐标为东经118°30′10″，北纬30°44′03″。泾县艺英轩宣纸工艺品厂创办于2000年，主要生产熟宣、半熟宣、色宣、水纹笺、虎皮宣等加工类纸，不生产宣纸与书画纸的原纸。2015年8月11日至2015年11月7日，调查组对泾县艺英轩宣纸工艺品厂进行过近10轮田野调查。截至调查时基础生产信息如下：泾县艺英轩宣纸工艺品厂有员工6人，加工纸类年产量4 000多刀，年销售额80多万元。

⊙1

琴溪镇位于泾县城东北7.5 km，青弋江与琴溪河交汇处，总面积93.12 km²，其中山场面积38 km²，耕地面积12.44 km²，辖8个村委会，192个村民组，1个居委会，3个居民小组，5 936户，19 955人。境内有琴高山隐雨石、炼丹洞、悬崖峭壁和历代名人摩石刻石像等景观，赤滩老街、马头明清建筑被列入安徽省级文物保护名录。

⊙1
"艺英轩"租用的生产车间——老粮库
An old granary, the production workshop rented by "Yiyingxuan"

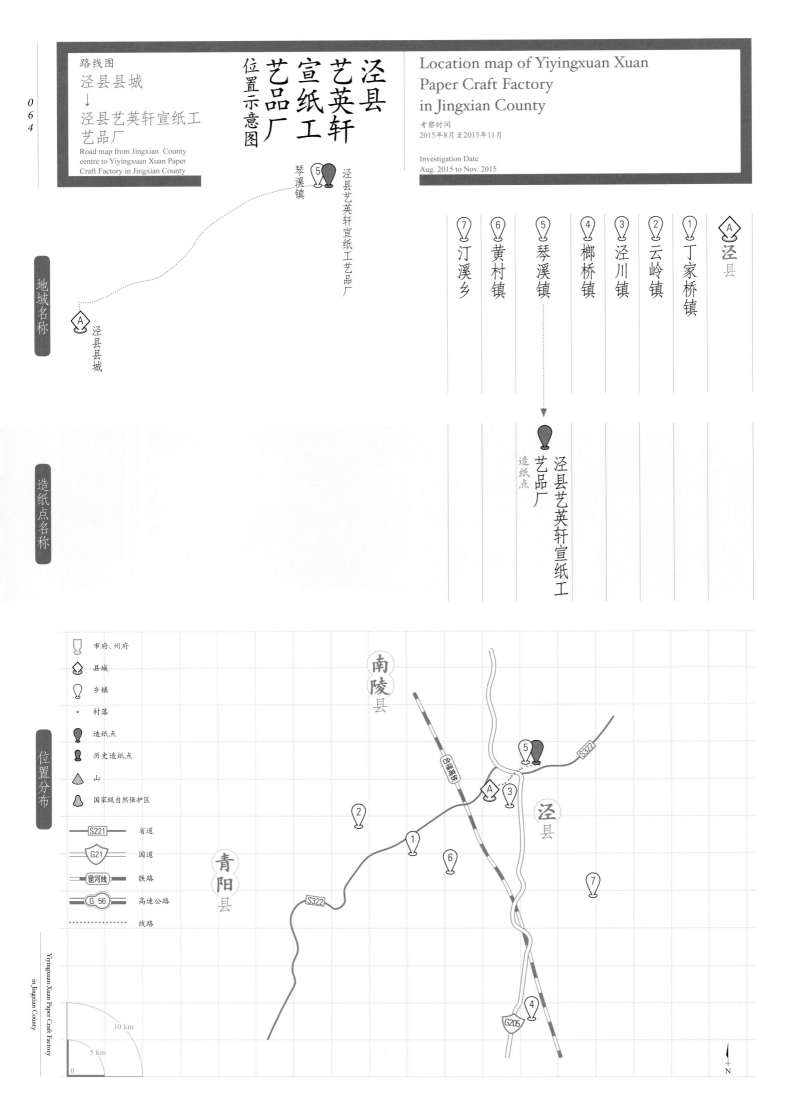

二

泾县艺英轩宣纸工艺品厂的
历史与传承情况

2

History and Inheritance of Yiyingxuan Xuan Paper Craft Factory in Jingxian County

泾县艺英轩宣纸工艺品厂注册地为安徽省泾县泾川镇。据泾县艺英轩宣纸工艺品厂创始人朱正海介绍，该厂前身为艺友堂宣纸工艺品厂，创办于2000年，以贴牌生产模式运营，厂址在原泾县宣纸二厂（生产"鸡球"牌宣纸）。2011年，朱正海的儿子朱大为大学毕业，子承父业，将艺友堂宣纸工艺品厂更名为艺英轩宣纸工艺品厂，并注册了"艺英轩"商标。2012年该厂在泾县开设直营店，同年12月底，朱大为在淘宝网上开设网店，借用淘宝网平台，以电子商务模式，直销"艺英轩"系列产品。2015年，因泾县宣纸二厂旧厂房拆迁，艺英轩宣纸工艺品厂将生产车间搬迁到了琴溪镇赤滩街道粮站内，县城仍留有经营门面。

朱正海，1961年7月出生，安徽宣城宣州区向阳镇裘村人。1978年考入合肥农业机械化学校机械化专业学习，1981年中专毕业后被分配到泾县农机厂。据朱正海在访谈中调侃式的回忆，当年视野与知识面都很窄，他在读中专之前根本不知道与宣州紧邻的泾县在哪里，当然更不知道宣纸。

1982年底，朱正海被抽调到泾县象山宣纸厂

1 老宣纸二厂所在地外景
Exterior of the Former Second Xuan Paper Factory

2 赤滩古镇一角
A corner of Chitan Ancient Town

（泾县宣纸二厂前身）从事办公室工作。1984年泾县象山宣纸厂更名为泾县宣纸二厂，并承担了一个国家级科研项目——宣纸抄纸新工艺。为了提高自己的专业技术能力，朱正海积极要求外出深造，1985年9月朱正海被泾县宣纸二厂派到南京林业大学进修制浆造纸工艺学，学习相关专业课并进行毕业论文设计及答辩，他开始从对造纸完全外行变为对制浆造纸基本工艺和原理有了系统的了解。完成学业后，朱正海到厂里专门从事工艺及质检工作。

朱正海回忆：当时泾县宣纸二厂厂长周乃空一直提倡搞技术的人要深入到实践中，要掌握整个操作过程。因周乃空的大力倡导，加上自己对宣纸工艺确有兴趣，便到每道制作工序上跟班蹲点，积累了整个宣纸传统制作工艺流程的经验。1992年他被抽调到泾县宣纸二厂机械化生产线，从事宣纸抄纸新工艺的技术改造工作。1997～1999年他又被抽调参加了泾县小岭宣纸厂（红旗宣纸厂）承担的安徽省重点项目——宣纸制浆"氧-碱法"新工艺。

朱正海于1997年下半年，在泾县宣纸二厂时首创泾县加工纸中的水纹笺配方。1999年底，泾县宣纸二厂基本停产后，作为技术骨干被借调到中国宣纸集团公司技术中心，从事"红星"牌宣纸生产过程和成品的检测工作。2002年，泾县宣纸二厂破产倒闭，朱正海与厂里解除劳动关系后只能自谋生路。

2003年6月至2012年5月，朱正海受邀担任私营合股的泾县双鹿宣纸有限公司技术主管。朱正海总结说，在泾县双鹿宣纸有限公司任主管的这一段生涯，是自己对宣纸工艺认识的一个系统总结和提升的阶段，收益很大。

2009年12月，朱正海被评为安徽省高级工艺美术师，2012年2月被评为第二届安徽省工艺美术大师。2012年10月他开始受邀担任泾县汪六吉宣纸有限公司技术顾问。从2000年开始，朱正海一边在多家宣纸企业从事宣纸工艺技术及质检工作，一边管理着自家的艺英轩宣纸工艺品厂的生产和销售工作。

1/2 朱正海正在调配原料
3 开网店的朱大为

三

泾县艺英轩宣纸工艺品厂的代表纸品及其用途与技术分析

3
Representative Paper and Its Uses and Technical Analysis of Yiyingxuan Xuan Paper Craft Factory in Jingxian County

（一）

代表纸品及其用途

据2015年8月11日至2015年11月7日多轮访谈确认的信息，艺英轩宣纸工艺品厂生产的代表产品为水纹纸（笺）、天然古法草木染色宣和熟煮捶纸。

水纹笺的名称源自唐宋时期，那时候的水纹笺指两种纸：一种指有纸的帘纹或者有水印的纸；一种指印染或绘制水纹的纸。水纹笺属于半生熟加工纸，是用多种染料或颜料配成染液，在染液中加入特制助剂，对宣纸进行表面染色制成的，色纸在风干过程中自然形成条纹。其原理是：纸页在未干前，由于助剂的作用，染料的微粒在纸面上自然絮聚成纹。从用途来说，水纹笺适用于书法中的小楷、行草、隶书，以及创作半工笔、半写意人物画及装饰。

天然古法草木染色宣采用宣纸中的净皮为原纸，利用植物染料纯手工表面染色，加工生产过程中不用任何化学原料，维持了宣纸本身的传统品质和内在纤维结构，在赋予纸张特有复古颜色的基础上，最大限度地保证了宣纸的寿命。植物染料浸染的色宣，纸张颜色久放不褪色，略微带五分熟，适合半工笔、半写意画以及速度慢的书写。

4
水纹笺的视觉效果
The visual effect of Shuiwen paper

5 / 7
"艺英轩" 天然古法草木染色宣
"Yiyingxuan" Vegetation Dyed Xuan paper in ancient natural ways

[1] 朱正海. 宣纸加工纸种类及色纹纸的制作[J]. 纸和造纸, 2004 (6): 35.

熟煮捶纸以正宗檀皮燎草为原料的净皮宣纸为底纸，按照传统煮捶加工工艺（糯米淀粉液+中药白芨水）对纸进行浆捶。浆捶后再对纸进行打蜡、研光等工序的处理。熟煮捶纸纸张绵软、表面细腻，书写流畅、墨色黑亮，纸张可以长期存放。熟煮捶纸适合于行书、草书、楷书、隶篆等书法创作。

（二）
代表纸品的技术分析

1. 代表纸品一："艺英轩"水纹纸（笺）

测试小组对采样自泾县艺英轩宣纸加工厂生产的水纹纸所做的性能分析，主要包括厚度、定量、紧度、抗张力、抗张强度、撕裂度、色度、吸水性等。按相应要求，每一指标都需重复测量若干次后求平均值，其中定量抽取5个样本进行测试，厚度抽取10个样本进行测试，抗张力抽取20个样本进行测试，撕裂度抽取10个样本进行测试，色度抽取10个样本进行测试，吸水性抽取10个样本进行测试。对"艺英轩"水纹纸进行测试分析所得到的相关性能参数见表6.2。表中列出了各参数的最大值、最小值及测量若干次所得到的平均值或者计算结果。

表6.2 "艺英轩"水纹纸相关性能参数
Table 6.2 Performance parameters of "Yiyingxuan" Shuiwen paper

指标	单位		最大值	最小值	平均值	结果
厚度	mm		0.097	0.081	0.088 4	0.088
定量	g/m²		—	—	—	30.5
紧度	g/cm³		—	—	—	0.347
抗张力	纵向	N	13.3	9.1	11.08	11.1
	横向	N	9.0	6.8	7.8	7.8
抗张强度	kN/m		—	—	—	0.630

续表

指标	单位		最大值	最小值	平均值	结果
撕裂度	纵向	mN	280	250	266	266
	横向	mN	450	360	410	410
撕裂指数	mN·m²/g		—	—	—	11.0
色度	%		78.51	75.29	77.39	77.4
吸水性	纵向	mm				19
	横向	mm				3

由表6.2中的数据可知，"艺英轩"水纹纸最厚约是最薄的1.198倍，经计算，其相对标准偏差为0.005，纸张厚薄较为一致。所测"艺英轩"水纹纸的平均定量为30.5 g/m²。通过计算可知，"艺英轩"水纹纸紧度为0.347 g/cm³。抗张强度为0.630 kN/m，抗张强度值较大。所测"艺英轩"水纹纸撕裂指数为11.0 mN·m²/g。

所测"艺英轩"水纹纸平均色度为77.4%。色度最大值是最小值的1.043倍，相对标准偏差为1.108，色度差异相对较小。吸水性纵横平均值为19 mm，纵横差为3 mm。

2. 代表纸品二："艺英轩"天然古法草木染色宣

测试小组对采样自泾县艺英轩宣纸加工厂生产的天然古法草木染色宣所做的性能分析，主要包括厚度、定量、紧度、抗张力、抗张强度、撕裂度、色度、吸水性等。按相应要求，每一指标都需重复测量若干次后求平均值，其中定量抽取5个样本进行测试，厚度抽取10个样本进行测试，抗张力抽取20个样本进行测试，撕裂度抽取10个样本进行测试，色度抽取10个样本进行测试，吸水性抽取10个样本进行测试。对"艺英轩"天然古法草木染色宣进行测试分析所得到的相关性能参数见表6.3。表中列出了各参数的最大值、最小值及测量若干次所得到的平均值或者计算结果。

表6.3 "艺英轩"天然古法草木染色宣相关性能参数
Table 6.3　Performance parameters of "Yiyingxuan" Vegetation Dyed Xuan paper in ancient natural ways

指标		单位	最大值	最小值	平均值	结果
厚度		mm	0.072	0.067	0.066	0.066
定量		g/m²	—	—	—	27.2
紧度		g/cm³	—	—	—	0.412
抗张力	纵向	N	15.7	10.7	12.8	12.8
	横向	N	10.9	9.2	8.8	8.8
抗张强度		kN/m	—	—	—	0.720
撕裂度	纵向	mN	160	150	156	156
	横向	mN	220	210	214	214
撕裂指数		mN·m²/g	—	—	—	6.6
色度		%	79.9	66.3	71.1	71.1
吸水性		mm	—	—	—	17

由表6.3中的数据可知，"艺英轩"天然古法草木染色宣最厚约是最薄的1.075倍，经计算，其相对标准偏差为0.001，纸张厚薄较为一致。所测"艺英轩"天然古法草木染色宣的平均定量为27.2 g/m²。通过计算可知，"艺英轩"天然古法草木染色宣紧度为0.412 g/cm³，抗张强度为0.720 kN/m，抗张强度值较大。所测"艺英轩"天然古法草木染色宣撕裂指数为6.6 mN·m²/g，撕裂度较大。

所测"艺英轩"天然古法草木染色宣平均色度为71.1%，色度最大值是最小值的1.21倍，相对标准偏差为4.509，色度差异相对较小。吸水性纵横平均值为17 mm。

3. 代表纸品三："艺英轩"熟煮捶纸

测试小组对采样自泾县艺英轩宣纸加工厂生产的熟煮捶纸所做的性能分析，主要包括厚度、定量、紧度、抗张力、抗张强度、撕裂度、色度、吸水性等。按相应要求，每一指标都需重复测量若干次后求平均值，其中定量抽取5个样本进行测试，厚度抽取10个样本进行测试，拉力抽取20个样本进行测试，撕裂度抽取10个样本进行测试，色度抽取10个样本进行测试，吸水性抽取10个样本进行测试。对"艺英轩"熟煮捶纸进行测试分析所得到的相关性能参数见表6.4。表中列出了各参数的最大值、最小值及测量若干次所得到的平均值或者计算结果。

表6.4 "艺英轩"熟煮捶纸相关性能参数
Table 6.4　Performance parameters of "Yiyingxuan" Shuzhuchui paper

指标		单位	最大值	最小值	平均值	结果
厚度		mm	0.096	0.080	0.090	0.090
定量		g/m²	—	—	—	33.1
紧度		g/cm³	—	—	—	0.368
抗张力	纵向	N	31.4	28.8	30.3	30.3
	横向	N	19.0	15.7	17.8	17.8
抗张强度		kN/m	—	—	—	1.603
撕裂度	纵向	mN	180	150	172	172
	横向	mN	220	210	214	214
撕裂指数		mN·m²/g	—	—	—	193
色度		%	70.8	69.7	70.2	70.2
吸水性		mm	—	—	—	4

由表6.4中的数据可知，"艺英轩"熟煮捶纸最厚约是最薄的1.20倍，经计算，其相对标准偏差为0.004，纸张厚薄较为一致。所测"艺英轩"熟煮捶纸的平均定量为33.1 g/m²。通过计算可知，"艺英轩"熟煮捶纸紧度为0.368 g/cm³，抗张强度为1.603 kN/m，抗张强度值较大。所测"艺英轩"熟煮捶纸撕裂指数为193 mN·m²/g，撕裂度较大。

所测"艺英轩"熟煮捶纸平均色度为70.2%，色度较高。色度最大值是最小值的1.016倍，相对标准偏差为0.352，色度差异相对较小。吸水性纵横平均值为4 mm。

四 "艺英轩"加工纸生产的原料、工艺流程与工具设备

4 Raw Material, Papermaking Techniques and Tools of "Yiyingxuan" Processed Paper

(一) "艺英轩"水纹纸

1. "艺英轩"水纹纸生产的原料

（1）主料：原纸。"艺英轩"水纹纸使用的原纸一般为手工捞制的纯龙须草书画纸。据朱正海介绍，"艺英轩"水纹纸原纸系从泾县当地书画纸生产厂家直接购买。

（2）辅料：染料和水源。"艺英轩"水纹纸加工时采用各种化工染料，系从泾县本地经销商处购买。

"艺英轩"水纹纸加工使用水源为当地井水，但由于加工水纹纸需要用软水（软水对染料不容易起反应），因此如果井水水性较硬时需要添加软水剂。据调查组成员在现场测试，"艺英轩"水纹纸所用的水源pH为7.09，呈碱性。

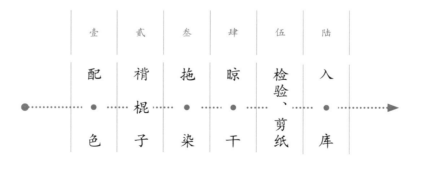

⊙1

2. "艺英轩"水纹纸生产的工艺流程

据朱正海描述，以及综合调查组多次在"艺英轩"加工纸车间现场的实地调查，归纳"艺英轩"水纹纸生产的工艺流程为：

壹	贰	叁	肆	伍	陆
配色	梢棍子	拖染	晾干	检验、剪纸	入库

染料
Dye ⊙1

壹 配色

1

水纹纸是将书画纸用表面染色的方式加工而成的，因此配色是关键环节。据朱正海介绍，"艺英轩"水纹纸一般按照10 kg的染液水进行计量配色，通常10 kg的混合液可以生产一刀纸（四尺100张）。仿古水纹纸具体配比为骨胶80 g、直接黄棕ND3G 3 g、直接深棕MM 1 g、直接耐晒黑G 8 g、助剂A 6 g、助剂B 18 g。再按照客户定制需要放入其他颜色染料，如米黄、橙红、淡绿、浅灰等颜色。需注意的是，染液存放不能超过8小时；pH在6.5～7.5范围内产生的水纹比较理想。

贰 楷棍子

2

将原纸的一个边刷上糨糊，使其可以粘上棍子，便于拖染及晾晒。

叁 拖染

3

将配过色的染料放入拖盆里面，两手拿着棍子的两端从拖盆的一头拖到另一头。拖染为制作水纹纸的重要步骤，因此一定要注意不能将水漫到纸面，否则会拖花；在拖染的过程中要均匀用力，不能操之过急，也不能过慢，要匀速地一气呵成，有一定的技巧要求。一般一天一个成熟的技工可以拖染6刀四尺水纹笺。

肆 晾干

4

将拖染过的纸拿到一边挂起来阴干，在阴干的过程中水纹会逐步出现，纸半干时水纹就全部出现了。挂纸时要注意纸与纸之间不能粘连在一起，车间室内湿度应大于65%，阴干的时间要根据季节和空气湿度决定，夏天一般为12小时。

伍 检验、剪纸 ⑧

将阴干后的水纹纸收下后,扒下粘在水纹纸上面的棍子,将其整理齐,挑出不合格的纸。然后将合格的纸按照客户需要进行裁剪,一般50张为一个刀口,压上石头,剪纸人站成箭步,持平剪刀一气呵成地剪下去。

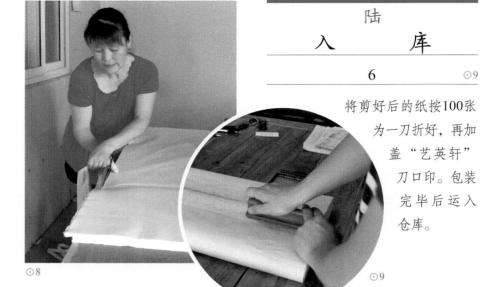

⊙8

陆 入库 ⑨

将剪好后的纸按100张为一刀折好,再加盖"艺英轩"刀口印。包装完毕后运入仓库。

⊙9

(二)

"艺英轩"天然古法草木染色宣

1. "艺英轩"草木染色宣生产的原料

(1) 主料:原纸。"艺英轩"天然古法草木染色宣使用的原纸一般为正宗的净皮四尺单宣纸。据朱正海介绍,原纸从泾县当地宣纸生产厂家直接订制。

(2) 辅料:染料。"艺英轩"天然古法草木染色宣加工时采用各种植物染料。据朱正海介绍,采用的植物染料有板栗壳、橡栗壳、槐米、藤黄等。

⊙10　　⊙11

⊙12

2. "艺英轩"天然古法草木染色宣生产的工艺流程

据朱正海介绍,以及综合调查组多次在泾县艺英轩宣纸工艺品厂的实地调查,归纳"艺英轩"天然古法草木染色宣生产的工艺流程为:

壹	贰	叁	肆	伍	陆	柒	捌	玖
泡料	过滤	调色	梢棍子	试色	拖染	阴干	检验、剪纸	入库

⊙8 剪纸　Cutting the paper
⊙9 盖章　Sealing the paper
⊙10/12 "艺英轩"天然古法草木染色宣加工使用的植物染料　Plant dye used for "Yiyingxuan" Vegetation Dyed Xuan paper in ancient natural ways

壹 泡料
1

根据客户需要,将所需要的植物染料原料放入大缸里,放水后封口浸泡3个月以上,或用水煮3~4小时。如果需要黄棕色宣,一般使用板栗壳;如果需要棕色宣,一般使用橡栗壳;如果需要黄色宣,一般使用槐米或者藤黄。

贰 过滤
2

将制作好的染料水取出,过滤出水中杂质后放在一边留用。

叁 调色
3

将过滤好的染料水进行调色,调试到需要的颜色,如加入胭脂红等染料。同时会根据客户需要适当加入白芨液(一种植物胶),可以防蛀而延长纸的寿命,同时增加纸的熟度。

肆 梢棍子
4

将原纸一头边上刷上糨糊,使其可以粘上棍子,便于拖染及晾晒。

伍 试色
5

用一张纸试一下颜色,如果色光满意就直接拖染,如果不满意需重新配色。

陆 拖染
6

将染料放入拖盆里面,拖染工艺与"艺英轩"水纹纸工序相同。

- 13 泡料 Soaking plant dye
- 14 白芨 Bletilla striata
- 15 梢棍子 Sticking the paper to the pole
- 16 试色 Testing the paper
- 17 拖染 Dyeing the paper

柒 阴干 7

将拖染好的纸挂在室内阴干，挂纸时要注意纸与纸之间不能粘连在一起。阴干晾晒的时间要根据季节和空气湿度决定，夏天干燥时控制在12小时左右，干燥过快纸面容易收缩起皱。

捌 检验、剪纸 8

基本操作工艺同"艺英轩"水纹纸。

玖 入库 9

将剪好后的纸按100张一刀折好，再加盖"艺英轩"刀口印。包装完毕后运入仓库。

（三）"艺英轩"熟煮捶纸

1. "艺英轩"熟煮捶纸生产的原料

（1）主料：原纸。"艺英轩"熟煮捶纸使用的原纸是正宗的净皮四尺单宣纸。据朱正海介绍，所用净皮原纸从泾县当地宣纸生产厂家按特制加工要求订购。

（2）辅料：糯米、中药白芨。"艺英轩"熟煮捶纸使用的糯米在泾县购买，2015年11月调查时的价格为7元/kg。一般先将糯米煮成稀饭状，过滤掉糯米渣后，用糯米浆与白芨混合后对原纸进行施胶。

"艺英轩"熟煮捶纸使用的白芨一般在泾县中药房购买，系安徽亳州中药材大市场进货的白芨，2015年调查时价格为500元/kg。白芨具有杀虫、抗水等作用，一般与糯米浆混合施胶的比例为1∶2（白芨∶糯米）。

2. "艺英轩"熟煮捶纸生产的工艺流程

据朱正海介绍，以及综合调查组多次在艺英轩宣纸工艺品厂的实地调查，归纳"艺英轩"熟煮捶纸生产的工艺流程为：

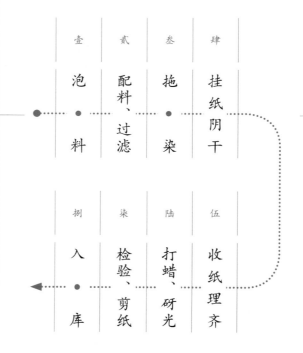

壹 泡料 → 贰 配料、过滤 → 叁 拖染 → 肆 挂纸阴干 → 伍 收纸理齐 → 陆 打蜡、砑光 → 柒 检验、剪纸 → 捌 入库

工艺流程

壹 泡料 1

根据客户需要,将所需要的植物染料原料放入大缸里,放水后封口浸泡3个月以上,或用水煮3~4小时。如果需要黄棕色宣,一般使用板栗壳;如果需要棕色宣,一般使用橡栗壳;如果需要黄色宣,一般使用槐米或者藤黄。

贰 配料、过滤 2

按照一刀四尺纸的量计算,每刀纸使用150~200 g糯米,将糯米用水蒸煮3小时,然后将白芨浸泡8小时以上(也可以蒸煮3小时以上),每刀纸使用150 g白芨。将糯米液和白芨液混合,过滤掉杂质,然后放在一边备用。

叁 拖染 3

将混合好的配料放入拖盆里面,两手拿着棍子的两端将原纸从拖盆的一头拖到另一头,让原纸浸湿浆液。

肆 挂纸阴干 4

将拖染好的纸挂在室内阴干。基本操作工艺同"艺英轩"天然古法草木染色宣。

伍 收纸理齐 5

将阴干好的纸收回,扒去棍子,整理齐。

陆 打蜡、砑光 6

将整理好的纸双面打上蜡,打蜡时用画圈的形式涂抹到纸的每个地方,然后用光滑平整的石头按照一致的方向磨纸。

⊙1

柒 检验、剪纸 7

将打蜡后的纸整理齐,挑出不合格的纸。然后将合格的纸按照客户需要进行裁剪。裁剪时,一般50张为一个单元进行裁剪。

捌 入库 8

操作工艺同"艺英轩"天然古法草木染色宣。

⊙2

⊙1 打蜡 Waxing
⊙2 砑光 Smoothing the paper

(四)
"艺英轩"加工纸生产的工具设备

壹 拖盆 1

用于盛装胶矾水,原纸在盆内进行表面施胶的工具。实测艺英轩宣纸工艺品厂的四尺拖盆尺寸为长108 cm,宽69 cm,高11 cm;六尺拖盆尺寸为长128 cm,宽72 cm,高11 cm。

⊙1

贰 棍子 2

用来拖纸和挂纸,实测艺英轩宣纸工艺品厂的四尺棍子尺寸为长98 cm,宽1.8 cm;六尺棍子尺寸为长121 cm,宽2 cm。

⊙2

叁 泡料缸 3

用来泡植物染料,实测泾县艺英轩宣纸工艺品厂的泡料缸尺寸为直径66 cm,高60 cm。

⊙3

肆 剪刀 4

检验后用来剪纸,剪刀口为钢制,其余部分为铁制。实测艺英轩宣纸工艺品厂所用的剪刀尺寸为长33 cm,最宽8.5 cm。

⊙4

五 泾县艺英轩宣纸工艺品厂的市场经营状况

5 Marketing Status of Yiyingxuan Xuan Paper Craft Factory in Jingxian County

泾县艺英轩宣纸工艺品厂在泾县县城内的稼祥北路设有销售店,在淘宝网上开设专营店,线下、线上的销售模式紧密协同,均由大学毕业回家中的朱大为负责运营,在淘宝网上进行电商销售已成为"艺英轩"重要的销售渠道。截至调查时的2015年秋天,艺英轩宣纸工艺品厂的"艺英轩"水纹纸市场价为260元/刀,草木染色宣市场价为2 500元/刀,熟煮捶纸市场价为2 200元/刀。

据朱正海介绍,2014年的年产量有4 000多刀,销售额80多万元,利润率在15%左右。目前泾县艺英轩宣纸工艺品厂采用以销售带动生产的模式,"艺英轩"水纹纸占全年销售额的60%~70%,天然古法草木染色宣和熟煮捶纸占全年销售额的10%。

⊙5

⊙6

⊙7

⊙5 店面经营图 Paper store
⊙6 产品价格表 Product price list
⊙7 黄淳为"艺英轩"题字的牌匾 A plaque for "Yiyingxuan" inscribed by Huang Chun

六 泾县艺英轩宣纸工艺品厂的品牌文化与民俗故事

6
Brand Culture and Stories of Yiyingxuan Xuan Paper Craft Factory in Jingxian County

1. "艺英轩"品牌的来历

据朱正海介绍,叫"艺英轩"实属意外。因为宣纸工艺品厂创建时,已选定的名字叫"艺友堂宣纸工艺品厂",寓意是要造出好纸,成为艺术家的朋友。但当时并未注册,到2011年朱大为回到厂里准备注册时,意外地发现"艺友堂"商标早已被他人注册。经过又一番推敲,选定"艺英轩"为宣纸工艺品厂的新品牌。"艺英轩"名称的灵感源自《美术报》上的一个栏目名称——"艺苑撷英",又因从事行业与宣纸加工有关,所以用了个谐音的"轩"字。

2. 朱正海:剪报与宣纸生涯

访谈中朱正海讲述了一段有趣的往事:朱正海出生于农村,直到1978年上中专离乡时,都不知道宣纸为何物,连近在咫尺的邻县泾县在哪里也不知道,是那个年代典型的孤陋寡闻的乡村少年。但是他从小就喜欢剪报,1980年上中专能够看见报纸了,有一期的《安徽日报》副刊《朝辉》上载有《最忆吴笺照墨光》的文章,写的是记者采访时任泾县宣纸厂(中国宣纸股份有限公司前身)厂长崔保来以及泾县宣纸厂发展情况,当时觉得写得特别好,很喜欢,就将这篇报道剪下贴在自己的练习本上并保存至今。

⊙ 1
《美术报·艺苑撷英》
China Art Weekly·Fine Arts

⊙ 2
朱正海收藏35年的《最忆吴笺照墨光》剪报
The Newspaper clipping on Recall the Calligraphy on Wujian Paper kept by Zhu Zhenghai for 35 years

七
泾县艺英轩宣纸工艺品厂的
传承现状与发展思考

7
Current Status of Business Inheritance and
Thoughts on Development of Yiyingxuan
Xuan Paper Craft Factory
in Jingxian County

泾县艺英轩宣纸工艺品厂加工纸坚持以传统加工工艺为主，在经历了从"贴牌"生产为主的企业生存阶段到以自己品牌为主的企业发展阶段后，又朝着多元化、多品牌、多渠道的循环发展之路前行。

访谈中朱正海表示，这几年加工纸市场较为混乱，价格战此起彼伏，但是在经营中艺英轩宣纸工艺品厂坚持传统，坚持不打价格战，闯出了自己的一片天空。谈及未来发展，朱正海说艺英轩宣纸工艺品厂在坚持恢复和发展传统工艺的同时，也会朝着多元化品种创新方向发展，希望能以此方式促进企业良性循环。

与此同时，朱大为在访谈中提及：泾县艺英轩宣纸工艺品厂在坚持原有销售模式（批发＋网

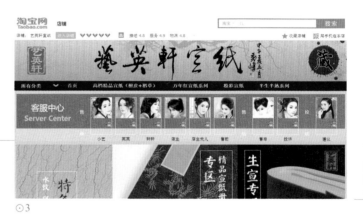

⊙3

店）基础上大力拓展销售渠道，如微信中的微拍等，在提高销售渠道的基础上增加产品的附加值和售后服务深度，以提高用户黏度和忠诚度。

⊙3
『艺英轩』淘宝网店截图
The screen shot of the Taobao online shop of "Yiyingxuan"

泾县艺英轩宣纸工艺品厂
加工纸

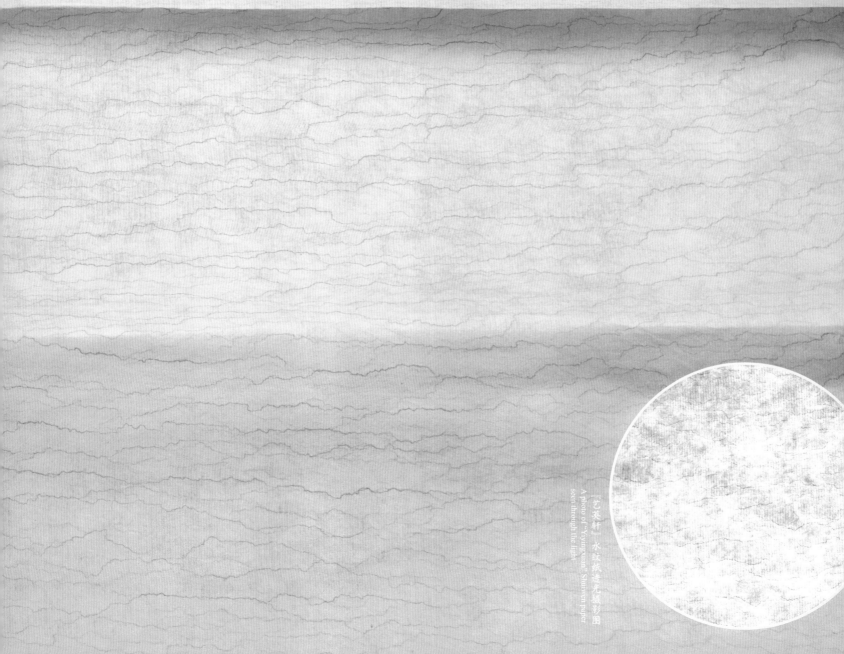

"艺英轩"水纹纸透光摄影图
A photo of "Yiyingxuan" Shuiwen paper seen through the light

加工纸

泾县艺英轩宣纸工艺品厂

Processed Paper of Yiyingxuan Xuan Paper Craft Factory in Jingxuan County

083

「艺英轩」草木染色纸透光摄影图
A photo of "Yiyingxuan" vegetation dyed paper seen through the light

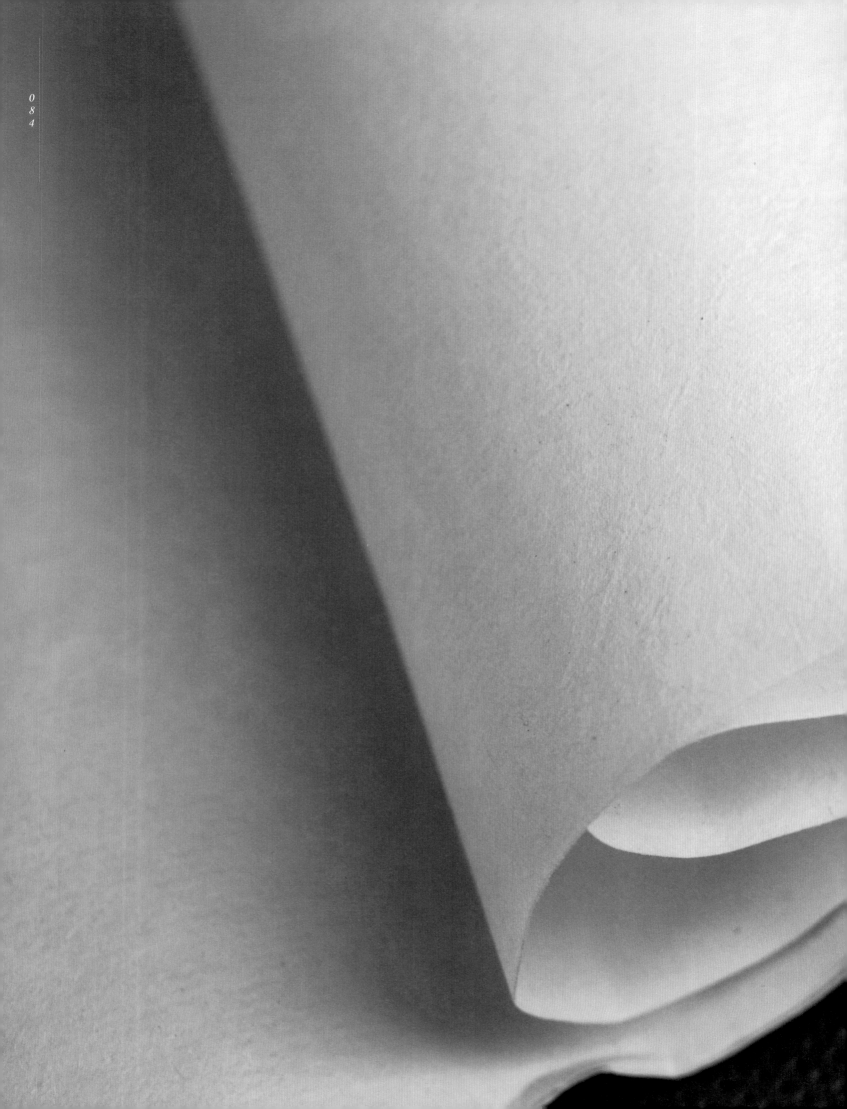

泾县
艺英轩宣纸工艺品厂
加工纸

「艺英轩」熟煮捶纸透光摄影图
A photo of "Yiyingxuan" Shuzhuichui paper seen through the light

第三节
泾县艺宣阁宣纸工艺品有限公司

安徽省
Anhui Province

宣城市
Xuancheng City

泾县
Jingxian County

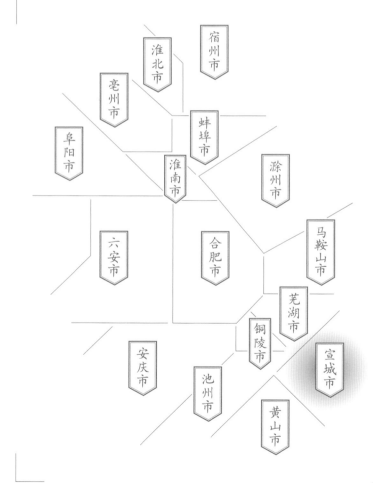

调查对象
泾川镇
泾县艺宣阁宣纸工艺品有限公司
加工纸

Section 3
Yixuange Xuan Paper Craft Co., Ltd. in Jingxian County

Subject
Processed Paper of Yixuange Xuan Paper Craft Co., Ltd. in Jingxian County in Jingchuan Town

一 泾县艺宣阁宣纸工艺品有限公司的基础信息与生产环境

1
Basic Information and Production Environment of Yixuange Xuan Paper Craft Co., Ltd. in Jingxian County

泾县艺宣阁宣纸工艺品有限公司位于泾县泾川镇城西工业集中区内,地理坐标为东经118°22′08″,北纬30°41′21″。泾县艺宣阁宣纸工艺品有限公司创办于1999年,是泾县最早从事宣纸加工纸制作的民营企业。主要生产原纸为宣纸与书画纸的加工纸,如粉蜡彩类、水印类、熟宣类等,使用"艺宣阁""源泉"两个商标,"艺宣阁"商标主要用于各类加工纸系列,"源泉"商标主要用于部分原纸的销售。

2015年8月8日、2016年4月16日和2017年6月2日,调查组3次前往泾县艺宣阁宣纸工艺品有限公司进行田野调查。2015年8月8日入厂调查时,现场据公司法人代表佘贤兵介绍,其基础信息如下:艺宣阁宣纸工艺品有限公司生产厂区有员工52人,年销售数量折合成四尺纸计算,年均加工量5万刀左右,2014年的销售额2 000余万元。

⊙1
艺宣阁宣纸工艺品有限公司厂区正门
Gate of Yixuange Xuan Paper Craft Co., Ltd.

Location map of Yixuange Xuan Paper Craft Co., Ltd. in Jingxian County

路线图
泾县县城 → 泾县艺宣阁宣纸工艺品有限公司

Road map from Jingxian County centre to Yixuange Xuan Paper Craft Co., Ltd. in Jingxian County

考察时间
2015年8月/2016年4月/2017年6月

Investigation Date
Aug. 2015/ Apr. 2016/ June 2017

地域名称
- A 泾县
- ① 丁家桥镇
- ② 云岭镇
- ③ 泾川镇
- ④ 榔桥镇
- ⑤ 琴溪镇
- ⑥ 黄村镇
- ⑦ 汀溪乡

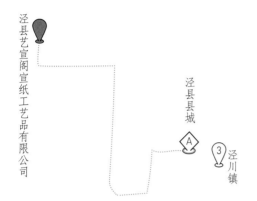

造纸点名称
泾县艺宣阁宣纸工艺品有限公司（造纸点）

位置分布

图例：
- 市府、州府
- 县城
- 乡镇
- · 村落
- 造纸点
- 历史造纸点
- 山
- 国家级自然保护区
- S221 省道
- G21 国道
- 皖赣线 铁路
- G56 高速公路
- ……… 线路

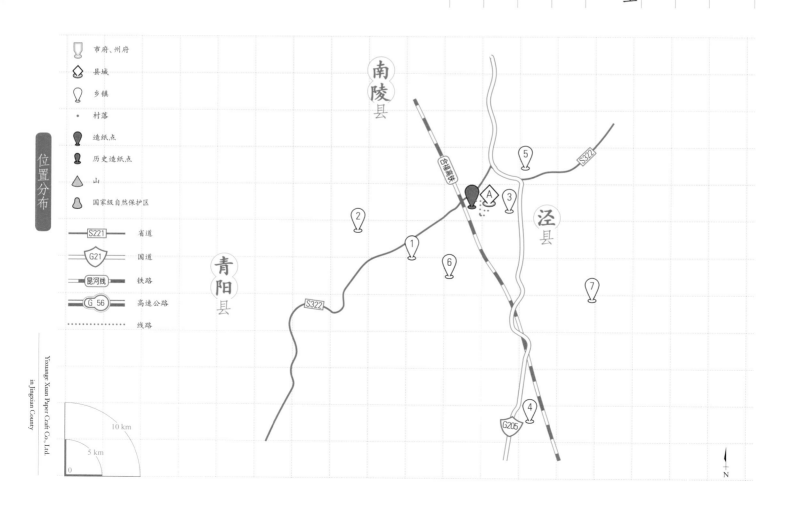

二 泾县艺宣阁宣纸工艺品有限公司的历史与传承情况

2 History and Inheritance of Yixuange Xuan Paper Craft Co., Ltd. in Jingxian County

泾县艺宣阁宣纸工艺品有限公司注册地为泾川镇城西工业集中区，创始人为余贤兵。访谈时余贤兵介绍的企业创立与沿革简史为：泾县艺宣阁宣纸工艺品有限公司创办于1999年，创办之初的定位就是生产加工纸。2001年，余贤兵结合自己在印刷行业积累的经验在泾县首创了粉彩系列加工纸。为保障原纸质量，泾县艺宣阁宣纸工艺品有限公司于2007年在小岭村的百岭坑村民组租用厂房创办了一个宣纸原纸的生产厂，设有2个纸槽，以生产"青檀皮+稻草"的四尺规格宣纸为主，同时也兼做一些特殊纸张试验，如古法、纯檀皮纸的试制。百岭坑的这个宣纸厂主要为泾县艺宣阁宣纸工艺品有限公司高端加工纸提供原纸。

2010年，泾县艺宣阁宣纸工艺品有限公司在泾县经济开发区租用了一个厂房专门进行较大批量的加工纸生产，2015年投入9台丝网自动烘干的跑板机。

1/2 百岭坑厂原址内外景观
Internal and external view of the former site of Bailingkeng Factory

3 泾县开发区厂区的车间内景
Internal view of the workshop in the factory area of Jingxian Development Zone

佘贤兵，1978年出生于泾县黄村镇（原安吴乡），从小就跟随家人从事宣纸原料的选草工作，供应给当年的小岭宣纸厂原料生产车间。16岁初中毕业后进入泾县南门老街的振裕印刷厂从事印刷工作，21岁开始自己办厂从事加工纸的生产。

振裕印刷厂是泾县老字号印刷厂，在中华人民共和国成立前就较有名，1956年实行公私合营，60年代中期转为地方国营[1]，80年代振裕印刷厂创始人的后代重新开办私营印刷厂，并使用"振裕"为厂名。该厂在印刷之余，也做一些土纸加工的订单。佘贤兵在振裕印刷厂工作期间，积累了一定的加工纸工艺经验。

佘贤兵在1999年创建泾县艺宣阁宣纸工艺品有限公司后，很注意加强与高等院校和科研机构的联系，借用"外脑"智慧，不断开发新产品、更新包装，使泾县艺宣阁宣纸工艺品有限公司在泾县加工纸行业中迅速崛起，截至2015年调查时，已成为泾县手工纸加工行业里产量最大、品种丰富度领先、包装更新领先的加工纸企业，其加工纸新产品、新包装成为泾县同行模仿的对象。

2008年，佘贤兵被评为安徽省民间文化传承人；2009年以来，参与承担国家文物局"指南针"等项目，潜心揣摩"澄心堂纸""金粟山写经纸""蠲纸""兰亭蚕纸""宣德贡笺"等历史名纸的加工工艺，并仿制性地恢复了历史名纸的生产。2014年，佘贤兵被评为安徽省工艺美术名人，同年，在佘贤兵的积极主张与申报下，"宣纸制品制作技艺"被列入安徽省第四批非物质文化遗产代表作名录；2015年，佘贤兵被评为安徽省第五批非物质文化遗产代表性传承人；2016年被评为第四批安徽省工艺美术大师。

⊙ 1 调查组成员在访谈佘贤兵
Researchers are interviewing She Xianbing
⊙ 2 省级工艺美术大师荣誉牌匾
Honor Plaque of Provincial Master of Craftsmanship and Art

[1] 泾县地方志编纂委员会.泾县志[M].北京：方志出版社，1996：206.

三 泾县艺宣阁宣纸工艺品有限公司的代表纸品及其用途与技术分析

3 Representative Paper and Its Uses and Technical Analysis of Yixuange Xuan Paper Craft Co., Ltd. in Jingxian County

（一）代表纸品及其用途

泾县艺宣阁宣纸工艺品有限公司加工纸种类相当丰富。特色产品有熟宣类、蝉翼类、印制类、四色套印类、染色类、植物颜色染色宣类、特制类、粉彩笺和蜡染笺类。当问及代表纸品时，佘贤兵的表述是：目前公司的代表纸品要推"艺宣阁"粉彩笺和蜡染笺两类。粉彩笺适用于书法创作，蜡染笺是粉蜡笺的简化，佘贤兵试验了粉蜡笺一年多的时间，研发出蜡染笺。蜡染笺既有粉蜡笺书写的效果，又比粉蜡笺便宜，最适合写小楷。

⊙3
书法家试用"艺宣阁"纸品
Calligrapher is testing the products of "Yixuange".

（二）代表纸品的技术分析

1. 代表纸品一："艺宣阁"粉彩笺

测试小组对采样自泾县艺宣阁宣纸工艺品有限公司生产的"艺宣阁"粉彩笺所做的性能分析，主要包括厚度、定量、紧度、抗张力、抗张强度、撕裂度、色度、吸水性等。按相应要求，每一指标都需重复测量若干次后求平均值，其中定量抽取5个样本进行测试，厚度抽取10个样本进行测试，抗张力抽取20个样本进行测试，撕裂度抽取10个样本进行测试，色度抽取10个样本进行测试,吸水性抽取10个样本进行测试。对"艺宣阁"粉彩笺加工纸进行测试分析所得到的相关性能参数见表6.5。表中列出了各参数的最大值、最小值及测量若干次所得到的平均值或者计算结果。

表6.5 "艺宣阁"粉彩笺相关性能参数
Table 6.5　Performance parameters of "Yixuange" Fencai paper

指标		单位	最大值	最小值	平均值	结果
厚度		mm	0.099	0.089	0.094	0.094
定量		g/m^2	—	—	—	34.4
紧度		g/cm^3	—	—	—	0.366
抗张力	纵向	N	15.3	11.2	14.1	14.1
	横向	N	10.2	8.2	9.1	9.1
抗张强度		kN/m	—	—	—	0.773
撕裂度	纵向	mN	340	305	313	313
	横向	mN	410	350	374	374
撕裂指数		mN·m^2/g	—	—	—	18.8
色度		%	77.3	71.9	73.9	73.9
吸水性		mm	—	—	—	15

由表6.5中的数据可知,"艺宣阁"粉彩笺最厚约是最薄的1.112倍,经计算,其相对标准偏差为0.003,纸张厚薄较为一致。所测"艺宣阁"粉彩笺的平均定量为34.4 g/m^2。通过计算可知,"艺宣阁"粉彩笺加工纸紧度为0.366 g/cm^3,抗张强度为0.773 kN/m,抗张强度值较大。所测"艺宣阁"粉彩笺撕裂指数为18.8 mN·m^2/g,撕裂度较大。

所测"艺宣阁"粉彩笺平均色度为73.9%,色度最大值是最小值的1.075倍,相对标准偏差为2.073,色度差异相对较小。吸水性纵横平均值为15 mm。

⊙1
"艺宣阁"粉彩笺

2. 代表纸品二:"艺宣阁"蜡染笺

测试小组对"艺宣阁"蜡染笺所做的性能分析,主要包括厚度、定量、紧度、抗张力、抗张强度、撕裂度、色度、吸水性等。按相应要求，每一指标都需重复测量若干次后求平均值，其中定量抽取5个样本进行测试，厚度抽取10个样本进行测试，抗张力抽取20个样本进行测试，撕裂度抽取10个样本进行测试，色度抽取10个样本进行测试，吸水性抽取10个样本进行测试。对"艺宣阁"蜡染笺进行测试分析所得到的相关性能参数见表6.6。表中列出了各参数的最大值、最小值及测量若干次所得到的平均值或者计算结果。

⊙2
"艺宣阁"蜡染笺
"Yixuange" Laran paper

表6.6 "艺宣阁"蜡染笺相关性能参数
Table 6.6 Performance parameters of "Yixuange" Laran paper

指标		单位	最大值	最小值	平均值	结果
厚度		mm	0.125	0.108	0.115	0.115
定量		g/m^2	—	—	—	54.7
紧度		g/cm^3	—	—	—	0.477
抗张力	纵向	N	14.8	12.5	13.3	13.3
	横向	N	9.4	7.9	8.4	8.4
抗张强度		kN/m	—	—	—	0.723
撕裂度	纵向	mN	360	300	330	330
	横向	mN	400	350	368	368
撕裂指数		mN·m^2/g	—	—	—	13.6
色度		%	59.0	57.2	58.1	58.1
吸水性		mm	—	—	—	很难吸水

由表6.6中的数据可知，"艺宣阁"蜡染笺最厚约是最薄的1.157倍，经计算，其相对标准偏差为0.006，纸张厚薄较为一致。所测"艺宣阁"蜡染笺的平均定量为54.7 g/m^2。通过计算可知，"艺宣阁"蜡染笺紧度为0.477 g/cm^3，抗张强度为0.723 kN/m，抗张强度值较大。所测"艺宣阁"蜡染笺撕裂指数为13.6 mN·m^2/g。

所测"艺宣阁"蜡染笺平均色度为58.1%。色度最大值是最小值的1.031倍，相对标准偏差为0.017，色度差异相对较小。很难吸水。

四 "艺宣阁"加工纸生产的原料、工艺流程与工具设备

4 Raw Materials, Papermaking Techniques and Tools of "Yixuange" Processed Paper

（一）"艺宣阁"粉彩笺

1. "艺宣阁"粉彩笺生产的原料

（1）主料：原纸。据佘贤兵介绍，"艺宣阁"粉彩笺使用的原纸是品质较好的书画纸，这种书画纸的原料必须是一级龙须草加青檀皮，而且是定点生产的原料。

（2）辅料：印染剂和水源。"艺宣阁"粉彩笺使用原纸加工时采用多种水性印染剂进行制作，这些印染剂都从泾县的相关商店购买，颜色与配比由佘贤兵自己试制后才正式批量投入生产。

加工纸的生产用水量很少，对水源要求也不如原纸生产高。泾县艺宣阁宣纸工艺品有限公司自建水塔，工厂使用的水源从水井中抽取后灌入水塔，随用随取。据调查组现场测试，"艺宣阁"加工纸所用水pH为6.23，呈弱酸性。

2. "艺宣阁"粉彩笺生产的工艺流程

⊙1 染料配制 Modulating the dye

据佘贤兵介绍，以及综合调查组2015年8月8日在泾县艺宣阁宣纸工艺品有限公司生产车间的实地调查，归纳"艺宣阁"粉彩笺生产的工艺流程为：

壹 套色印染 → 贰 检验、剪纸 → 叁 成品包装

壹 套色印染

1 ⊙2～⊙4

首先按照桶计量，一般一桶需要高岭土粉3 kg，胶水1.2 kg，防渗增稠剂0.6 kg，胺水0.05 kg和适当水性颜料混合，放在一边备用；其次将原纸放在板上，再把丝网放在原纸上，用刮刀将这些染料涂在原纸上。涂好后的纸放在一边阴干，一般第二天即可收回，如遇到阴雨天则需要2～3天。

⊙2

⊙3

⊙4

贰 检验、剪纸

2 ⊙5⊙6

将已阴干收回的粉彩笺进行品质检验，遇到不合格的纸取出另行处理，合格的纸按照市场需求或者客户要求的规格，在裁纸机上进行裁剪或者使用美工刀进行裁剪。

⊙5

⊙6

⊙2 上纸 Pasting the paper
⊙3 浇染料 Watering the dye
⊙4 套色印染 Overprinting the paper
⊙5 检验 Checking the quality
⊙6 准备裁剪纸 Preparing to cut the paper

叁 成品包装

将裁剪好的纸按照客户需求分好，再加盖"艺宣阁"的加工纸章，包装完毕后运入仓库。

⊙7

（二）"艺宣阁"蜡染笺

1. "艺宣阁"蜡染笺生产的原料

（1）主料：原纸。据佘贤兵介绍，"艺宣阁"蜡染笺的原纸为楮皮纸，在试验阶段使用的是"红星"牌棉料夹宣纸。现在批量生产的楮皮纸为纯楮皮纸，在泾县载元堂工艺厂定制。

（2）辅料：印染剂和水源。"艺宣阁"蜡染笺使用原纸加工时采用各种水性印染剂。据佘贤兵介绍，这些水性印染剂均从泾县当地的代理商处购买。

⊙8
⊙9

2. "艺宣阁"蜡染笺生产的工艺流程

据佘贤兵介绍，以及综合调查组2015年8月8日在泾县艺宣阁宣纸工艺品有限公司生产车间的实地调查，归纳"艺宣阁"蜡染笺生产的工艺流程为：

壹 印刷 涂蜡 → 贰 检验 剪纸 → 叁 成品包装

⊙7 包装 Packing the paper
⊙8 购自载元堂工艺厂的楮皮原纸 Mulberry bark paper bought from Zaiyuantang Craft Factory
⊙9 水性印染剂浆料 Waterborne printing agent paste

壹 印刷、涂蜡

1 ⊙10~⊙12

"艺宣阁"蜡染笺是在粉彩配方的基础上，再加上水溶性蜡进行配比的。工序为：先用丝网在楮皮纸上印上图案打底后阴干，将阴干的纸放在房间压一天，主要是压平回潮，将回潮后的纸用羊毛刷染料收光后，再用光滑的研石和蜡块进行交错研光、打蜡。

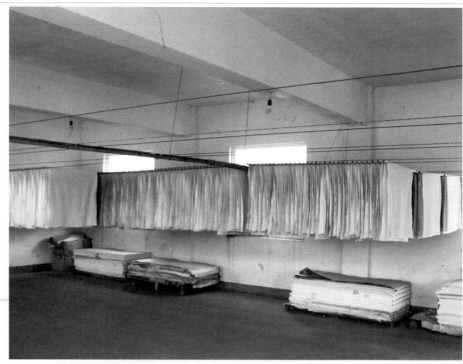

⊙10

贰 检验、剪纸

2 ⊙13

将涂好蜡的蜡染笺进行检验，遇到不合格的纸取出另行处理，合格的纸按照市场需求或者客户要求，在裁纸机上进行裁剪或者使用美工刀进行裁剪。

⊙11　　⊙12

叁 成品包装

3 ⊙14

将剪好的纸按照客户需求分好，再加盖"艺宣阁"的加工纸章，包装完毕后运入仓库。

⊙13

⊙14

⊙10 阴干 Drying the paper in the shadow
⊙11/⊙12 收光与交错研光 Smoothing and polishing the paper
⊙13 美工刀裁剪 Cutting the paper by art knife
⊙14 盖章 Sealing the paper

(三)"艺宣阁"粉彩笺和蜡染笺生产的工具设备

壹 丝网板 1

加工纸所用工具,调查时据佘贤兵介绍,一般在泾县当地购买。实测制作"艺宣阁"粉彩笺和蜡染笺的丝网板尺寸为内长181 cm,内宽88 cm;外长186 cm,外宽92 cm。

⊙1

贰 刮刀 2

在丝网上刮印染剂所用,刀柄为木头,刀头为橡胶。实测制作"艺宣阁"粉彩笺和蜡染笺所用刮刀尺寸为长78 cm,宽12.5 cm,其中橡胶长3.5 cm。

⊙2

叁 研光石 3

为"艺宣阁"蜡染笺打磨涂蜡所用,长方形,外表十分光滑。实测制作"艺宣阁"蜡染笺所用研光石尺寸为长19.5 cm,宽10 cm,高3.5 cm。

⊙3

肆 裁纸机 4

用来裁剪制作好的加工纸,机器控制,可以精准地裁剪出所需要的尺寸。

⊙4

五 泾县艺宣阁宣纸工艺品有限公司的市场经营状况

5 Marketing Status of Yixuange Xuan Paper Craft Co., Ltd. in Jingxian County

泾县艺宣阁宣纸工艺品有限公司目前厂址占地约6 000 m²，厂区投资共1 000万元左右。截至2017年6月2日第三次调查时，年销售额1 500万元，年产量约2万刀，利润率在20%左右。其中雅韵粉彩笺销售最好，年销售量1万刀，单价336元/刀。销售模式主要是经销商销售，截至2017年，全国大部分一、二线城市都有其经销商分布，共约200多家经销商，其中北京和郑州经销商的销售量占全年总销量的20%左右。泾县艺宣阁宣纸工艺品有限公司在洛阳还开有一家直营店，并且在2015年由佘贤兵的妻子沈丽红开始在淘宝网上开设专营店，销售量占全年销售量的10%左右。

2016年9月泾县艺宣阁宣纸工艺品有限公司与中国邮政集团公司合作，为其邮票制作手工纸册页10 000套。此前还一直与大英博物馆合作，定制古籍古画的复制、打印和印刷照片所用的加厚夹宣。

⊙5

⊙6

值得注意的是，佘贤兵与厂里工人从2012年开始就不断研发新产品，希望拓宽销售渠道。截至2017年6月调查时，泾县艺宣阁宣纸工艺品有限公司开始正式量产专用于打印的手工纸，与传

⊙5 精致粉彩泥金泥银卷筒纸（厂家供图）
Delicate colored gold and silver paper (provided by the factory)

⊙6 中国邮政集团邮票册页
The stamp album released by China Post Group

⊙1

⊙2

统手工纸不同的是,这种手工纸更换了涂层涂料和原材料,使着墨的还原度更加真实。据佘贤兵介绍,当初萌发研发这一新纸品是因为一个朋友在无意中提及,但是自己后来坚持研发了5年之久,直到现在才可以开始量产。目前打印使用的纸价格定为40~50元/刀。

泾县艺宣阁宣纸工艺品厂除了维持正常的生产经营外,还不断加强与外界的联系;除挖掘并恢复传统失传的加工纸生产外,不断尝试跨界探索,在加工纸的产品包装等方面也领先于同行业。如首创了粉彩笺,改进了粉蜡笺,并形成了蜡染笺这一新加工门类,引起了市场的关注。在传统加工纸的挖掘方面,推出了清代宫廷龙纹纸等产品。目前,泾县艺宣阁宣纸工艺品有限公司销售渠道已经覆盖到全国一线城市和大部分二线城市,主要以国内经销为主。截至调查时的2017年初,"艺宣阁"粉彩笺销售价为400元/刀,"艺宣阁"蜡染笺为1 600元/刀。

⊙1 龙纹纸 Paper with dragon decorations
⊙2 纸库 Warehouse of paper

六
泾县艺宣阁宣纸工艺品有限公司的品牌文化与民俗故事

6
Brand Culture and Stories of Yixuange Xuan Paper Craft Co., Ltd. in Jingxian County

⊙3
韩美林题写的"艺宣阁"
The plaque of "Yixuange" inscribed by Han Meilin

"艺宣阁"品牌的由来比较偶然。创始人佘贤兵初中毕业后即到振裕印刷厂工作,长期接触泾县地方图书、资料的印制,受到乡土文化感染较深。在筹划创办宣纸加工厂时,在气象部门工作的舅爷是泾县较为知名的乡土文化学者。据佘贤兵回忆,舅爷当年认为,宣纸是国宝,是为中国书画艺术提供服务的宝贝,办企业赚钱重要,但是要为中国艺术做好服务才有境界,办厂的人如果没有境界就不会有大发展。所以舅爷提示佘贤兵:要创办一个为书画艺术服务的机构,便为其取名"艺术宣纸研制中心"。

不过佘贤兵在办理工商登记时,由于厂子当时的规模很小,而泾县城里有"宣纸大厦"这样的品牌,不免心里惶惑,自己的企业充其量只是泾县造纸行业里的一个小房间、小阁楼,便将舅爷起的"艺术宣纸研制中心"简化为相对低调的"艺宣",联系自身定位问题,将企业名称组合成"艺宣阁"。

也正是源于舅爷的提示,佘贤兵一直坚持"产品为艺术服务"的理念,坚持从挖掘传统、拓展应用这两端出发推陈出新,逐步成为泾县加工纸领域的领头羊。随着企业规模的不断扩大,社会美誉度与关注度也越来越高,先后有杨仁恺、韩美林、张海、苏士澍等一批书画名人为企业题名题匾。

七 泾县艺宣阁宣纸工艺品有限公司的传承现状与发展思考

7 Current status of Business Inheritance and Thoughts on Development of Yixuange Xuan Paper Craft Co., Ltd. in Jingxian County

作为创始人兼泾县颇有影响的造纸企业家,佘贤兵的发展理念较为清晰,即一方面抓住机会开放性学习专业机构的宝贵知识,另一方面抓住机会借力现代技术和创新工艺,并挖掘古代加工名纸的技艺传统,拓展加工纸当代适用的消费领域,实现高端产品和大宗消费产品的并行区隔发展。传统加工纸行业与古典木板水印、丝网版印制等技术融合紧密,佘贤兵结合自己在印刷行业多年积累的经验,探索加工纸的恢复性传承和拓展性创新,为"艺宣阁"的发展打造了一条富有特色的路,从而成为泾县乃至全国加工纸行业的代表性厂家。

访谈中据佘贤兵介绍,2009年以来,作为参与承担国家文物局"指南针"项目等工作的民营加工纸企业,泾县艺宣阁宣纸工艺品有限公司通过与国家级研究单位的紧密合作,经过反复的试

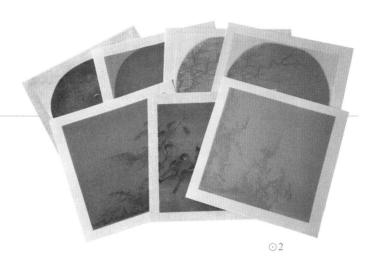

⊙1 蜡染笺禅意手札(厂家供图)
Laran letter paper of Zen style (offered by the factory)

⊙2 仿古蜡染笺(厂家供图)
Antique Laran paper (offered by the factory)

验探索,仿制历史上名声极大的一代名纸"澄心堂纸""金粟山写经纸""兰亭蚕纸""宣德贡笺""蠲纸"等,摸索、提炼了历史名纸加工的材料、工艺并试验性地恢复生产,其努力得到国家文物局、故宫博物院、北京大学、中国科学院

自然科学史研究所等的认可。

在2016年4月的访谈中，佘贤兵对于加工纸行业的现状以及泾县艺宣阁宣纸工艺品有限公司的发展发表了如下看法：

第一，目前加工纸市场比较乱，鱼龙混杂，多数企业的产品档次不高，一些纸企、纸商用价格和品质都处于低端的机械书画原纸加工后充当中高端加工纸出售，以次充好；而普通消费者辨识能力不足，这已经较严重地扰乱了市场，给包括艺宣阁宣纸工艺品有限公司在内的规范生产厂家带来了冲击，迫切需要有市场监管的力量进行干预。

第二，高端加工纸市场在刚进入21世纪的那几年一直都很繁荣，近2～3年受到礼品市场清理和规范等因素的影响，加上网上营销渠道的火爆，价格要素变得相当敏感，市场接受能力和空间都突然收缩，给不少做高端加工纸的企业带来了产能过剩的压力，这是需要关注的行业新问题。

第三，加工纸行业发展到今天，已经可以通过科技手段对品质和工艺进行精准分析与提升。但作为泾县实力最强的加工纸企业，艺宣阁宣纸工艺品有限公司没有专业的检测设备，基本上只是凭操作者的感觉和经验进行加工，实际上已经限制了产品优化发展的空间。佘贤兵表示：他正在积极寻找专业的科研机构进行合作，在挖掘传统加工工艺基础上更好地借助科技力量来创新，希望能率先建立加工纸行业的科学化发展模式。

⊙3
"艺宣阁"富有特色的门楣
The featured lintel of "Yixuange".

加工纸

泾县艺宣阁宣纸工艺品有限公司
Processed Paper of Yixuange Xuan Paper Craft Co., Ltd. in Jingxian County

「艺宣阁」粉彩笺透光摄影图
A photo of "Yixuange" Fencai paper seen through the light

泾县艺宣阁宣纸工艺品有限公司
加工纸

Processed Paper of Yixuange Xuan Paper Craft Co., Ltd. in Jingxian County

［艺阁］蜡染笺透光摄影图
A photo of "Yixuange" Laran paper seen through the light

第四节
泾县宣艺斋宣纸工艺厂

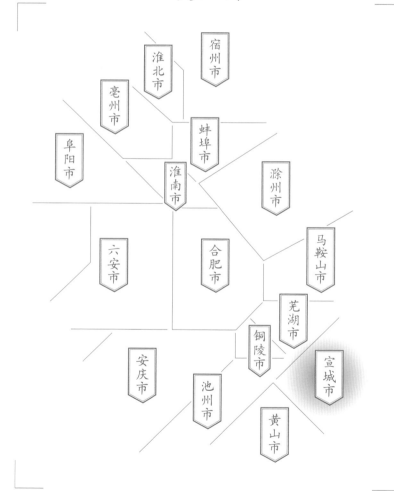

安徽省
Anhui Province

宣城市
Xuancheng City

泾县
Jingxian County

调查对象
泾川镇
泾县宣艺斋宣纸工艺厂
加工纸

Section 4
Xuanyizhai Xuan Paper Craft Factory in Jingxian County

Subject
Processed Paper of Xuanyizhai Xuan Paper Craft Factory in Jingxian County in Jingchuan Town

一 泾县宣艺斋宣纸工艺厂的基础信息与生产环境

1
Basic Information and Production Environment of Xuanyizhai Xuan Paper Craft Factory in Jingxian County

泾县宣艺斋宣纸工艺厂坐落于泾县泾川镇城西工业集中区内的坊林大道旁（原太园乡园林村），地理坐标为东经118°22′12″，北纬30°41′24″。泾县宣艺斋宣纸工艺厂创建于1987年10月，注册商标为"宣艺斋"，主要从事书画纸和宣纸加工纸的生产。2015年8月13日和2016年4月16日，调查组先后两次对泾县宣艺斋宣纸工艺厂进行了田野调查和工艺访谈。

2015年8月13日，调查组入泾县宣艺斋宣纸工艺厂生产厂区调查获得的基础信息如下：泾县宣艺斋宣纸工艺厂共有2个生产点，合计占地约4 667 m²，厂房建筑面积约2 000 m²，两个生产区相距约500 m。其中一个为原纸的生产区，配置了4帘槽的生产设备，从事书画纸原纸的生产。调查时，该厂区共有2个院落，原纸生产流程中的各车间依北面的院落分展布局，捞纸槽位有4个，分别为六尺槽1个、尺八屏槽1个、"一改二"（四尺）槽1个、六尺与尺八屏共用槽1个。调查当天，只有"一改二"槽在生产。现场据厂长张金泉介绍，由于宣纸加工成本昂贵，宣艺斋宣纸工艺厂目前只捞制以龙须草浆板为主料的书画纸。

另一个厂区为加工纸的制作场所，是泾县宣艺斋

⊙1

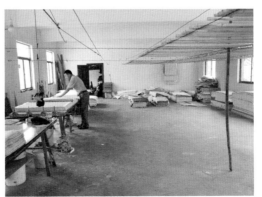
⊙2

1 泾县宣艺斋宣纸工艺厂大门
The main entrance of Xuanyizhai Xuan Paper Craft Factory in Jingxian County
2 加工纸厂区车间内景
Internal view of the workshop of the processed paper factory

泾县宣艺斋宣纸工艺厂位置示意图

Location map of Xuanyizhai Xuan Paper Craft Factory in Jingxian County

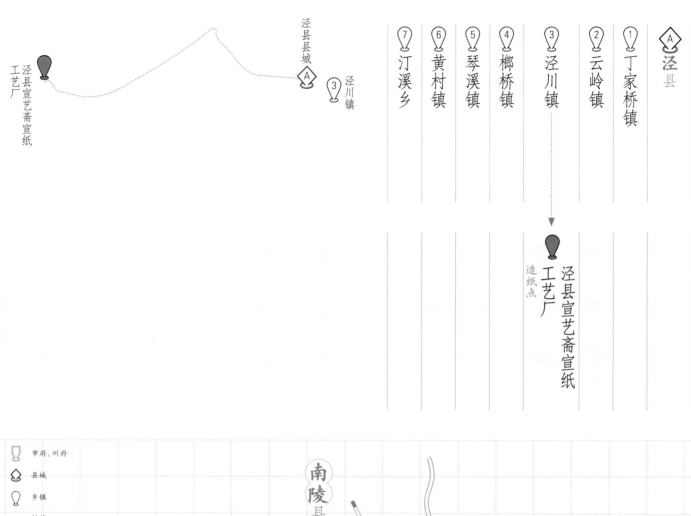

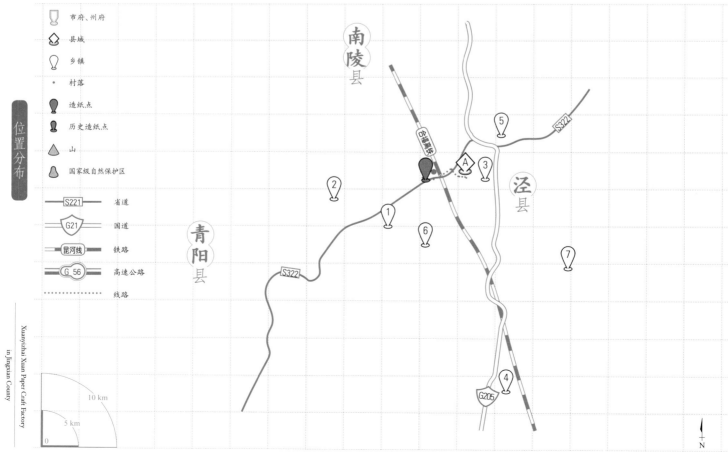

○1

宣纸工艺厂的主厂区，空间内包括各类加工纸制作、纸厂办公和产品销售展陈。该厂区布局呈四合院的形式，北面沿省道的厂房已改成销售商铺，东面和西面的厂房主要为加工车间，南面的厂房为成品库和办公区。

据张金泉介绍，泾县宣艺斋宣纸工艺厂近年的年产量约为30 000刀，其中原纸年产量约为15 000刀。"宣艺斋"纸品中粉彩纸笺、洒银纸笺、洒金纸笺为其代表产品，其中，泾县宣艺斋宣纸工艺厂生产的加工宣纸系列产品入选2003年9月的"中国质量万里行"消费者放心产品。

二 泾县宣艺斋宣纸工艺厂的历史与传承情况

2
History and Inheritance of Xuanyizhai Xuan Paper Craft Factory in Jingxian County

访谈中张金泉回忆：1987年他在园林村创办泾县宣艺斋宣纸工艺厂时，启动资金约为2万元，因资金少，故从事的业务只是单纯的宣纸加工。原纸都从当年的泾县宣纸厂（中国宣纸股份有限公司前身）、泾县小岭宣纸厂、泾县百岭坑宣纸厂（泾县双鹿宣纸有限公司前身）购买。但购买原纸压缩了利润空间，为降低生产成本，1992年，张金泉咬牙开设了2帘槽的原纸生产，以收购价格相对便宜的宣纸加工边角料为原料。1994年，纸厂业务规模不断扩大，

○2 张金泉
Zhang Jinquan

○1 "中国质量万里行"颁发的证书
Certificate issued by "China Association for Quality Promotion"

张金泉等人在老厂不远处兴建了新的厂区，即为调查时生产加工纸兼办公区的主厂区。

张金泉，1964年生，1983年高中毕业后，前往当时泾县泾川镇园林村村办企业百岭轩宣纸工艺厂从事宣纸染色加工工作，一直持续到泾县宣艺斋宣纸工艺厂建立的1987年，这期间积累了宣纸加工的经验。

据张金泉回忆，1986年左右，泾县乡镇企业兴起，大多为私营企业，张金泉所在的村办企业百岭轩宣纸工艺厂也受其影响，于是自己动了创

角料为原料生产自用宣纸。随着使用龙须草原料在泾县造纸行业内的流行，泾县宣艺斋宣纸工艺厂也开始使用龙须草浆板制作书画纸原纸，部分产品直接销售，部分产品自己加工后营销。随着市场的拓展，泾县宣艺斋宣纸工艺厂逐渐扩充到4帘"一改二"的产能。调查组2016年第二次入厂时，该厂维持2帘槽的生产。

张金泉在加工纸上的探索让他获得不少的社会褒奖，比较突出的如2001年被泾县县委和泾县人民政府评为"加工状元"，这在原纸及加工纸

办企业的念头，希望将所学之技与实践经验用起来。1987年，年仅22岁的张金泉筹集了2万元的启动资金，创办了泾县宣艺斋宣纸工艺厂。

纸厂起初生产宣纸印谱、册页、木版水印等加工纸品，直接销往全国部分省会城市的文房四宝专卖店，这种销售模式持续约5年。1992年，受福建商人启发，张金泉开始尝试品牌的广告推广，便大胆地在《中国书画报》等专业媒体上做平面广告宣传，市场自此逐步扩大，销售模式由直销转为订单销售，外部订单也逐渐增多。

1992年纸厂开设了2帘槽生产原纸，以收购中国宣纸集团公司劳动服务公司等厂的宣纸加工边

生产的集聚地的泾县是不容易的事。

张金泉有两个女儿：大女儿张琼2016年30岁，主要从事泾县宣艺斋宣纸工艺厂的实体和网络销售工作。一个比较值得提出的关系是，张琼与泾县双鹿宣纸有限公司的法人代表张先荣的儿子结婚，这层关系加深了两家纸厂在原纸、加工纸制作和销售上的合作。小女儿目前在韩国留学，未从事与宣纸相关行业的工作。

⊙1 厂区一角 A corner of the factory
⊙2 生产现场 Production site
⊙3 张金泉所获的"加工状元"荣誉证书 "Top in the processed paper industry" certificate of honor received by Zhang Jinquan

三 泾县宣艺斋宣纸工艺厂的代表纸品及其用途与技术分析

3 Representative Paper and Its Uses and Technical Analysis of Xuanyizhai Xuan Paper Craft Factory in Jingxian County

（一）

代表纸品及用途

泾县宣艺斋宣纸工艺厂从事书画纸和加工纸的生产，其中加工纸为主营业务。调查中据张金泉介绍，代表纸品为粉彩纸笺、色宣和洒金洒银纸笺。粉彩纸笺采用双层宣纸，熟度为三分或五分，共有10种颜色，分别为瓦灰色、宫廷黄、石青色、翠绿色、浅仿古色、米黄色、深仿古、佛教黄色、明紫色、檀香色。为增加外观美感，宣艺斋宣纸工艺厂的粉彩纸笺绘有描金云龙、云鹤、冷金，描银则为梅、兰、竹、菊。截至调查时在售的主要有四尺、六尺和八尺3种规格（具体长度、宽度见表6.7），其中八尺规格的纸为十色冷金宣，适用于书法创作。

色宣为经过染色后的宣纸，主要为单层生纸，共有10种颜色，分别为仿古色、佛教黄、淡黄、银灰、血青、浅蓝、浅绿、肉红、米色、白色。截至调查时在售的主要有四尺、六尺、八尺和尺八屏4种规格的色纸（具体长度、宽度见表6.7），主要用于书法和绘画用纸、裱托用纸、拓碑用纸。据张金泉介绍，销量较好的纸品为四尺全熟仿古色宣，系著名书法家曹宝麟专门定制的纸品，主要用于临摹画、工笔画、硬笔书法、小楷书法。

⊙4

⊙5　⊙6

⊙4 「宣艺斋」粉彩宣（厂家供图）"Xuanyizhai" Fencai Xuan paper (offered by the factory)

⊙5 「宣艺斋」色宣（厂家供图）"Xuanyizhai" Dyed Xuan paper (offered by the factory)

⊙6 「宣艺斋」仿古色宣 "Xuanyizhai" antique Dyed Xuan paper

洒金洒银纸笺为泾县宣艺斋宣纸工艺厂另一代表纸品，共有生宣洒金和熟宣洒金、洒银两个系列。生宣洒金为单层生纸，共有5种颜色，分别为米色、深仿古色、白色、佛教黄色、浅仿古色。截至2015年8月在售规格仅有四尺，主要用于书法练习。半熟宣洒金、洒银为单层生纸，熟度为五分，共有10种颜色，分别为瓦灰色、宫廷黄、石青色、翠绿色、浅仿古色、米黄色、深仿古、佛教黄色、明紫色、檀香色。截至2015年8月在售规格有四尺、六尺、八尺和尺八屏4种，适用于书法练习。

⊙1

表6.7 泾县宣艺斋宣纸工艺厂主要纸品规格（截至2015年8月在生产的纸品）
Table 6.7 Main paper specifications of Xuanyizhai Xuan Paper Craft Factory in Jingxian County
(Paper products produced up to August 2015)

品名	规格	尺寸（长×宽，cm）
粉彩宣	四尺	136×66
	六尺	176×95
冷金宣	八尺	260×66
色宣	四尺	138×70
	六尺	180×97
	尺八屏	234×53
全熟仿古色宣	四尺	133×66
洒金、洒银 （五分熟）	四尺	133×66
	六尺	174×92
	八尺	260×66
	尺八屏	234×53
洒金（生宣）	四尺	138×70

（二）代表纸品的技术分析

"宣艺斋"代表加工纸品种虽然很丰富，但原纸的同质性强，因此选择米黄洒银加工纸一种进行分析。

测试小组对"宣艺斋"米黄洒银加工纸所做的性能分析，主要包括厚度、定量、紧度、抗张力、抗张强度、撕裂度、色度、吸水性等。按相应要求，每一指标都需重复测量若干次后求平均值，其中定量抽取5个样本进行测试，厚度抽取10个样本进行测试，抗张力抽取20个样本进行测试，撕裂度抽取10个样本进行测试，色度抽取10个样本进行测试，吸水性抽取10个样本进行测试。对"宣艺斋"米黄洒银加工纸进行测试分析所得到的相关性能参数见表6.8。表中列出了各参数的最大值、最小值及测量若干次所得到的平均值或者计算结果。

表6.8 "宣艺斋"米黄洒银加工纸相关性能参数
Table 6.8 Performance parameters of "Xuanyizhai" Cream-colored Sayin processed paper

指标		单位	最大值	最小值	平均值	结果
厚度		mm	0.103	0.080	0.095	0.095
定量		g/m²	—	—	—	32.5
紧度		g/cm³	—	—	—	0.342
抗张力	纵向	N	16.3	10.8	13.9	13.9
	横向	N	10.1	8.2	9.2	9.2
抗张强度		kN/m				0.770
撕裂度	纵向	mN	420	370	392	392
	横向	mN	500	460	476	476
撕裂指数		mN·m²/g	—	—	—	13.1

⊙1
洒金纸笺

续表

指标	单位	最大值	最小值	平均值	结果
色度	%	59.1	58.8	59.0	59.0
吸水性	mm	—	—	—	较难吸水

由表6.8中的数据可知，"宣艺斋"米黄洒银加工纸最厚约是最薄的1.288倍，经计算，其相对标准偏差为0.007，纸张厚薄较为一致。所测"宣艺斋"米黄洒银加工纸的平均定量为32.5 g/m²。通过计算可知，"宣艺斋"米黄洒银加工纸紧度为0.342 g/cm³，抗张强度为0.770 kN/m，抗张强度值较大。所测"宣艺斋"米黄洒银加工纸撕裂指数为13.1 mN·m²/g。

所测"宣艺斋"米黄洒银加工纸平均色度为59.0%。色度最大值是最小值的1.005倍，相对标准偏差为0.116，色度差异相对较小，较难吸水。

四 "宣艺斋"洒银纸生产的原料、工艺流程和工具设备

4
Raw materials, Papermaking Techniques and Tools of "Xuanyizhai" Sayin Paper

（一）"宣艺斋"洒银纸生产的原料

1. 主料

（1）青檀树皮。泾县宣艺斋宣纸工艺厂每年11月过后前往泾县当地农户家中收购毛皮。2015年8月，檀皮收购价格为850元/50 kg，出浆率约为25%。据张金泉介绍，2012年后由于受到环保政策的影响，纸厂不再自行加工檀皮浆，所收购而来的毛皮送至泾县双鹿宣纸有限公司加工成檀皮浆再运回纸厂混合。

(2) 龙须草。截至调查时，泾县宣艺斋宣纸工艺厂所生产的洒银纸原纸为书画纸。原纸为该厂自行生产，采用龙须草浆板加青檀皮捞制。据张金泉介绍，纸厂所采用的龙须草都从泾县丁家桥镇浆板代理商处购买，据代理商介绍，其龙须草浆板原先来自湖北丹江口一带，受到南水北调工程影响，现从河南郑州采购龙须草漂白浆板居多，成本约为11 500元/t。

2.辅料

(1) 纸药。制作"宣艺斋"洒银纸使用的纸药为化学纸药聚丙烯酰胺。

(2) 胶。因手工纸的原纸纤维孔粗大，将胶涂在纸上能起到阻墨的作用，加工纸有很多品种都需要用胶来对纸张进行加工。泾县宣艺斋宣纸工艺厂所生产的洒银纸熟度约为三分至五分之间，因此在洒银之前需使用胶。据张金泉介绍，该厂所使用的胶为骨胶或糨糊水，购买于泾县当地经销商，1刀四尺纸所需的骨胶成本约为4元，糨糊水成本约为3元。

(3) 铝箔。据张金泉介绍，"宣艺斋"洒银纸上的银片实为铝箔，购买于泾县当地经销商，1刀四尺纸所需的铝箔成本约为10元。

(4) 水。泾县宣艺斋宣纸工艺厂造纸时所使用的水源来自当地地下水，经调查时现场测量，其pH为6.29，呈偏酸性。

① 堆放着的青檀皮原料 Stacked raw materials of *pteroceltis tatarinowii* maxim. bark
② 龙须草浆板原料 Raw materials of eulaliopsis binata pulp board
③ 盛在盆中的"胶" "Glue" in the basin
④ 铝箔原料 Raw materials of aluminum foil
⑤ 地下水 Groundwater

(二)
"宣艺斋"洒银纸生产的工艺流程

据张金泉描述,以及综合调查组2015年8月13日和2016年4月16日在泾县宣艺斋宣纸工艺厂对生产工艺的实地调查,归纳"宣艺斋"洒银纸生产的工艺流程为:

壹	贰	叁	肆	伍	陆	柒	捌
浸泡	挑选	打浆	配浆	捞纸	压榨	晒纸	检验、剪纸

玖	拾	拾壹	拾贰	拾叁	拾肆	拾伍	拾陆	拾柒
搭棍子	上胶	洒银	压纸	悬挂风干	收纸、理纸	掸银	裁剪	装箱打包

工艺流程

壹 浸泡 1

在打浆之前，需先将购买来的龙须草浆板放在池中浸泡，约浸泡24小时，等浆板软化后再进行下步打浆。

贰 挑选 2

将在双鹿宣纸厂加工的漂白檀皮运至宣艺斋宣纸工艺厂后，需人工进行挑选，挑出其中的黑皮和杂质。同时在龙须草浆板浸泡过程中也需要人工剔选出黑色浆料和杂物。

叁 打浆 3

皮料和草料分开打浆。皮料制浆是将挑好的皮料放入打浆机中即可，而草料制浆则需先将浸泡后的浆板堆放一段时间，等水分沥干后再放入打浆机中进行打浆。

⊙1

肆 配浆 4

据张金泉介绍，洒银纸的原纸为含皮量比较高的书画纸，这种书画纸中檀皮所占比为30%~35%，其余为龙须草浆。如果遇到别人定做，有的纸檀皮含量最高可达50%。张金泉还特别表示：实际上檀皮含量过高容易导致纸张匀度不足，反而不便画家创作，如客户坚持要高檀皮配比，厂里虽然会委婉提示但也接受制作。将配浆池配好后的浆料作为纸浆运至捞纸车间进行捞纸。

伍 捞纸 5

（1）和浆。通过配比的混合浆进入捞纸车间后，工人按照自己的生产需求将混合浆放入纸槽中，注入水，用扒头将混合浆搅拌均匀，再按照一定比例放入化学纸药后搅匀。捞纸过程中工人会根据自己的经验增加混合浆和水。

⊙2

（2）捞纸。从事捞纸的工人共分两种岗位，一为掌帘，一为抬帘，操作办法与泾县的宣纸、手工书画纸操作基本一致。

⊙3

⊙4

（3）放纸。掌帘工会将有湿纸的一面朝下，按照已放好的湿纸边际放置。放置时，会从左边倾斜放置纸帘，然后沿着下方湿纸的边际慢慢落帘，当确保整张纸帘放置完毕后，再从额头向梢部迅速将帘揭起。

陆 压榨 6

捞纸工下班后,帮槽工先将湿纸帖盖上盖纸帘,帘上方盖木板,木板上架上榨杆、木枕,木枕上方加千斤顶,通过千斤顶将纸帖榨干。

柒 晒纸 7

(1)烘帖。将榨干的纸帖当天晚上送到纸焙上,利用纸焙余温进行烘烤。次日晒纸工上班时,将纸帖架在纸焙上方继续烘烤。

(2)浇帖。将烘烤好的纸帖平放在浇帖架上,用水慢慢将其润湿。

(3)鞭帖。将浇好的纸帖架在纸架上,用鞭帖板抽打纸帖,使其反弹后便于晒纸。

(4)做额。鞭帖后,用木棍或手指甲在纸的上方从一侧划至另一侧,疏松每张纸上方的边,方便下一步揭纸。

(5)揭纸。晒纸工在纸的左上角起头,将纸从左至右揭下。

(6)晒纸。晒纸工人先将揭下来的纸张上方贴在晒纸墙上,然后用刷子将其从上到下刷满,让其完全浮贴在晒纸墙上,并确保其表面平整。

(7)收纸。在晒纸墙贴满后,再根据晒纸时的上墙顺序依次揭下。

捌 检验、剪纸 8

将晒好的纸进行逐张检验。检验工遇到不合格的纸立即取出或者做上记号,积压到一定数量的残次品或废品后,将残次品或废品回笼打浆或者低价出售;检验合格的纸整理好,数好数,一般50张为一个刀口,压上石头,剪纸人站成箭步,按照一定的尺寸持平剪刀一气呵成地剪下去,至此环节,原纸制作工序完成。

玖 搭棍子 9

将原纸某侧约4 cm的范围内刷上糨糊,取一根木棍放在纸的某一端,这一端俗称为"纸头";然后将纸沿着木棍侧面卷起,此过程中需保证与木棍接触的纸面内均已刷上糨糊,卷好经过按压后纸张便可"挂"在木棍上,方便后期晾晒。洒银开始前,将搭好棍子的纸带有棍子的一侧放在桌子上,其余部分平摊在桌子下,便于压纸时取纸。

⊙5 搭棍子 Pasting the paper with sticks

拾 上　胶
10

据张金泉介绍，泾县宣艺斋宣纸工艺厂所使用的纸张熟度为三四分熟，因此需要在纸张表面上胶。上胶之前，工人师傅会按照5 kg水与0.25 kg胶的比例将"胶"调好，大部分情况下原料为骨胶。然后将搭好棍子的纸平铺在桌面上，用排笔从盆中蘸取胶后，从某一侧开始从

⊙6

上至下沿着一定的顺序向另一侧刷纸，直至将纸刷满胶水，使纸变熟。时间长短取决于纸张的尺寸大小，整个过程中纸张仅刷一遍即可。

拾壹 洒　银
11

上完胶后，工人拿起一个小瓶子，瓶中装有铝箔和一个石头，铝箔即为纸上的"银片"，石头方便工人洒银摇晃瓶身时将铝箔打碎。瓶子一端为带孔的盖子，便于铝箔洒出。访谈中据张金泉介绍，铝箔不宜放太满，每次仅放至瓶身的一半即可，一般1瓶铝箔可洒2刀

⊙7

纸左右。

洒银过程中，师傅常从纸张左上侧开始，沿着一定顺序上下左右均匀摇晃瓶身倾洒铝箔，直至洒满整张纸。洒银只需从左到右进行一次即可，时间依据纸张大小不同，以四尺纸来计，一般一个熟练的师傅一次洒银约1分钟。

拾贰 压　纸
12

洒完银后，铝箔往往并不服帖，因此工人会将下一张待加工的纸张放置在刚洒完银的纸张上，放置过程中，上下两层纸边需尽量对齐，保证所有的铝箔均可受力。然后再用棕刷沿着纸张的一侧从上至下、从左到右将纸张刷满。利用棕刷传递的力量和纸张

⊙8

的覆盖面让铝箔服帖在纸上，这一过程便成为"压纸"。

拾叁 悬挂风干
13

上完胶后的纸比较潮湿，因此压纸工序结束后，利用纸张一端所搭的棍子将其悬挂在铁丝上风干。风干所需时间取决于天气的干湿程度，如遇晴天1天便可，如遇雨天则需2天左右。

拾肆
收 纸、理 纸
14

当纸张悬挂风干后进入收纸环节，每次以一刀为计量单位收纸。收下后将纸头原先所搭的棍子抽出，然后放在工作平台上理齐放好，收纸后的这一过程被称为理纸。

拾伍
掸 银
15

虽然经过压纸环节，但是依然还有一些铝箔未完全紧贴在纸上，师傅称之为"没有压扁的铝箔"。因此理完纸后，师傅需用掸子一张一张地掸去纸面上没有压扁的铝箔，这一过程称为"掸银"。

拾陆
裁 剪
16

掸银过后，纸张送往剪纸车间裁去用于搭棍子的"纸头"和周边不规整的部分。以四尺为例，宣艺斋宣纸工艺厂生产的四尺洒银纸尺寸为133 cm×66 cm（长×宽），而原纸的尺寸为138 cm×70 cm（长×宽），因此大约需剪去5 cm×4 cm（长×宽）的大小。整理好后，一般50张为一个刀口，压上石头，剪纸人站成箭步一气呵成地剪下去。

拾柒
装 箱 打 包
17

剪好后的纸按100张一刀折好，再加盖"宣艺斋"、尺寸、品种等的刀口印。一般根据纸张规格以7～11刀装一箱（件），包装完毕后运入仓库。

9 悬挂风干 Hanging and drying
10 刚理好的纸 Just finished paper
11 掸银 Removing the redundant aluminum foil
12 待裁剪的纸 Paper to be cut
13 包装车间 Packaging workshop

(三)
"宣艺斋"洒银纸生产的工具设备

壹 打浆机 1

用来制作浆料的机械，自动搅拌。

⊙1

贰 捞纸槽 2

盛浆工具，宣艺斋宣纸工艺厂的捞纸槽为水泥浇筑，调查时实测"一改二"捞纸槽尺寸为长350 cm，宽94 cm，外立面高80 cm，内深50 cm；尺八屏捞纸槽尺寸为长294 cm，宽223 cm，外立面高80 cm，内深50 cm；六尺捞纸槽尺寸为长244 cm，宽224 cm，外立面高80 cm，内深50 cm；八尺和六尺合用捞纸槽尺寸为长352 cm，宽225 cm，外立面高80 cm，内深50 cm。

⊙2

叁 纸帘 3

用于捞纸，竹丝编织而成，表面很光滑平整，帘纹细而密集。实测宣艺斋宣纸工艺厂生产时所使用的"一改二"纸帘尺寸为长315 cm，宽103 cm。

⊙3

肆 帘床 4

捞纸时放置纸帘的架子，多由竹子或木头制成。

⊙4

伍 扒头 5

捞纸前用于和浆的工具，值得注意的是调查时单槽的扒头已经更新为电动扒头，节约了大量的人力，主要起搅拌作用。

⊙5

陆 浇帖架 6

用于浇帖时盛放纸的木架。

⊙6

柒 刷子 7

晒纸时将纸刷上晒纸墙，刷柄为木制，刷毛为松针。

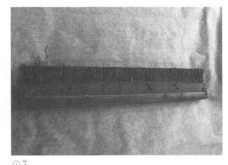
⊙7

捌 纸焙 8

用来晒纸，由两块长方形钢板焊接而成，中间贮水加热，双面墙，可以两边晒纸。

⊙8

玖 剪刀 9

检验后用来剪纸，剪刀口为钢制，其余部分为铁制。

⊙9

拾
排　笔
10

由平列的一排毛笔构成，用于上胶、绘画、裱糊、粉刷等，笔毛材质主要是羊毛，购买于当地的毛笔厂。实测宣艺斋宣纸工艺厂生产时所使用的排笔尺寸为长34 cm，宽17 cm。

⊙10

拾壹
洒银筒
11

盛装铝箔的筒，一端制成蜂窝孔状便于铝箔片洒出。

⊙11

拾贰
棕　刷
12

由棕榈树生长的棕丝和剑麻纤维制成的各种刷类，用于压纸时刷纸。实测宣艺斋宣纸工艺厂生产时所使用的棕刷尺寸为长21 cm，宽15 cm。

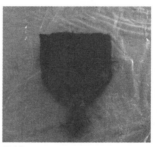

⊙12

拾叁
掸金（银）把子
13

掸金（银）时掸走不服帖铝箔的工具。实测宣艺斋宣纸工艺厂生产时所使用的掸子尺寸为长73 cm。

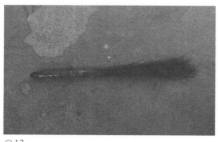

⊙13

五 泾县宣艺斋宣纸工艺厂的市场经营状况

5 Marketing Status of Xuanyizhai Xuan Paper Craft Factory in Jingxian County

据访谈时张金泉提供的销售数据，泾县宣艺斋宣纸工艺厂近年来年销售额300万~400万元，原纸和加工纸销量约各占1/2，而粉彩色宣系列加工纸则约占总销量的1/3，为销量最大的纸品。

2015年8月入厂调查时，泾县宣艺斋宣纸工艺厂已创立28年，其间经历了多次为适应消费潮流转变的销售渠道转变：

开始创业的3~5年，直接销售市场基本上在泾县本地。20世纪90年代张金泉希望市场走出泾县，虽打通了经销商通路，但是价格自主权受到一定限制；在网络销售未诞生的环境下，张金泉看中了邮购这一销售渠道，于是1992年宣艺斋宣纸工艺厂相继在《书法报》《中国书画报》《中国宣纸》等书法类或纸类专业性报纸或期刊上刊登产品广告，开启了经销商和邮购的双渠道经营模式。其中1997年的双渠道销售成效最明显，该年销售额约为100万元。

随着邮政购物逐渐衰落，网络购物成为新流行的销售渠道，张金泉应时开启了这一销售渠道，于2009年在网上开设专营店。2015年8月调查时，宣艺斋宣纸工艺厂一直维持经销商和网络销售的双渠道模式，其销售额各占1/2，其中线下的经销商达30多家，均分布在地级市以上的城市。

泾县宣艺斋宣纸工艺厂成立之初，由于资金不足而缺乏相应的工艺设备，主要以销售原纸和加工耗费流动资金小的洒金纸笺、册页为主。随着资金与技术能力的优化，制版技术随之增强，引进了丝网印制技艺和设备，纸品类型不断丰富。2004年后，纸厂开始陆续引进丝网加工设备，色宣等产品应运而生并逐渐成为主流产品，粉彩宣、花粉宣、描金等进入市场，打造了色宣系列作为优势主导产品占销售量1/3的局面。此后，纸厂又推出了木版水印、水纹宣等产品，形成色宣系列一枝独秀、20多种加工纸品环绕的产品群销售格局。

⊙14
泾县宣艺斋宣纸工艺厂厂牌
Plate of Xuanyizhai Xuan Paper Craft Factory in Jingxian County

张金泉在访谈中谈的一点体会是："现在消费者更加注重产品的外观，一种较为花哨的产品会满足人们的新鲜感，会更加受到青睐。"因此，泾县宣艺斋宣纸工艺厂近年来不断加强对加工纸外观的创新探索，尝试生产出品种更多、形式更美的纸品。

不过，虽然追求纸品的丰富多彩，但是内在原则依然是有区别的。张金泉特别表示：泾县宣艺斋宣纸工艺厂生产的洒金洒银类加工纸，以及裱手绢、扇叶等用途的纸品为传统手工产品；色宣、描金、方格等纸品是在现代工艺（例如丝网）基础上保持部分手工制作的产品，因此后者制作规模较传统手工制品要大得多。

不仅如此，原纸材料也在发生变化。创立之初至1992年，泾县宣艺斋宣纸工艺厂生产加工纸的原纸主要为宣纸中的棉料类，张金泉回忆当时的价格为60～70元/刀（四尺）。后来随着宣纸价格的不断上涨，使用价格昂贵的宣纸做原纸已成为一种奢侈，于是到了1992年，宣纸原纸逐渐被龙须草浆板为主原料的书画纸取代，而且这种替代并非宣艺斋宣纸工艺厂一家，泾县大多数加工纸厂家迫于成本压力，都已选择了书画纸为原纸。不过在交流时张金泉特别声明：这并不意味着以宣纸为原纸的加工纸已消失，如果有客户需求，泾县宣艺斋宣纸工艺厂自然会生产此类加工纸，只是现阶段不再批量生产。

据调查时泾县宣艺斋宣纸工艺厂所提供的价

目表，截至2015年8月，该厂生产的"宣艺斋"品牌代表性纸品中，粉彩宣售价约为255元/刀（四尺）、510元/刀（六尺），色宣售价约为156元/刀（四尺）、312元/刀（六尺）和280元/刀（尺八屏），洒金洒银（熟宣）售价约为235元/刀（四尺，此次调查组采样分析的纸品）、470元/刀（六尺）和330元/刀（尺八屏）。

2015年8月调查组入厂调查时，纸厂共有工人25人，其中一半以上为女工，这与加工纸无需凌晨即起、体力消耗小等工作特性有关。据张金泉

① 融现代工艺的"万年红"纸
②/③ 丝网加工

⊙4

介绍，厂内员工大部分为临时工，因为泾县各类纸厂众多，纸工可不断比较各家薪酬选择跳槽，因此员工稳固性较差；尤其是原纸生产环节，劳动强度大，纸工稳定程度在纸厂内最低，相比之下加工纸生产环节中的员工则相对固定。

在泾县宣艺斋宣纸工艺厂，工人每天工作时间约为12小时，以捞纸工人为例，每天凌晨4点多上班，下午4点多下班。工作量上，不同工种任务不同，以洒银师傅算，一天可洒3～4刀纸；以捞纸师傅算，每天1槽2名工人的捞纸量，以"一改二"规格计算为10多刀，以六尺规格计算约为10刀。工人每星期工作6天，每月约为25天，除去过年期间1个月的假期和周末放假，一年实际生产不足10个月。

六 泾县宣艺斋宣纸工艺厂的品牌文化与民俗故事

6
Brand Culture and Stories of Xuanyizhai Xuan Paper Craft Factory in Jingxian County

1. "宣艺斋"的名称来历

当问及公司名称的由来时，张金泉回忆道："一开始真不知道取什么名称，但是'荣宝斋'名号在文房四宝和书画业界十分响亮，成为许多纸厂纷纷效仿的对象，自己一开始也就不自觉地往上面去想了。"张金泉受到百年老号"荣宝斋"名号的启发，也希望自己的纸厂可如"荣宝斋"一样兴旺多年；同时，身边所见的众多纸厂偏爱采用"斋""堂""轩"等字，于是随大流采用"宣艺"+"斋"的形式。之所以取名"宣艺"，张金泉的说法是因为公司创立之初生产内容为"宣纸加工工艺"，取其中的"宣"和"艺"二字，自己感觉既显得文雅，也直接表明了纸厂的主营产品。

2. 曹宝麟与"宣艺斋"的试纸缘

著名书法家曹宝麟与泾县宣艺斋宣纸工艺厂

关系甚密，不仅为纸厂题字"中国安徽宣艺斋牌宣纸"，还专门定制了"宣艺斋"的熟仿古宣纸为书法用纸。曹宝麟擅长书写米芾体而又有自己变化的小字，用过多种纸来试，但一直苦于寻找不出适合自己笔路的纸。一次偶然试笔过程中，曹宝麟用了"宣艺斋"的纸，发现颇为接近自己所期望的纸，但依然有不顺心顺手处。于是，渴望用到贴心纸的曹宝麟不断将试纸感受和要求反馈给"宣艺斋"，张金泉等人根据他的要求不断改进纸品，最终形成曹宝麟满意的个人定制熟仿古宣纸。

⊙1 曹宝麟为泾县宣艺斋宣纸工艺厂题字（厂家供图）

⊙2 曹宝麟用自己的定制纸写赠张金泉的书法（厂家供图）

七 泾县宣艺斋宣纸工艺厂的传承现状与发展思考

7
Current Status of Business Inheritance and Thoughts on Development of Xuanyizhai Xuan Paper Craft Factory in Jingxian County

⊙3
忙活着的加工纸车间
Processed paper workshop in production

1. 建立固定客户群的深耕策略

访谈中印象比较深的一点是，泾县宣艺斋宣纸工艺厂作为以生产加工纸为主的纸厂，面对2014～2016年行业普遍认为的市场寒冬不以为然。张金泉的说法是：泾县宣艺斋宣纸工艺厂这三年销量保持良好，未出现明显波动，这与销售终端拥有固定的批量客户群关系密切。固定客户群不仅可以维持纸厂原有利益，在一定程度上排除其他竞争对手，也可通过优质客户自身的影响与能量帮助纸厂开发和引进潜在客户群，滚动式扩大市场。

张金泉表示：造纸的企业都知道固定客户群好，但绝不是想有就会有的，固定客户群建立在企业与客户之间多年诚信交易和量身满足需求的基础上。比如，泾县宣艺斋宣纸工艺厂一直通过各种机会搜集和了解客户的需求，全心全意为优质客户制作满足需求的纸品，同时也不断更新散户大众较为偏爱的纸品，加深客户群的稳定性，多年的付出才有困难时期的回报。

2. 眼光独到的广告精准投放策略

据张金泉分析，泾县宣艺斋宣纸工艺厂固定客户群的建立与当初特立独行的广告宣传和与时俱进的销售模式关系甚密。泾县宣艺斋宣纸工艺厂（实际应是张金泉本人）偏爱目标明确、精准投放的广告宣传，例如20世纪90年代在《书法报》《中国书画报》《中国宣纸》等书画类或纸类专业性报刊上登产品广告，这些平台的订阅者也是宣纸需求者，因此广告的目标群体明确；而且，在网络没有发展起来的20世纪90年代，张金泉眼光独到地看中了邮购这一销售渠道，比其他纸厂先人一步的邮购宣传方式，开启了在当地邮购销售模式上的辉煌，同时越早被吸引的客户演变为固定客户的概率越高。

130

加工纸

泾县宣艺斋宣纸工艺厂

Processed Paper of Xuanyizhai Xuan Paper Craft Factory in Jingxian County

131

"宣艺斋"米黄洒银加工纸
透光摄影图
A photo of "Xuanyizhai" Cream-colored Sayin processed paper seen through the light

第五节
泾县贡玉堂宣纸工艺厂

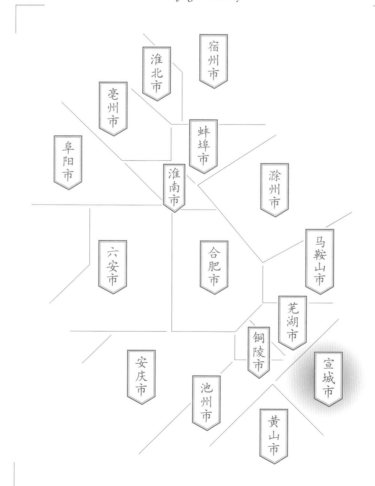

安徽省
Anhui Province

宣城市
Xuancheng City

泾县
Jingxian County

调查对象
泾县贡玉堂宣纸工艺厂
黄村镇
加工纸

Section 5
Gongyutang Xuan Paper Craft Factory in Jingxian County

Subject
Processed Paper of Gongyutang Xuan Paper Craft Factory in Jingxian County in Huangcun Town

一 泾县贡玉堂宣纸工艺厂的基础信息与生产环境

1 Basic Information and Production Environment of Gongyutang Xuan Paper Craft Factory in Jingxian County

泾县贡玉堂宣纸工艺厂坐落于泾县黄村镇紫阳行政村沙丰村民组，地理坐标为东经118°17′18″，北纬30°35′29″。贡玉堂宣纸工艺厂创办于1990年，是一家从事书画纸原纸生产及后续加工产品的民营企业。黄村境内拥有闻名全国的旅游胜地"江南第一漂"，漂流所经地段山清水秀、风光旖旎，在当地素有"小漓江"的誉称。

2015年8月12日和11月下旬，调查组两次入厂现场考察交流，并分别对"贡玉堂"的现负责人王学兵以及已退休的创始人司绍先进行了访谈。

2015年8月12日，在调查组第一次实地调查期间，由于天气等原因，原纸生产处于停产整顿阶段，所见为加工纸生产现场。

1 贡玉堂宣纸工艺厂指示牌
Road sign towards Gongyutang Xuan Paper Craft Factory

2 工人加工墨流宣现场
Workers processing Moliu Xuan paper

3 黄村镇风景
Scenery of Huangcun Town

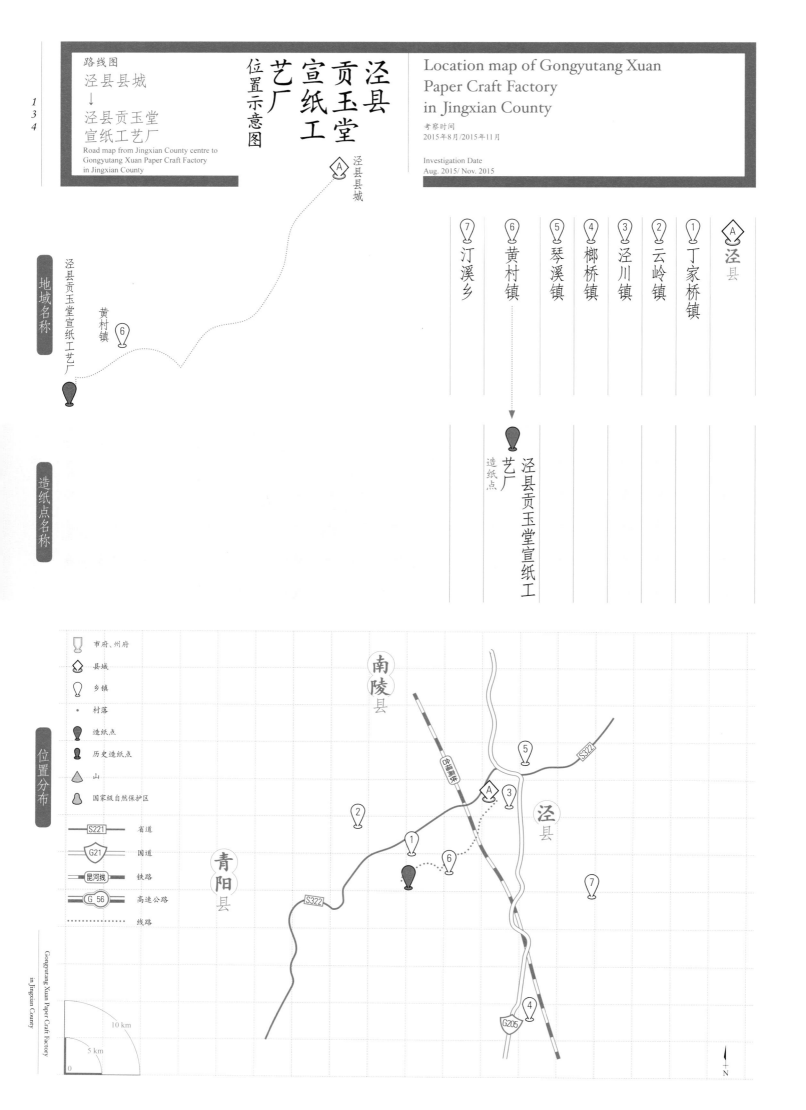

二 泾县贡玉堂宣纸工艺厂的历史与传承情况

2 History and Inheritance of Gongyutang Xuan Paper Craft Factory in Jingxian County

2015年8月12日访谈时，据王学兵介绍，泾县贡玉堂宣纸工艺厂的创始人为司绍先，他本人是在2011年才接手贡玉堂宣纸工艺厂的。

泾县贡玉堂宣纸工艺厂创办于1990年，主要产品是宣纸和书画纸加工产品，创建以来的10年左右一直以外销为主，产品出口到日本、韩国等东亚国家，出口量占总加工量的90%左右，其加工的纸品在东亚地区具有一定的知名度。2012年以来，由于市场情况的变化，"贡玉"加工纸开始转向以国内市场销售为主。

2002年，泾县贡玉堂宣纸工艺厂在从事宣纸和书画纸加工的同时，开始生产原纸，原纸生产的负责人为王学兵与其弟。据王学兵介绍，当时共投资15万元左右，利用自家宅基地修建厂房，购买并安装了造纸生产设备，共耗时2个月完成所有准备工作，当年9月份投入生产，主要生产书画纸，被命名为兴潭造纸厂。

王学兵，1968年生于泾县，1985年初中毕业后，由其亲戚刘荣林（1979~1984年任泾县宣纸厂党委副书记兼小岭宣纸厂总支书记，后任泾县轻工业局局长兼小岭宣纸厂总支书记，1985~1988年任中国宣纸公司副经理，1988年任泾县宣纸工业管理局局长，直至1993年初退休时止。刘荣林也是司绍先和王学兵之间关于贡玉堂宣纸工艺厂转让的中间介绍人）推荐进入泾县小岭宣纸厂西山车间从事晒纸工作9年，由于小岭宣纸厂改制为红旗宣纸有限责任公司后不久停产，便进入泾县宣纸二厂（生产"鸡

⊙1 王学兵 Wang Xuebing

球"牌宣纸）上班。

1996年，紧邻青弋江的泾县宣纸二厂因遭受洪灾停产，在灾后清理期间，因王学兵是临时工，享受不到停工期间的工资待遇，便转到泾县宣纸厂（中国宣纸股份有限公司前身）工作。1998年辞职，到宣城市宣州区高桥镇一私人创办的纸厂作为工艺专家指导生产。1999年受朋友邀请到北京帮朋友开文房四宝店，2002年返乡创办兴潭造纸厂。2004年在北京琉璃厂开办自己的文房四宝店，除了销售自己生产的书画纸外，还销售毛笔、墨汁、砚台等文房用品。

2015年11月，调查组对泾县贡玉堂宣纸工艺厂创始人司绍先老人进行了访谈，据司绍先介绍，他原在泾县宣笔厂担任书记一职，1986年调任原泾县宣纸二厂工作，为扩大就业范围，并解决宣纸厂非规格纸张的再利用问题，泾县宣纸二厂成立了浣月轩，主要进行宣纸加工，司绍先担任浣月轩的主要负责人。在其任职的4年里，浣月轩从无到有，并发展到28个人的规模。

1990年，司绍先离开泾县宣纸二厂回到原来工作的泾县宣笔厂所在地安吴乡（现为黄村镇）紫阳村从事宣纸加工工作。司绍先出生于1940年，2017年已77岁高龄，由于年事已高，子女也有稳定工作，无人愿意转行从事宣纸加工业，老人力不从心之下，便将厂子转让给了王学兵，并在厂里帮助指导生产管理2年，直到2013年回到泾县县城居家养老。

⊙ 1 访谈司绍先老人 Interviewing Si Shaoxian
⊙ 2 工人加工粉蜡笺现场 A worker processing Fenla paper
⊙ 3 紫阳村 Ziyang Village

三 泾县贡玉堂宣纸工艺厂的代表纸品及其用途与技术分析

3 Representative Paper and Its Uses and Technical Analysis of Gongyutang Xuan Paper Craft Factory in Jingxian County

（一）代表纸品及其用途

泾县贡玉堂宣纸工艺厂创始人司绍先是一名技术较为全面的老艺人，能够根据客户提供的纸样，自己选配料，加工出能达到纸样要求的纸，同时对市场的判断也比较准确，生产的多种加工纸均走俏市场。调查时，泾县贡玉堂宣纸工艺厂主要从事书画纸与宣纸深加工系列产品的生产，可根据客户需求自行调制各种规格、颜色的生、熟书画纸与宣纸染色产品。王学兵认为，有代表性的加工纸品是虎皮宣、墨流宣、手工洒金以及染色笺。

在调查中，王学兵较详细地介绍了"贡玉"虎皮宣与墨流宣的制作工艺及其流程。

虎皮宣是传统宣纸原纸加工纸，泾县贡玉堂宣纸工艺厂加工虎皮宣所需的原纸大多由兴潭造纸厂生产，少量的原纸根据客户需要外购不同的宣纸或书画纸品种。王学兵表示，他在2011年初接手贡玉堂宣纸工艺厂时，大部分的原纸使用书画纸原纸，但后来逐渐转向部分使用加檀皮的高档书画纸为原纸进行加工。

泾县贡玉堂宣纸工艺厂加工虎皮宣使用的书画纸较多是龙须草加10%～15%的檀皮，纯手工捞制，每刀纸的重量为3.5 kg左右。生产的虎皮宣系列产品由于绚烂多彩，类似于虎皮花纹，主要用于商品的设计包装以及标签制作，较少用于书法绘画，王学兵觉得虎皮宣未来在装饰用纸中很有希望能占有一片市场。

墨流宣，又称流沙纸，是一种半生半熟的加工纸，纸面上呈现出千变万化的图案，用途广泛，

⊙4 不同颜色的虎皮宣
Different colors of Hupi Xuan paper

⊙5 司绍先展示墨流宣
Si Shaoxian showing Moliu Xuan paper

可用于书法、装饰、纸制工艺品和包装等。司绍先表示：制作墨流宣不仅需要制作者掌握熟练的技艺，还需要选择火候添加2种试剂，这在泾县的加工纸厂中少有做到的，为避免同行竞争，泾县贡玉堂宣纸工艺厂这一制作技艺处于保密状态中。

⊙1 / 2
墨流宣及特写效果
Moliu Xuan paper and its close-up effect

（二）代表纸品的技术分析

1. 代表纸品一："贡玉"虎皮宣

测试小组对采样自泾县贡玉堂宣纸工艺厂生产的"贡玉"虎皮宣所做的性能分析，主要包括厚度、定量、紧度、抗张力、抗张强度、撕裂度、色度、吸水性等。按相应要求，每一指标都需重复测量若干次后求平均值，其中定量抽取5个样本进行测试，厚度抽取10个样本进行测试，抗张力抽取20个样本进行测试，撕裂度抽取10个样本进行测试，色度抽取10个样本进行测试，吸水性抽取10个样本进行测试。对"贡玉"虎皮宣进行测试分析所得到的相关性能参数见表6.9。表中列出了各参数的最大值、最小值及测量若干次所得到的平均值或者计算结果。

表6.9 "贡玉"虎皮宣相关性能参数
Table 6.9 Performance parameters of "Gongyu" Hupi Xuan paper

指标		单位	最大值	最小值	平均值	结果
厚度		mm	0.092	0.077	0.095	0.095
定量		g/m²	—	—	—	29.1
紧度		g/cm³	—	—	—	0.306
抗张力	纵向	N	20.0	11.3	15.8	15.8
	横向	N	11.1	8.7	9.7	9.7
抗张强度		kN/m	—	—	—	0.850
撕裂度	纵向	mN	200	190	196	196
	横向	mN	280	215	247	247
撕裂指数		mN·m²/g	—	—	—	7.2
色度		%	71.5	69.2	70.6	70.6
吸水性		mm	—	—	—	很难吸水

由表6.9中的数据可知,"贡玉"虎皮宣最厚约是最薄的1.195倍,经计算,其相对标准偏差为0.005,纸张厚薄较为一致。所测"贡玉"虎皮宣的平均定量为29.1 g/m²。通过计算可知,"贡玉"虎皮宣紧度为0.306 g/cm³,抗张强度为0.850 kN/m,抗张强度较小。所测"贡玉"虎皮宣撕裂指数为7.2 mN·m²/g。

所测"贡玉"虎皮宣平均色度为70.6%。色度最大值是最小值的1.033倍,相对标准偏差为0.850,色度差异相对较小。很难吸水。

2. 代表纸品二:"贡玉"墨流宣

测试小组对采样自泾县贡玉堂宣纸工艺厂生产的"贡玉"墨流宣所做的性能分析,主要包括厚度、定量、紧度、抗张力、抗张强度、撕裂度、色度、吸水性等。按相应要求,每一指标都需重复测量若干次后求平均值,其中定量抽取5个样本进行测试,厚度抽取10个样本进行测试,抗张力抽取20个样本进行测试,撕裂度抽取10个

⊙3

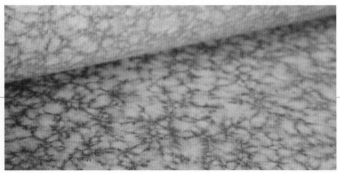

⊙4

样本进行测试,色度抽取10个样本进行测试,吸水性抽取10个样本进行测试。对"贡玉"墨流宣进行测试分析所得到的相关性能参数见表6.10。表中列出了各参数的最大值、最小值及测量若干次所得到的平均值或者计算结果。

表6.10 "贡玉"墨流宣相关性能参数
Table 6.10 Performance parameters of "Gongyu" Moliu Xuan paper

指标		单位	最大值	最小值	平均值	结果
厚度		mm	0.085	0.074	0.080	0.080
定量		g/m²	—	—	—	34.6
紧度		g/cm³	—	—	—	0.433
抗张力	纵向	N	13.4	11.4	12.5	12.5
	横向	N	7.2	6.2	6.6	6.6
抗张强度		kN/m	—	—	—	0.637
撕裂度	纵向	mN	260	220	244	244
	横向	mN	300	260	290	290
撕裂指数		mN·m²/g	—	—	—	22.1
色度		%	71.5	69.2	70.6	70.6
吸水性		mm	—	—	—	17

由表6.10中的数据可知，"贡玉"墨流宣最厚约是最薄的1.149倍，经计算，其相对标准偏差为0.005，纸张厚薄较为一致。所测"贡玉"墨流宣的平均定量为34.6 g/m²。通过计算可知，"贡玉"墨流宣紧度为0.433 g/cm³，抗张强度为0.637 kN/m，抗张强度较小。所测"贡玉"墨流宣撕裂指数为22.1 mN·m²/g。

所测"贡玉"墨流宣平均色度为70.6%。色度最大值是最小值的1.033倍，相对标准偏差为0.855，色度差异相对较小。所测"贡玉"墨流宣吸水性纵横平均值为17 mm。

⊙1
车间里现场悬挂的墨流宣
Moliu Xuan paper hanging in the workshop

四 "贡玉"加工纸生产的原料、工艺流程与工具设备

4 Raw materials, Papermaking Techniques and Tools of "Gongyu" Processed Paper

（一）"贡玉"加工纸生产的原料

1. 主料

原纸。泾县贡玉堂宣纸工艺厂加工纸使用的原纸以书画纸为主，书画纸主要由兴潭造纸厂生产，少量的特殊品种在泾县当地购买。有时根据客户需要，也使用不同规格的宣纸加工，泾县贡玉堂宣纸工艺厂所用的宣纸原纸会到泾县定点的宣纸厂购买。

2. 辅料

（1）染料。"贡玉"加工纸使用的辅料主要是各种颜色的染料，根据染料属性可分为植物染料、矿物染料、化工染料3种。植物染料是最传统的染料，其原料一般自己采集或在中药店里购买，根据配比成分自行配置并提取，但通常提取时间长，工序复杂，价格也相对昂贵；矿物染料和化工染料都从当地文房四宝商店里购买，回厂勾兑加工成所需的颜色。据王学兵介绍，现在泾县当地加工纸使用植物染料越来越少，如果客户有特别要求，就使用植物染料。

（2）水。泾县贡玉堂宣纸工艺厂加工纸用水量小，对水质要求没有原纸生产高。"贡玉"使用的水是来自本地的地下水，其水质的混浊度低，杂质含量少，水的温度较低，适合加工纸用。经调查组成员现场取样观测，"贡玉"加工纸用水的pH为6.51，呈弱酸性。

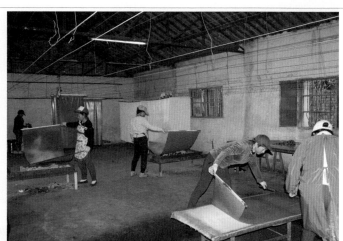

⊙2
染料（因涉及保密对方不愿告知）
Dye (not willing to inform because of the confidentiality involved)

⊙3
制作过程
Working process

（二）"贡玉"加工纸生产的工艺流程

1. 虎皮宣生产的工艺流程

调查组成员于2015年8月12日对贡玉堂宣纸工艺厂的生产工艺进行了实地调查和访谈，归纳其代表纸品虎皮宣生产的工艺流程为：

壹 选纸 → 贰 搭棍子 → 叁 染色 → 肆 烘烤 → 伍 撒花 → 陆 二次烘烤 → 柒 检验、剪纸 → 捌 成品包装

工艺流程

壹　选纸

1

一般选用书画纸作为虎皮宣制作的原纸，当客户有特别需求时选用更高质量的宣纸作为原纸，一般尺寸为70 cm×138 cm。

贰　搭棍子

2　⊙1

用糨糊将原纸的一端与木棍粘在一起，主要为后期悬挂晾晒方便。据王学兵介绍，贡玉堂宣纸工艺厂平均一个人2小时可以完成3刀原纸的搭棍子工作。

⊙1

叁　染色

3　⊙2～⊙4

提前将需要的虎皮宣的底色颜料调配好，盛于拖纸盆中，将原纸逐一缓慢均匀拖曳通过拖纸盆颜料表面进行染色，然后晾挂在绳索搭建的架子上。当拖完第20张纸时即进入烤火环节，首先从第一张开始，20张纸一循环，刚刚染色完的纸由于含有大量颜料中的水分，不能立即烘烤，必须自然晾干一段时间。以20张纸循环染色的间隔时间作为晾晒时间。

⊙2

⊙3

⊙4

⊙ 1
搭完棍子的原纸
Raw paper pasted with sticks
⊙ 2/3
拖染
Drag the paper through the surface of the pigment for dyeing
⊙ 4
拖染后自然晾干
Drying naturally after dyeing

肆 烘烤

4 ⊙5～⊙7

在烘烤前还需要在染色完的纸的另一端搭一根棍子，烘烤时由一人两手各抬纸的一端，平行放置在火盆上方，来回均匀受热，使其颜料水分蒸发20%后，首次烘烤结束。

⊙5

⊙6

首次烘烤的作用是避免纸中含有过多水分，以免影响后面撒花的效果。如果烤得过干，缺少水分，花纹也会显示不出来。水分蒸发量靠工人的工作经验去判断，得用眼睛去观察水汽与纸面的情况；同时，也受火盆的温度和来回速度的影响，一张纸首次烘烤大概需要几十秒的时间。烘烤也是以20张纸作为一个循环单位。

⊙7

⊙5 烘烤的炭盆 Charcoal brazier for baking
⊙6/7 烘烤 Baking

伍 撒花

5

撒虎皮斑纹之前要首先调制好能够受热扩散形成花纹的调和液体，泼洒调和的液体主要通过一些特殊物质原料配备（涉及机密未透露），用过滤网过滤后使用。过滤网主要起到排出调和液中颗粒成分的作用。用撒把沾染已过滤的调和液体向纸张上泼洒，最后形成一个个斑点花纹。花纹大小和力度是相关联的，需来回泼洒多次，四尺的1张纸最少要泼洒100多下，时间约1分钟。

⊙8

陆 二次烘烤

6

泼洒完成后立即用炭火烘烤。烘烤时，撒上去的粉点在高温下会膨胀炸裂开来，迅速扩散形成虎皮状斑纹。烘烤至纸张水分蒸发80%左右时结束，然后卸下一根棍子，用另一根悬挂纸张，使其自然阴干。纸张一般悬挂在没有剧烈空气流动的屋里，用炭火加温烘烤一夜，让水分完全蒸发，第二天早上就完成了整个加工过程。有时候受天气因素影响，空气湿度大时，会延长制作周期。

调查期间由于夏天天气炎热，火盆温度高，考虑到工人的工作环境与身体状况，没有进行生产。完成上述工序通常需要5个工人轮流，一天大概能够生产200多张纸。

⊙9

⊙10

⊙8/9 撒花 Speckling the paper
⊙10/11 第二次烘烤 Second baking

⊙11

柒
检 验、剪 纸
7

将晾晒干后的纸张运到剪纸车间进行裁剪。剪纸前需要对纸张进行检查，将有破损或瑕疵的纸张剔除出去。然后将检验好的纸张放在纸台上，工人用特制大剪刀一气呵成进行裁剪，裁剪出来的纸张要求四边整齐。除人工剪纸外，也有用切纸机进行切割的。

捌
成 品 包 装
8

⊙12

⊙12

剪完之后，根据不同规格分装好，盖上印有"贡玉"商标的加工纸印章，再用特定包装纸包装方算完成。

⊙12 纸品仓库的存货架
Stock shelf in paper warehouse

2. 墨流宣生产的工艺流程

由于墨流宣在最初的选纸、搭棍子，以及后期晾干、检验、包装等环节与普通半熟加工纸基本相似，其保密的部分在于生产者拖染的过程中添加的试剂不对外告知。根据司绍先口述及其提供的纸样，调查组成员查找相关文献资料了解到墨流宣生产的工艺流程大致为：

壹 配料 → 贰 拖染 → 叁 晾干

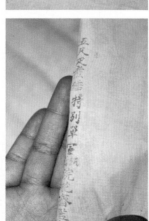

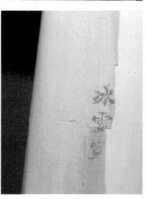

⊙1
司绍先展示不同历史时期的宣纸加工纸
Si Shaoxian showing the processed Xuan paper in different historical periods

壹 配料 1

在墨汁、颜料或者水溶性国画颜料中添加一种扩散防混合剂，同时在拖染盆的水中添加一种悬浮剂，使得颜料入水后漂浮在水面上而不溶解于水。然后将各种颜色用毛笔沾染点在水面上，各种颜色的染料在水中迅速扩散并且不相互融合。

贰 拖染 2

对事先搭好棍子的原纸，让一个人两手拿着棍子的两端从拖盆的一头拖到另一头。拖染时纸从水面上缓缓移动，依靠水的流动性使颜料吸附在纸上，并且形成流沙流动的效果或者云彩漂浮的形状；在拖染的过程中一定要均匀用力，注意技巧，一气呵成。当拖染若干张纸后，再往盆中添加颜料，什么时候添由拖染者自己把握。

叁 晾干 3

将拖染好的湿纸悬挂阴干，注意纸与纸之间要保持一定的距离，防止风吹相互沾染，待纸干后收集整理好，去除棍子等待检验、剪纸。

(三)

"贡玉"加工纸生产的工具设备

壹 棍子 1

用于悬挂纸张以及烘烤时手持纸张的棍子，主要用糨糊粘在纸张两端。实测泾县贡玉堂宣纸工艺厂所用的棍子长约112 cm。

⊙2

贰 撒把 2

用来向纸张泼洒调和液的工具，材质主要是长竹签。

⊙3

叁 拖纸盆 3

用来盛放染料，给纸张染色的长方体木盆。实测泾县贡玉堂宣纸工艺厂所用拖纸盆尺寸为长156 cm，宽88 cm，高12 cm。

⊙4

肆 工作台 4

用于制作洒金、洒银平铺纸张的桌台。实测泾县贡玉堂宣纸工艺厂所用工作台尺寸为长262 cm，宽122 cm。

⊙5

伍 刷胶刷子 5

羊毛材质，用于涂刷胶纸材料的工具，实测泾县贡玉堂宣纸工艺厂所用刷胶刷子有2种尺寸。刷六尺纸的大刷为35排齿，长35 cm，宽17.5 cm；刷四尺纸的小刷为30排齿，长29 cm，宽16.5 cm。

⊙6

陆 丝网 6

用于丝网印刷纸的网状工具。实测泾县贡玉堂宣纸工艺厂所用四尺丝网尺寸为长181 cm，宽87 cm。

⊙7

柒 刮刀 7

用于丝网印刷纸涂抹染料的工具，主要由橡胶刀片和木质刀把组成。实测泾县贡玉堂宣纸工艺厂所用四尺刮刀尺寸为长78 cm，宽12.5 cm 六尺刮刀尺寸为长101 cm，宽12.5 cm。

⊙8

五 泾县贡玉堂宣纸工艺厂的市场经营状况

5 Marketing Status of Gongyutang Xuan Paper Craft Factory in Jingxian County

司绍先在经营泾县贡玉堂宣纸工艺厂期间，生产的产品主要出口到日本等国家。2011年转让给王学兵后，在维持原有国外市场的前提下，从2012年开始逐步拓展国内市场。调查时企业的出口量达到30%～40%，品种达到几百种，规格各异，其产品在日本加工纸行业已具有一定的品牌认同。

王学兵2004年在北京经营第一家文房四宝店后，目前又经营了第二家直营店铺，产品主要以泾县贡玉堂宣纸工艺厂生产的产品为主。王学兵表示，他完全是出于对宣纸的兴趣爱好才承接下泾县贡玉堂宣纸工艺厂的，从那以后他的工作重心从北京转移到泾县黄村镇，主要从事宣纸加工纸品的研发与应用。北京直营店的销售生意完全交给其妻子和儿子在经营。

⊙1
泾县贡玉堂宣纸工艺厂大门
Main entrance of Gongyutang Xuan Paper Craft Factory in Jingxian County

除了直营店，目前泾县贡玉堂宣纸工艺厂主要采取代理商的模式进行纸品销售，另外在网络平台还开有一家网店。起初几年企业效益不好，调查时情况已在逐年好转，年销售额最高可达200万元。用王学兵的话说，做传统宣纸行业利润较低，只是想在原来的基础上把传统的技艺传承下去，希望丰富纸类品种，继续将宣纸加工工艺发扬光大。

六
泾县贡玉堂宣纸工艺厂的品牌文化

6
Brand Culture of Gongyutang Xuan Paper Craft Factory in Jingxian County

据司绍先回忆,"贡玉堂"名称得自文房四宝书籍上记载的诗句,源于唐代大诗人杜甫创作的一首七言古诗《寄韩谏议注》:"美人胡为隔秋水,焉得置之贡玉堂。"

七
泾县贡玉堂宣纸工艺厂的传承现状与发展思考

7
Current Status of Business Inheritance and Thoughts on Development of Gongyutang Xuan Paper Craft Factory in Jingxian County

在调查期间,王学兵感慨道:其从事宣纸制造工作30多年,从最初的宣纸生产技术工人,到自己开造纸厂进行全程生产和宣纸产品的市场销售,再到最后的宣纸加工纸品的生产,基本上经历了宣纸的各个工作环节。访谈时,王学兵反复强调对于宣纸的热爱,最大的体会就是如今宣纸传统手工技艺随着时代的进步逐渐被人们遗忘,他觉得很痛心。相比当代若干新兴的行业,当前从事传统工艺生产收益低,制作流程繁琐,生产条件艰苦。王学兵表示:最大的期望是在传承传统工艺的同时,能通过潜心研究、大胆探索,不断开发工艺新品,享受发明发现的快乐。

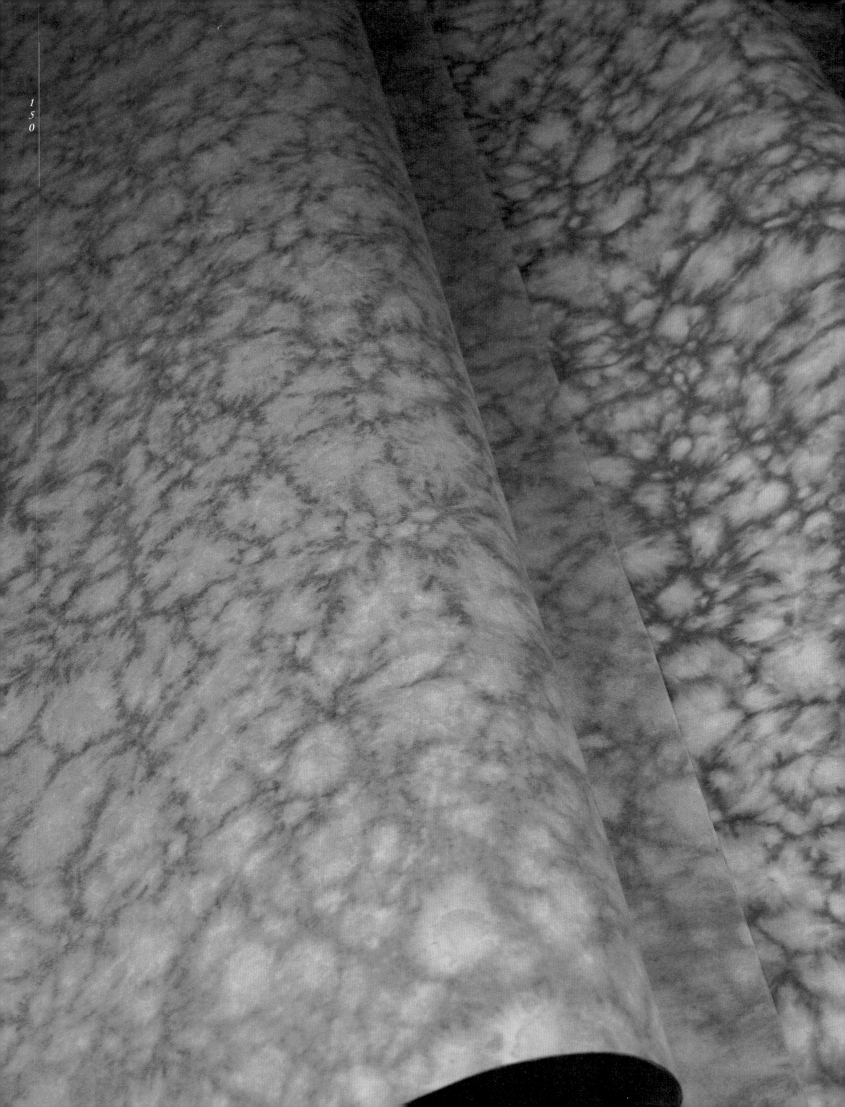

泾县贡玉堂宣纸工艺厂 加工纸

Processed Paper of Gongyutang Xuan Paper Craft Factory in Jingxian County

"贡玉"虎皮宣透光摄影图
A photo of "Gongyu" Hupi Xuan paper seen through the light

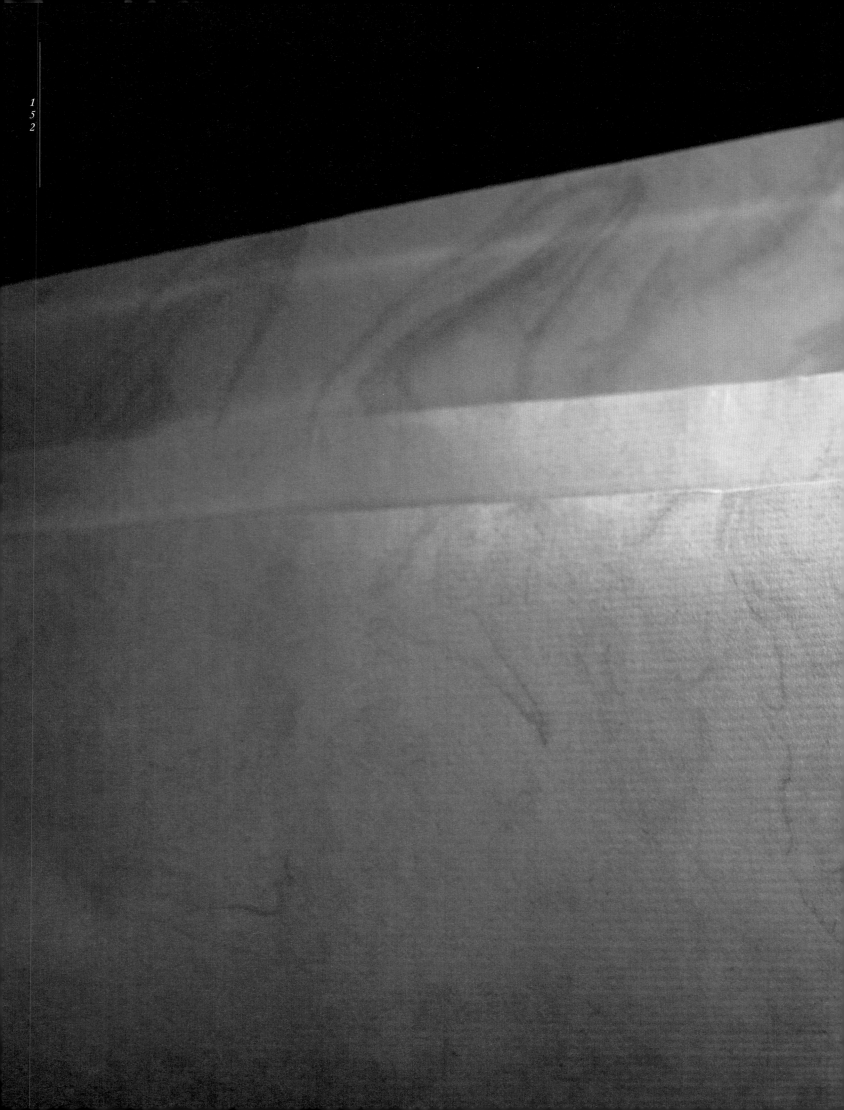

加工纸

泾县贡玉堂宣纸工艺厂

Processed Paper of Gongyutang Xuan Paper Craft Factory in Jingxian County

「贡玉」墨流宣透光摄影图
A photo of "Gongyu" Moliu Xuan paper seen through the light

第六节
泾县博古堂宣纸工艺厂

Section 6
Bogutang Xuan Paper Craft Factory
in Jingxian County

Subject
Processed Paper of Bogutang Xuan Paper Craft Factory
in Jingxian County
in Dingjiaqiao Town

调查对象
丁家桥镇
泾县博古堂宣纸工艺厂
加工纸

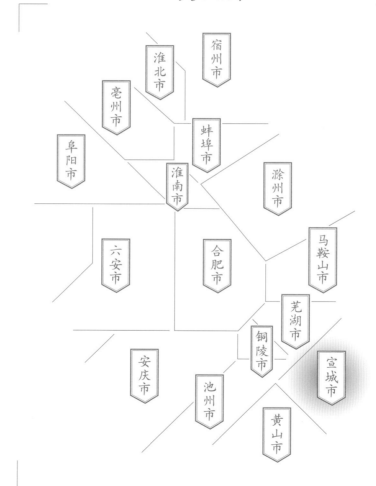

一

泾县博古堂宣纸工艺厂的基础信息与生产环境

1
Basic Information and Production Environment of Bogutang Xuan Paper Craft Factory in Jingxian County

泾县博古堂宣纸工艺厂位于泾县丁家桥镇小岭村，现厂址建立在小岭宣纸厂旧址上，地理坐标为东经118°18′50″，北纬30°41′20″。泾县博古堂宣纸工艺厂创办于2006年，主要生产书画纸和加工纸，代表加工纸品种有粉彩、泥金、泥银、万年红等。

2015年7月22日和2016年4月21日，调查组前往泾县博古堂宣纸工艺厂进行田野调查获得的基础信息如下：泾县博古堂宣纸工艺厂占地超过1 000 m²，动态保持有员工20多人，有2条自动烘干套色机生

产线、4个丝网印刷及加工纸的工作台。据泾县博古堂宣纸工艺厂厂长曹迎春介绍，泾县博古堂宣纸工艺厂近3年平均年销售额300万~400万元。

小岭村位于泾县丁家桥镇北部，西与云岭镇、东与泾川镇相接，有"九岭十三坑"之称。据调查时小岭村委会主任曹晓晖介绍，小岭村面积22.5 km²，目前常住人口数1 749左右，约576户。

⊙1 "博古堂"生产厂区 "Bogutang" Xuan Paper Factory
⊙2 "博古堂"堂名题写原件 The original inscription of the factory name "Bogutang"
⊙3/5 小岭村 Xiaoling Village

Location map of Bogutang Xuan Paper Craft Factory in Jingxian County

路线图
泾县县城 → 泾县博古堂宣纸工艺厂

Road map from Jingxian County centre to Bogutang Xuan Paper Craft Factory in Jingxian County

考察时间
2015年7月 / 2016年4月

Investigation Date
July 2015 / Apr. 2016

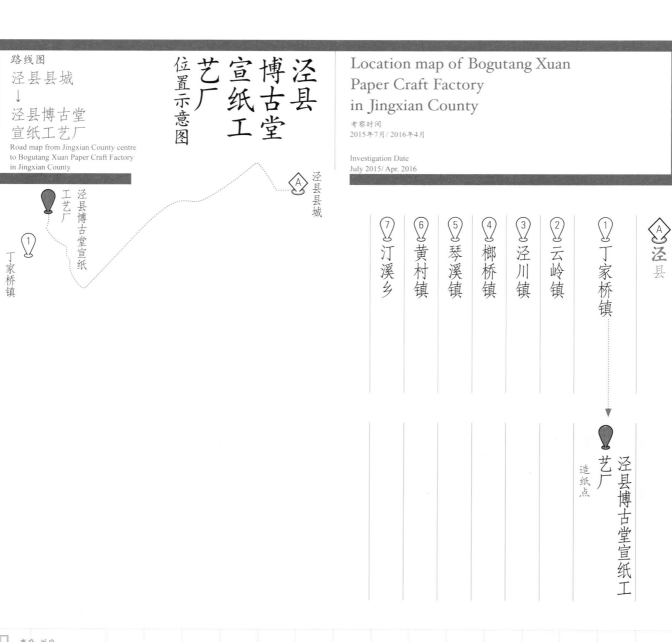

地域名称：泾县
造纸点名称：泾县博古堂宣纸工艺厂

1. 丁家桥镇
2. 云岭镇
3. 泾川镇
4. 榔桥镇
5. 琴溪镇
6. 黄村镇
7. 汀溪乡

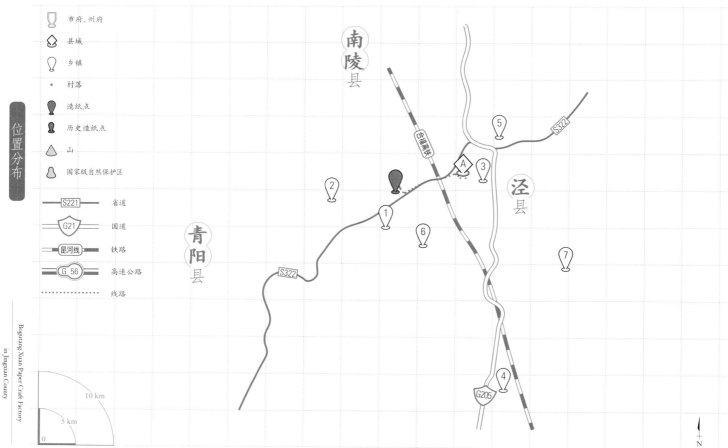

位置分布

图例：
- 市府、州府
- 县城
- 乡镇
- 村落
- 造纸点
- 历史造纸点
- 山
- 国家级自然保护区
- S221 省道
- G21 国道
- 昆河线 铁路
- G56 高速公路
- 线路

二 泾县博古堂宣纸工艺厂的历史与传承情况

2 History and Inheritance of Bogutang Xuan Paper Craft Factory in Jingxian County

① 1
泾县小岭宣纸厂旧址标志牌
The signboard of former Xiaoling Xuan Paper Factory in Jingxian County

① 2
调查组成员采访曹迎春
A researcher interviewing Cao Yingchun

泾县博古堂宣纸工艺厂注册地为泾县丁家桥镇小岭村。据泾县博古堂宣纸工艺厂法人代表曹迎春介绍，泾县博古堂宣纸工艺厂创办于2006年，租用原小岭宣纸厂的"氧-碱法"生产车间，每年租金2万~3万元，主要生产加工纸。创办初期厂里只有3个人，随着生产的不断发展，为了保障加工纸的质量，2011年创办了一个书画纸厂，为泾县博古堂宣纸工艺厂的分厂，分厂员工有13~15人，而目前加工纸车间有20人左右。2014年底，泾县博古堂宣纸工艺厂从浙江一家服装厂引进自动烘干套色设备2台，是泾县最早引进该设备的3家企业之一（其他2家为源流加工纸厂和兴文斋加工纸厂），2015年初开始投入生产。

曹迎春，1978年出生，泾县小岭村人。父亲曹康贵，1943年出生，在小岭宣纸厂担任捞纸工直到小岭宣纸厂改制后停产回家；母亲曹国英，1949年出生，曾在小岭宣纸厂担任原料选检工直到小岭宣纸厂改制后停产回家。外公曹滋生，1949年10月~1951年10月在许湾担任最初的小岭宣纸厂负责人，也是泾县宣纸联营处早期的领导成员之一，负责过宣纸生产、销售管理工作。

由于从小耳濡目染，曹迎春1998年开始在小岭一家私人宣纸厂（代牌生产，无厂名）学晒纸，2002年前往浙江、上海等地企业打工。2006年3月回到家乡小岭村开始筹建泾县博古堂宣纸工艺厂，2006年6月正式投入生产，启动资金7 000元，最初生产洒金、描金等加工纸。2011年在云岭镇靠山村创办泾县博古堂宣纸工艺厂分厂，生产书画纸，分厂由曹迎春岳父肖先兰和内兄肖必峰负责，主要生产半自动喷浆书画纸和手工书画纸，共有2个手工捞纸槽和6个半自动喷浆口，书画纸的原料都是从泾县小岭代理商处购买，由代理商负责帮其调配原料。曹迎春的大姐曹仙红在西安从事文房四宝销售，二姐曹金莲在泾县博古堂宣纸工艺厂务工。

三 泾县博古堂宣纸工艺厂的代表纸品及其用途与技术分析

3 Representative Paper and Its Uses and Technical Analysis of Bogutang Xuan Paper Craft Factory in Jingxian County

(一) 代表纸品及其用途

调查组2015年7月22日调查所获信息如下：泾县博古堂宣纸工艺厂加工纸种类繁多，品种规格各异，如色宣、粉彩、素彩、雅光、套色、泥金、泥银及信笺。据曹迎春介绍，目前泾县博古堂宣纸工艺厂代表产品为精品套色加工纸，属于丝网漏网印刷，约90%的购买人群用来练习和创作书法，10%的人用来画写意画。

(二) 技术分析

测试小组对采样自泾县博古堂宣纸工艺厂生产的粉彩套色笺所做的性能分析，主要包括厚度、定量、紧度、抗张力、抗张强度、撕裂度、色度、吸水性等。按相应要求，每一指标都需重复测量若干次后求平均值，其中定量抽取5个样本进行测试，厚度抽取10个样本进行测试，抗张力抽取20个样本进行测试，撕裂度抽取10个样本进行测试，色度抽取10个样本进行测试，吸水性抽取10个样本进行测试。对博古堂宣纸工艺厂粉彩套色笺进行测试分析所得到的相关性能参数见表6.11。表中列出了各参数的最大值、最小值及测量若干次所得到的平均值或者计算结果。

表6.11 "博古堂"粉彩套色笺相关性能参数
Table 6.11 Performance parameters of "Bogutang" Fencai Taose paper

指标	单位	最大值	最小值	平均值	结果
厚度	mm	0.118	0.108	0.110	0.110
定量	g/m^2	—	—	—	43.8
紧度	g/cm^3	—	—	—	0.398

续表

指标		单位	最大值	最小值	平均值	结果
抗张力	纵向	N	16.2	14.4	15.2	15.2
	横向	N	11.5	10.5	11.1	11.1
抗张强度		kN/m	—	—	—	0.876
撕裂度	纵向	mN	400	340	380	380
	横向	mN	430	390	412	412
撕裂指数		mN·m²/g	—	—	—	9.0
色度		%	51.8	51.3	51.49	51.5
吸水性		mm	—	—	—	10

由表6.11中的数据可知，"博古堂"粉彩套色笺最厚约是最薄的1.09倍，经计算，其相对标准偏差为0.003，纸张厚薄较为一致。所测"博古堂"粉彩套色笺的平均定量为43.8 g/m²。通过计算可知，"博古堂"粉彩套色笺紧度为0.398 g/cm³，抗张强度为0.876 kN/m，抗张强度较小。所测"博古堂"粉彩套色笺撕裂指数为9.0 mN·m²/g，撕裂度较小。

所测"博古堂"粉彩套色笺平均色度为51.5%。色度最大值是最小值的1.009 7倍，相对标准偏差为0.166，色度差异相对较小。所测"博古堂"粉彩套色笺吸水性纵横平均值为10 mm。

四
"博古堂"精品套色加工纸生产的原料、工艺流程与工具设备

4
Raw Materials, Papermaking Techniques and Tools of "Bogutang" Fine Taose Processed Paper

（一）
"博古堂"精品套色加工纸生产的原料

1. 主料

原纸。"博古堂"精品套色加工纸使用的原纸一般为半自动喷浆书画纸。据曹迎春介绍，"博古堂"精品套色加工纸原纸一般由博古堂宣纸工艺厂分厂提供，但需求量大时也从小岭当地私人厂购买。

2. 辅料

（1）印染剂。"博古堂"精品套色加工纸

使用原纸加工时，采用各种水性印染剂进行套色制作，印染剂一般从杭州、义乌等地购买。

（2）水。"博古堂"精品套色加工纸使用的是小岭村下游山涧水。据调查组在现场的测试，"博古堂"精品套色加工纸制作所用的山涧水pH为7.59，呈弱碱性。

（二）
"博古堂"精品套色加工纸生产的工艺流程

据曹迎春介绍，以及综合调查组2015年7月22日在博古堂宣纸工艺厂的实地观察，归纳"博古堂"精品套色加工纸生产工艺流程为：

壹 套色 → 贰 检验、裁纸 → 叁 装箱打包

壹 套色

⊙2～⊙6

（1）先将印染剂按照一定比例混合，放在一边等套色时使用。

（2）将跑板机液压表开到4个液压，待跑板机升温到450 ℃左右后，工人将原纸放在自动烘干套色生产线上，另一个工人紧随其后，将丝网板放在原纸上进行第一遍打

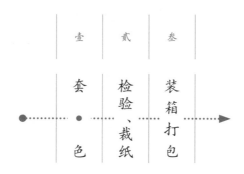

⊙1 印染剂 Dye
⊙2 液压表 Hydraulic pressure gauge
⊙3 套色 Color register

底，一般使用浅色印染剂。按照桶计量，一般一个桶需要高岭土粉3 kg、化学试剂1.8 kg、胺水0.05 kg，再按照需要根据经验放置适量的色浆。一般一个桶的剂量可以制作2～3刀"博古堂"精品套色加工纸，等到印染剂烘干后，一般不超过1分钟。

（3）再将丝网板放在打底后的纸上，在纸四周根据需要刷上泥金、泥银或者粉彩，这个步骤是制作用户落款使用的花色，等到纸烘干约需5分钟。

（4）在烘干的纸上放置丝网板，在纸的中间或者所有位置刷上所需要的图案，等到全部烘干后即可收回，约烘10分钟就可以将纸收回放置一边。"博古堂"精品套色加工纸每张纸套色加工需要15～20分钟。使用过的丝网一般用洗衣粉进行清洗。

贰 检验、裁纸

将收回的套色加工纸进行检验，遇到不合格的纸取出另行处理，合格的纸按照市场需求或者客户要求在裁纸机上进行裁剪。

叁 装箱打包 3 ⊙9 ⊙10

将剪好后的纸按客户需要包好,再加盖"博古堂"的加工纸章,包装完毕后运入仓库。

⊙9 包纸 Wrapping the paper
⊙10 打包 Packaging

工具设备

(三) "博古堂"精品套色加工纸生产的工具设备

壹 自动烘干套色机 1

也叫跑板机,因工人一直拿着丝网板在生产线上染色而得名。用环保蒸汽发生器烘干,锅炉通气,工作时温度控制在50～60 ℃。机器上还有卡子,专门定位使用,方便工人更好地套色印染。

贰 丝网板 2

加工纸所用工具,据曹迎春介绍,在泾县当地购买,"博古堂"自己设计图案,泾县当地制作商制作丝网板。四尺丝网购买价为300元/张,使用寿命为2年左右。

实测"博古堂"四尺丝网板尺寸为外长181 cm,外宽87 cm,内长175 cm,内宽81 cm;六尺丝网板尺寸为外长224 cm,外宽116 cm,内长216 cm,内宽108 cm;八尺丝网板尺寸为外长320 cm,外宽160 cm,内长312 cm,内宽152 cm;丈二丝网板尺寸为外长440 cm,外宽184 cm,内长430 cm,内宽174 cm。

⊙11

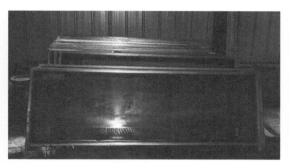

⊙12

⊙13

叁
刮 刀
3

用来在丝网上刮印染剂，刀柄为木头所制，刀头为塑料所制。实测"博古堂"四尺丝网板所用刮刀长77 cm，六尺丝网板所用刮刀长104 cm，八尺丝网板所用刮刀长148 cm，丈二丝网板所用刮刀长170 cm。

⊙14

肆
裁纸机
4

用来裁剪制作好的加工纸，机器控制，可以精准裁剪出所需要的尺寸。

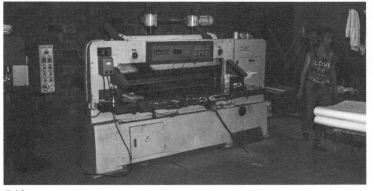

⊙15

⊙ 11／13 丝网板 Screen board
⊙ 14 刮刀 Scraper
⊙ 15 裁纸机 Paper cutter

五 泾县博古堂宣纸工艺厂的市场经营状况

5 Marketing Status of Bogutang Xuan Paper Craft Factory in Jingxian County

访谈中曹迎春表示：泾县博古堂宣纸工艺厂在发展运营过程中一直坚持以市场为主导，同时也关注从别的领域吸取可供借鉴的工艺，在加工纸领域不断进行技术改革探索。目前泾县博古堂宣纸工艺厂销售渠道已经覆盖到全国一线城市和大部分二线城市，主要以国内经销为主。截至调查组调查时为止，"博古堂"四尺精品套色加工纸的出厂价为700~800元/刀。据曹迎春介绍，博古堂宣纸工艺厂员工有20多人，工人两班倒，每班9小时。2015年年销售收入为300万~400万元，利润率约为5%。

⊙1

⊙2

⊙1 生产车间 Production workshop
⊙2 仓库 Warehouse

六
泾县博古堂宣纸工艺厂的品牌文化

6
Brand Culture of Bogutang Xuan Paper Craft Factory in Jingxian County

当问到为什么会取"博古堂"这个名字时,曹迎春介绍说,他在一开始创业时考虑到纸的深加工行业发展潜力会比白纸行业潜力大,因此他想让自己的企业产品不仅学习古人的深厚,而且希望做得比古时候更加博大精深,所以取名为"博古堂",并将其注册了商标。

⊙3
产品上印制的"博古堂"品牌商标
"Bogutang" brand trademark printed on the product

七
泾县博古堂宣纸工艺厂的传承现状与发展思考

7
Current Status of Business Inheritance and Thoughts on Development of Bogutang Xuan Paper Craft Factory in Jingxian County

在泾县加工纸行业里,调查组发现有好几家加工纸厂在探索跨界创新,泾县博古堂宣纸工艺厂亦是如此。例如,在考察了服装印刷厂后,曹迎春就敏锐地察觉到将服装印染行业早已不新奇的自动烘干印染设备引入到加工纸行业将有大的优势,尽管这2条自动烘干套色生产线投入生产不到一年,但据曹迎春介绍,它所带来的变化是始料不及的:不仅解决了加工纸时常受到天气影响的作业时间局限问题,还明显提高了效率,销量也随之大幅度提高。当然,对于跨界工艺对产品的复杂影响,业界也有不同的看法。2016年4月再次回访时,曹迎春兴致很高地说:"在刚刚结束的北京全国文房四宝博览会上,尽管全国纸的市场有些疲软,但是消费者对创新产品仍有着较高的兴趣。"

泾县博古堂宣纸工艺厂 加工纸

Processed Paper of Boguitang Xuan Paper Craft Factory in Jingxian County

「博古堂」粉彩套色笺（冷金）透光摄影图
A photo of "Boguitang" Fencai laose paper seen through the light

第七节
泾县汇宣堂宣纸工艺厂

调查对象
泾川镇
泾县汇宣堂宣纸工艺厂
加工纸

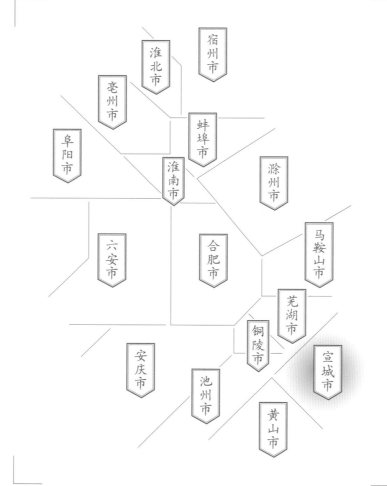

Section 7
Huixuantang Xuan Paper Craft Factory in Jingxian County

Subject
Processed Paper of Huixuantang Xuan Paper Craft Factory in Jingxian County in Jingchuan Town

一

泾县汇宣堂宣纸工艺厂的基础信息与生产环境

1

Basic Information and Production Environment of Huixuantang Xuan Paper Craft Factory in Jingxian County

泾县汇宣堂宣纸工艺厂位于泾县泾川镇曹家行政村，地理坐标为东经118°21′50″，北纬30°41′11″。厂房坐落于S322省道路旁，交通便利，周边自然环境优美。泾县汇宣堂宣纸工艺厂创办于2010年，是一家股份合作制民营企业。2005年，在原红宝龙宣纸工艺厂基础上，经历家族企业分割、引进战略投资者的过程后，最终于2010年正式成立泾县汇宣堂宣纸工艺厂。2014年在泾县榔桥镇新建了规模更大的分厂，纸厂本部现占地面积约600 m²，分厂占地面积约1 200 m²。

2015年7月21日与2016年4月20日，调查组两次前往泾县汇宣堂宣纸工艺厂进行田野调查，截至2016年4月时的调查信息如下：纸厂有员工20人，年销售额约100万元。泾县汇宣堂宣纸工艺厂产品主要有加工纸和各类宣纸制品，其中加工纸包括泥金纸、蜡染笺、复古笺、对联（万年红、黑、蓝）和各类色宣，宣纸制品包括册页、印谱、卡纸（硬卡和软卡）、手卷、折页、镜片和扇面等。

⊙1 在S322省道上的路口指示牌
Road sign on the S322 provincial road

⊙2 汇宣堂宣纸工艺厂外围厂名标牌
Name tag outside the Huixuantang Xuan Paper Craft Factory

据2016年的泾县泾川镇人民政府网站介绍，泾川镇地处皖南山区，是泾县县委、县政府所在地，是全县政治、经济、文化、交通中心。全镇总面积114.22 km²，辖14个村和8个社区居委会，总人口6.8万多人，其中农业人口2.1万人。境内G205国道、S322省道、青弋江穿镇而过，自古水陆交通便捷。

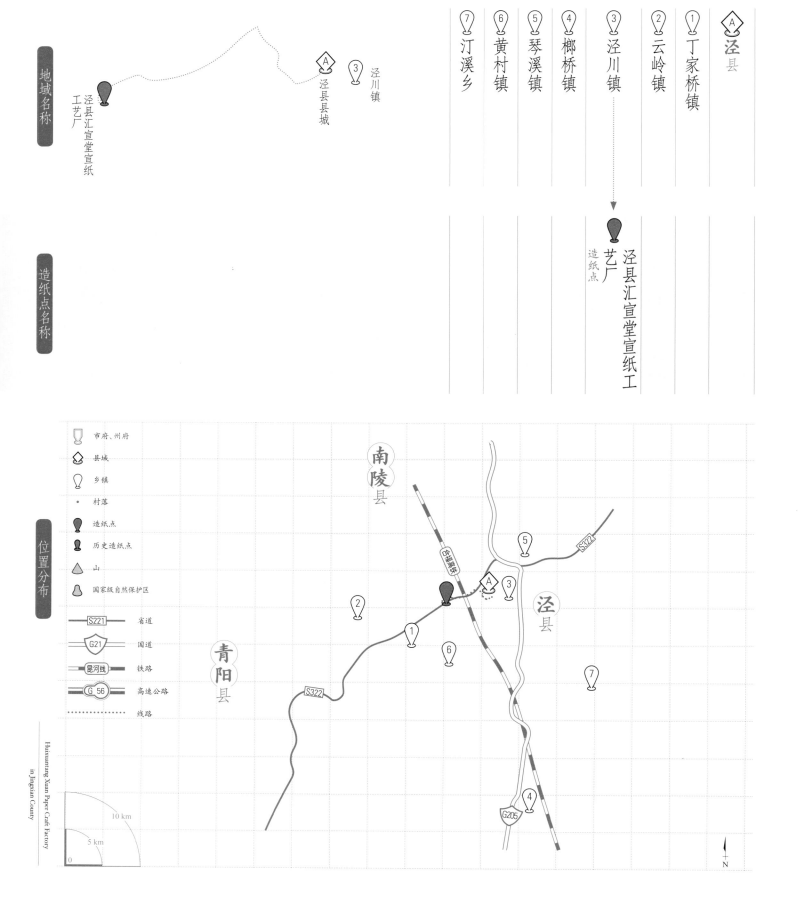

二 泾县汇宣堂宣纸工艺厂的历史与传承情况

2
History and Inheritance of Huixuantang Xuan Paper Craft Factory in Jingxian County

泾县汇宣堂宣纸工艺厂注册地为泾县泾川镇，据参股人涂振华介绍，泾县汇宣堂宣纸工艺厂创办于2010年，企业法人为他的姨父汤贵宝，两人各占公司50%的股份。该厂于2010年5月15日正式投入生产，企业商标为"汇宣堂"。

访谈中涂振华提供的企业历史信息为：2001年，汤贵宝与他的兄弟在泾县丁家桥街道合伙成立了泾县红宝龙宣纸工艺厂。随着加工纸市场需求量的增长以及家族产业的分割，老厂泾县红宝龙宣纸工艺厂全部由汤贵宝的弟弟汤龙平管理。2010年汤贵宝联合涂振华成立泾县汇宣堂宣纸工艺厂，成立之初启动资金约20万元，占地面积约600 m²。调查时的总厂由涂振华管理生产，汤贵宝负责技术指导和运营统筹。2014年，为扩大生产规模，在泾县榔桥镇街道建立了分厂，厂名仍使用"汇宣堂宣纸工艺厂"，占地面积约1 200 m²，就近招聘了10多个员工，交由涂振华的父亲涂来胜管理，2015年初正式投产，调查期间尚未实现满负荷生产。之所以在相隔较远的地方设立分厂，涂振华解释是因为泾川镇厂房不好找，宣纸生产企业数量多，大一点的厂房都被别人租完了，只好将分厂设在比较熟的榔桥镇了。

关于纸厂的技艺传承情况，涂振华表示：他和他父亲都是从汤贵宝那里学习宣纸工艺的，父亲原本不做纸，祖辈也没有什么造纸传承。

汤贵宝是泾县章渡镇（现已归入泾县云岭镇）人，2000年，汤贵宝受表兄肖海冲启发开始涉足宣纸制造行业，2001年先后前往浙江省江山县、山东省威海市的服装厂学习丝网印刷技术，并将其与宣纸加工相结合；2001年下半年至2003年借安吴三兔宣笔厂的厂房生产"红宝龙"牌宣纸，2003年迁至泾县丁桥镇。2008年汤贵宝开设汇宣堂宣纸用品店，主营宣纸厂家生产用品（宣纸包装、化工材料），2010年联合涂振华成立泾县汇宣堂宣纸工艺厂。调查时，汤贵宝仍在泾县红宝龙宣纸工艺厂和泾县汇宣堂宣纸工艺厂担任技术指导，工艺与技术经验丰富。

⊙1 调查组成员访谈涂振华
A researcher interviewing Tu Zhenhua

⊙2 调查组成员现场请教汤贵宝
Researchers consulting Tang Guibao

⊙3 汤贵宝
Tang Guibao

三 泾县汇宣堂宣纸工艺厂的代表纸品及其用途与技术分析

3 Representative Paper and Its Uses and Technical Analysis of Huixuantang Xuan Paper Craft Factory in Jingxian County

（一）代表纸品及其用途

泾县汇宣堂宣纸工艺厂加工纸产品种类繁多，包括粉彩类、信笺印谱类、蜡染笺类、泥金泥银类、复古笺类、万年红蓝黑类、手卷类、卡纸类，规格有四尺、六尺、四尺对开、六尺对开和各种特定规格等。调查时据涂振华介绍，泾县汇宣堂宣纸工艺厂的代表纸品主要有泥金纸、蜡染笺、复古笺和卡纸，对联和手卷与其他厂家产品差别不大，代表纸品以泥金纸最为突出，泾县目前能做出与"汇宣堂"泥金纸类似效果的几乎没有。泾县汇宣堂宣纸工艺厂自身不生产原纸，各类加工纸原纸都从泾县当地厂家直接购买。

泥金纸原纸要求纯手工、做工精细、拉力强的熟宣，宣纸和书画纸原纸种类选择根据客户要求来定，宣纸加工自然高档一点，但价格也贵，适合写小楷、画工笔画、抄经、画佛像等。

蜡染笺原纸跟泥金纸要求类似，原纸半熟。据涂振华介绍，蜡染笺是唐代的一种很高档的书法用纸，用汰白粉和改性动物蜡等繁琐的古代工艺配制而成，配方中又加入中草药，

⊙1 蜡染笺 Laran paper
⊙2 粉彩笺 Fencai paper
⊙3 整刀的泥金纸 A set of Nijin paper
⊙4 精品宣 High-quality Xuan paper

⊙5

防虫防蛀，表面光滑，具有书写顺滑不损毫、字迹显黑发亮有神等特点。蜡染笺主要用于书法（楷书）创作。

复古笺原纸要求用生宣，润墨效果好，主要用于书法创作和绘画。卡纸以硬卡为特色，原纸要求纯手工，使用生宣还是熟宣则根据客户要求，大多用生宣，主要用途也是绘画和书法。

（二）
技术分析

1. 代表纸品一："汇宣堂"泥金纸

测试小组对采样自泾县汇宣堂宣纸工艺厂生产的泥金纸所做的性能分析，主要包括厚度、定量、紧度、抗张力、抗张强度、撕裂度、色度等（由于所测纸张很难吸水，没有测试吸水性）。

⊙6

按相应要求，每一指标都需重复测量若干次后求平均值，其中定量抽取5个样本进行测试，厚度抽取10个样本进行测试，抗张力抽取20个样本进行测试，撕裂度抽取10个样本进行测试，色度抽取5个样本进行测试。对"汇宣堂"泥金纸进行测试分析所得到的相关性能参数见表6.12。表中列出了各参数的最大值、最小值及测量若干次所得到的平均值或者计算结果。

表6.12 "汇宣堂"泥金纸相关性能参数
Table 6.12 Performance parameters of "Huixuantang" Nijin paper

指标		单位	最大值	最小值	平均值	结果
厚度		mm	0.095	0.090	0.092	0.092
定量		g/m^2	—	—	—	45.2
紧度		g/cm^3	—	—	—	0.491
抗张力	纵向	N	16.2	11.1	13.8	13.8
	横向	N	11.8	9.1	10.9	10.9
抗张强度		kN/m	—	—	—	0.823
撕裂度	纵向	mN	160	150	157	157
	横向	mN	200	180	189	189
撕裂指数		mN·m^2/g	—	—	—	3.72
色度		%	69.6	62.2	65.2	65.2

⊙5 复古笺 / Xuan paper made by traditional ways
⊙6 涂振华带领调查组成员参观车间 / Tu Zhenhua leading researchers to visit the workshop

由表6.12中的数据可知,"汇宣堂"泥金纸最厚约是最薄的1.056倍,经计算,其相对标准偏差为0.001 48,纸张厚薄较为一致。所测"汇宣堂"泥金纸的平均定量为45.2 g/m²。通过计算可知,"汇宣堂"泥金纸紧度为0.491 g/cm³。抗张强度为0.823 kN/m,抗张强度较小。所测"汇宣堂"泥金纸撕裂指数为3.72 mN·m²/g,撕裂度较小。

所测"汇宣堂"泥金纸平均色度为65.2%。色度最大值是最小值的1.119倍,相对标准偏差为2.86,色度差异相对较大。

续表

指标	单位	最大值	最小值	平均值	结果
紧度	g/cm³	—	—	—	0.508
抗张力	纵向 N	19.3	15.1	17.6	17.6
	横向 N	15.3	12.7	14.2	14.2
抗张强度	kN/m	—	—	—	1.060
撕裂度	纵向 mN	230	190	208	208
	横向 mN	240	200	220	220
撕裂指数	mN·m²/g	—	—	—	4.5
色度	%	80.7	76.2	78.5	78.5
吸水性	mm	—	—	—	很难吸水

2. 代表纸品二:"汇宣堂"泥银纸

测试小组对采样自汇宣堂宣纸工艺厂生产的泥银纸所做的性能分析,主要包括厚度、定量、紧度、抗张力、抗张强度、撕裂度、色度、吸水性等。按相应要求,每一指标都需重复测量若干次后求平均值,其中定量抽取5个样本进行测试,厚度抽取10个样本进行测试,拉力抽取20个样本进行测试,撕裂度抽取10个样本进行测试,色度抽取10个样本进行测试,吸水性抽取10个样本进行测试。对"汇宣堂"泥银纸进行测试分析所得到的相关性能参数见表6.13。表中列出了各参数的最大值、最小值及测量若干次所得到的平均值或者计算结果。

表6.13 "汇宣堂"泥银纸相关性能参数
Table 6.13 Performance parameters of "Huixuantang" Niyin paper

指标	单位	最大值	最小值	平均值	结果
厚度	mm	0.095	0.086	0.092	0.092
定量	g/m²	—	—	—	46.8

由表6.13中的数据可知,"汇宣堂"泥银纸最厚约是最薄的1.105倍,经计算,其相对标准偏差为0.003,纸张厚薄较为一致。所测"汇宣堂"泥银纸的平均定量为46.8 g/m²。通过计算可知,"汇宣堂"泥银纸紧度为0.508 g/cm³,抗张强度为1.060 kN/m,抗张强度较大。所测"汇宣堂"泥银纸撕裂指数为4.5 mN·m²/g,撕裂度较大。

所测"汇宣堂"泥银纸平均色度为78.5%,色度最大值是最小值的1.06倍,相对标准偏差为1.794,色度差异相对较小。由于泥银纸表面附有铜粉,故吸水性较差,很难吸水。

四 "汇宣堂"泥金纸生产的原料、工艺流程与工具设备

4
Raw Materials, Papermaking Techniques and Tools of "Huixuantang" Nijin Paper

（一）"汇宣堂"泥金纸生产的原料

1. 主料

"汇宣堂"泥金纸生产的主料是原纸、铜金粉。原纸包括宣纸和书画纸，根据客户需要，选择不同的熟宣，原纸均从泾县当地购买。铜金粉是进口产品，用于配置涂料，国内有代理商，泾县汇宣堂宣纸工艺厂主要从浙江义乌、温州等地购买。

2. 辅料

"汇宣堂"泥金纸加入的辅料包括增稠剂、黏合剂、骨胶和中草药。增稠剂、黏合剂和骨胶主要是增加涂料的黏性，便于涂料的印刷；中草药是纸厂自己配制的，从周边山上采集而来，其作用是防虫防蛀。

制造"汇宣堂"泥金纸使用的水是地下井水。涂振华特别强调：水不能使用自来水，自来水中含有漂白净化物质，对涂料会有影响。调查组成员通过实测得出，泾县汇宣堂宣纸工艺厂所用水pH为6.5，呈弱酸性。

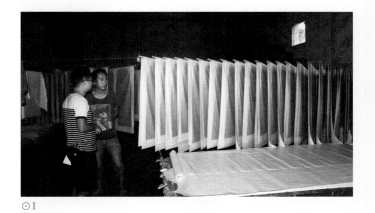

⊙1

（二）"汇宣堂"泥金纸生产的工艺流程

据涂振华较为详细介绍，归纳"汇宣堂"泥金纸生产的工艺流程为：

壹 配料 → 贰 丝网印刷 → 叁 裁剪、印花 → 肆 包装

⊙1
涂振华在介绍工序
Tu Zhenhua introducing the papermaking procedures

工艺流程

壹 配料

"汇宣堂"泥金纸拥有自身独有的配方。配涂料时将进口铜金粉按照配方配比，一般是20 kg的配料桶加入2 kg的地下井水，在搅拌过程中加入一定比例的增稠剂、黏合剂、骨胶和中草药，不断搅拌直到涂料变得浓稠为止。

⊙1

贰 丝网印刷

泥金纸的丝网印刷工序较为复杂，大体可分为丝网印刷、阴干、收下、平整、丝网印刷、阴干、平整这几个环节。

（1）在印刷台上，将原纸放在丝网正下方，刮刀沾上涂料后沿丝网边用力来回刷两遍，涂料会随着丝网的细孔渗透到纸张表面。

⊙3

（2）印刷完涂料后用木棍从纸张背面的中间轻轻挑起，放在晾纸车间进行自然阴干，一般，晴天2天左右即可晾干，阴雨天用时长点。

⊙4

⊙5

（3）收集自然阴干后的纸张，使其自然平整，一般用时2~3天；自然平整后的纸张需要用丝网再次印刷，印刷后重复前面的自然阴干和自然平整过程。总的来说，"汇宣堂"泥金纸需要经过2次丝网印刷，有时根据客户要求需要印刷3次。

⊙1 泥金涂料
Coating material of Nijin paper
⊙2/5 丝网印刷环节
Printing with screen

⊙6

叁 裁剪、印花

3　⊙6

丝网印刷后的泥金纸可根据客户需要定制，需要用丝网印花的，可在泥金纸上印制各种图案或花纹，工序与上一环节丝网印刷相同，主要是丝网选择不同；不需要印花的则可直接裁剪成各种规格。涂振华表示：汇宣堂宣纸工艺厂的泥金纸之前是人工裁剪的，裁剪工具是大剪刀，现在则使用切纸机进行裁剪。

肆 包　装

4　⊙7～⊙10

裁剪后的泥金纸需要人工进行包装，100张一刀，放入成品仓库等待销售。

⊙7

⊙9

⊙10

(三)

"汇宣堂"泥金纸生产的工具设备

壹 丝网 1

丝网根据用途不同分为不同品种，泥金纸根据需要有时会用到丝网印花。制作"汇宣堂"泥金纸使用的丝网都从泾县丝网生产商处购买，实测现场正在使用的丝网尺寸为长177 cm，宽89 cm。

⊙1

⊙2

⊙3

贰 刮刀 2

刮刀主要用来涂刮涂料，木制，实测尺寸为长80 cm，宽11 cm。

⊙4

叁 印刷台 3

长方形木制刷纸台，用来放置丝网，印刷工在上面完成丝网印刷工作，实测尺寸为长265 cm，宽126 cm。

⊙5

肆 切纸机 4

主要用来裁剪泥金纸，按照规格切成不同尺寸。

⊙6

伍 配料桶 5

用来配置泥金涂料，塑料桶，一桶可装10 kg涂料。

⊙7 ⊙8

陆
搭纸棍
6

用来晾晒丝网印刷后的泥金纸，挂在晾晒房内用于自然阴干纸张。

⊙9

⊙9
晾晒纸张
Drying the paper

五
泾县汇宣堂宣纸工艺厂的市场经营状况

5
Marketing Status of Huixuantang Xuan Paper Craft Factory in Jingxian County

据涂振华介绍，自己的加工纸大约从2000年开始生产，经过十几年的发展，加工纸市场行情依然较旺。泾县汇宣堂宣纸工艺厂在老厂的基础上建立了分厂，生产规模不断扩大，产品日益丰富。截至调查时，泾县汇宣堂宣纸工艺厂年销售额约100万元，年利润率约10%。

"汇宣堂"加工纸产品价格根据产品种类、规格的不同而不同。调查时的品种与价格数据如下：粉彩类四尺粉彩批发价为160元/刀，六尺粉彩为320元/刀；对联类四尺万年蓝、万年黑批发价为220元/刀，六尺万年蓝、万年黑批发价为440元/刀，四尺万年红批发价为180元/刀，六尺万年红批发价为360元/刀；蜡染笺类四尺蜡染笺批发价为400元/刀，六尺蜡染笺批发价为800元/刀；复古笺类四尺复古笺批发价为260

元/刀，六尺复古笺批发价为520元/刀；卡纸类规格多样，有24 cm×27 cm、33 cm×33 cm、50 cm×26 cm、38 cm×38 cm、50 cm×30 cm等，其中50 cm×30 cm的泥金硬卡、银硬卡纸批发价为9.5元/张；四尺泥金纸批发价约为600元/刀。

与泾县众多宣纸制造厂家类似，泾县汇宣堂宣纸工艺厂的产品销售渠道以代理商为主，80%的代理商为泾县代理商，全国省会城市基本都有经销商。据汤贵宝介绍，"汇宣堂"品牌纸以国内销售为主，不直接进行出口，卖给国内的外贸供应商的倒有一部分。纸厂目前建立了官方网站，以产品宣传推广为主，还没有建立自己的网店，网络销售主要是通过泾县附近开淘宝店的经营者来厂进货销售。

⊙1

⊙2

六 泾县汇宣堂宣纸工艺厂的品牌文化

6
Brand Culture of
Huixuantang Xuan Paper Craft Factory
in Jingxian County

调查中据汤贵宝介绍，"汇宣堂"泥金纸采取独家配方，选料精细、做工精心、质量精良、使用效果佳、特色显著，超出了泾县其他纸厂生产的泥金纸，建立了自身品牌。郑州书画家黄春增曾试用"汇宣堂"泥金纸即兴题字，评价其为"汇宣堂宣纸纸中极品"，虽有溢美之嫌，但也可视为其质量优良的一种证明。

⊙1 晾晒中的『汇宣堂』加工纸
"Huixuantang" processed paper drying in the air
⊙2 书画家的赞誉
Appreciation of the painter

七 泾县汇宣堂宣纸工艺厂的传承现状与发展思考

7 Current Status of Business Inheritance and Thoughts on Development of Huixuantang Xuan Paper Craft Factory in Jingxian County

⊙3

丝网印刷机器
Screen printer

泾县汇宣堂宣纸工艺厂经过几年的发展，品牌逐渐得到传播，产品销路也在不断拓宽。然而，纸厂发展也面临着若干考验。涂振华表示：近几年泾县加工纸厂快速发展，从事加工纸品生产或销售的人越来越多，而加工纸市场尚不规范，纸品种类鱼目混珠，良莠不齐，已经初步呈现较明显的恶性竞争态势。另外，网店销售产品价格低廉，商家经常通过压低价格来吸引顾客，打价格战，这对纸厂的实体店销售模式造成一定的冲击。

汇宣堂宣纸工艺厂虽然创办时间不长，现有规模不大，但一些创新性举措值得肯定：

第一，泾县汇宣堂宣纸工艺厂在保证产品种类丰富度的基础上，专注于研发和生产若干具有特色性和典型性的纸品，如泥金纸、硬卡、复古笺和蜡染笺。除了上面介绍的泥金纸，"汇宣堂"硬卡的特色在于不易变形，复古笺和蜡染笺的特色在于书写效果佳。泾县汇宣堂宣纸工艺厂以量为基础、以质为保障的生产定位，确实支撑了其在激烈的市场竞争中保有若干产品优势。

第二，泾县汇宣堂宣纸工艺厂目前正在试验丝网印刷机械化，厂里已经购置了一台机器在进行生产性试验。丝网印刷机械化一旦试验成功，将有效提高产量，节约人工成本。泾县汇宣堂宣纸工艺厂这种将新技术工具引入传统丝网印刷过程中，改进传统生产工艺的尝试效果值得期待。

加工纸

泾县汇宣堂宣纸工艺厂

Processed Paper of Huixuantang Xuan Paper Craft Factory in Jingxian County

「汇宣堂」泥金纸透光摄影图
A photo of "Huixuantang" Nijin paper seen through the light

加工纸

泾县汇宣堂宣纸工艺厂

Processed Paper of Huixuantang Xuan Paper Craft Factory in Jingxian County

185

"汇宣堂"泥银纸透光摄影图
A photo of "Huixuantang" Niyin paper seen through the light

第八节
泾县风和堂宣纸加工厂

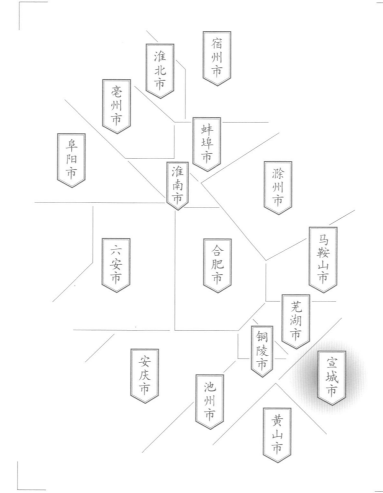

安徽省
Anhui Province

宣城市
Xuancheng City

泾县
Jingxian County

调查对象
泾川镇
泾县风和堂宣纸加工厂
加工纸

Section 8
Fenghetang Xuan Paper Factory
in Jingxian County

Subject
Processed Paper of Fenghetang Xuan Paper Factory
in Jingxian County
in Jingchuan Town

一 泾县风和堂宣纸加工厂的基础信息与生产环境

1
Basic Information and Production Environment of Fenghetang Xuan Paper Factory in Jingxian County

泾县风和堂宣纸加工厂与泾县风和堂文化艺术中心名称互用,法人代表均为翟春平。泾县风和堂宣纸加工厂创办于1996年,注册商标为"风和堂",加工厂房是翟春平家的老宅,也是该厂的注册地,位于泾县县城西南方6 km处的泾川镇五星村翟家村民组,地理坐标为东经118°20′45″,北纬30°40′7″。翟家村民组北侧是泾县第一大河流——青弋江,村内有泾县唯一存世的"南坛三圣"庙,旧日曾是当地人"过会"[1]时首要祭拜的场所。

泾县风和堂文化艺术中心注册地是安徽省泾县泾川镇荷花塘城市文化广场北区,地理坐标为东经118°24′7″,北纬30°41′38″,是风和堂加工纸厂展示产品与洽谈业务之地。两地相距约10 km,有县道、村村通公路连接,交通较便利。

2015年7月21~22日与2018年8月4~5日,调查组两次前往风和堂宣纸加工厂与风和堂文化艺术中心进行现场调查。截至第二次入厂调查时的2018年8月,泾县风和堂宣纸加工厂与泾县风和堂文化艺术中心固定从业人员只有郑智源、翟春平夫妇二人,每年的销售收入约100万元。泾县风和堂宣纸加工厂的产品主要有精制加工纸和宣纸延伸制品。加工纸包括为书画家定制的粉蜡笺、泥金纸、蜡笺等系列;宣纸延伸制品有印谱、手卷等。调查中了解到的信息如下:"风和堂"加工纸大都由郑智源亲手制作,每天的加工量在5~6张,不同纸的价格差别很大,低端在100~200元,高端可达10 000元。翟春平空闲时会回村里的纸坊协助郑智源做些纸品加工,但主要以泾县风和堂

⊙1 翟家村内的"南坛三圣"庙
"Nantan Sansheng" Temple in Zhaijia Village

⊙2 风和堂宣纸加工厂大门
Main entrance of Fenghetang Xuan Paper Factory

[1] 传统民间习俗,城隍出巡、庙会等民间节日的一种集体游艺活动。

泾县风和堂宣纸加工厂位置示意图

Location map of Fenghetang Xuan Paper Factory in Jingxian County

路线图
泾县县城 → 泾县风和堂宣纸加工厂

Road map from Jingxian County centre to Fenghetang Xuan Paper Factory in Jingxian County

考察时间 2015年7月 / 2018年8月

Investigation Date July 2015 / Aug. 2018

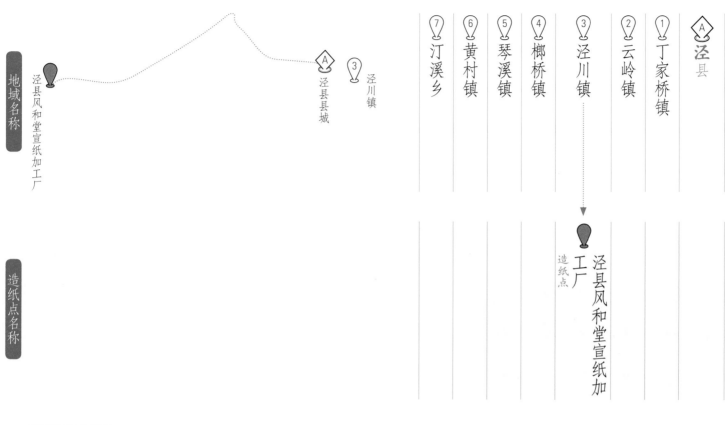

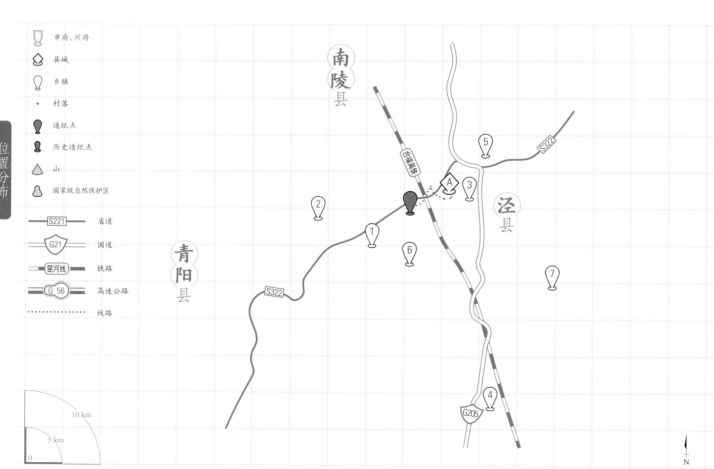

① 1

② 2

1 青弋江翟家村民组段自然环境
Natural environment of Zhaijia Village near the Qingyi River

2 风和堂艺术中心外景
Exterior view of Fenghetang Art Center

艺术中心为平台来经营，一种方式是通过与书画家交流，采用以纸换等值书画作品实现利润，另一种方式是直接出售加工纸。

二 泾县风和堂宣纸加工厂的历史与传承情况

2 History and Inheritance of Fenghetang Xuan Paper Factory in Jingxian County

调查组在前后两次深入调查中得知，尽管泾县风和堂宣纸加工厂的法人代表是翟春平，但从事加工纸及书画艺术品交易的主要是郑智源。

郑智源，1969年生于泾县，家族世居五星村。其父郑长根1986年与亲戚沈维正、张金泉、沈小泉等人在当时的百元乡（调查时已并入泾川镇）五星村创办了振兴宣纸工艺厂，主要从事宣纸深加工，主要工人除张金泉、沈小泉外，还有郑智源的大哥郑智铭、二哥郑智奇及若干村里的回乡青年。其时，沈维正尚在泾县百岭轩宣纸厂工作,其他人也因各种原因不能直接出面开办企业,所以该厂法人代表就由郑长根担任。

访谈时据郑智源介绍,1988年,初中毕业后的他到百岭轩宣纸厂师从沈维正学习宣纸加工技艺。在做学徒的2年时间里,掌握了拖矾、打蜡、研光、染色、染潢、木版水印等加工工艺,对传统的宣纸加工有了较为深入的理解。1990年,学徒期满后他便回自家宣纸工艺厂,一边从事宣纸加工工作,一边外出联系业务。一次,在与泾县文华宣笔厂厂长汤良骥一起出差途中,因汤良骥身体不好,带了很多药却又不识字,郑智源根据汤良骥提供的药方帮助汤良骥按顿配拿药品,给汤良骥留下了较深印象。见多识广的汤良骥敏锐地发现了泾县尚无私人裱画的市场空白,建议并介绍郑智源到位于北京高碑店的当年故宫博物院裱画场所学习。不久后,郑智源便到了故宫博物院裱画场所学习,由于郑智源有幸拜了名师,虽然时间不太长,但掌握了较系统的装裱技术。据郑智源回忆:1993年10月,在泾县举办首届宣纸艺术节的前夕,颇为思乡的他回到了泾县,而继续留京的同批5~6名学员则大多被"荣宝斋"等机构录用。

回到泾县后的郑智源在父亲主持的厂里从事宣纸加工和字画裱装工作。其间,因百岭轩宣纸厂业务每况愈下,沈维正已正式脱离百岭轩宣纸厂,并抽回与郑智源父亲等人合股的泾县振兴宣纸工艺厂股份,创办了泾县载元堂工艺厂;张金泉独立创建了泾县宣艺斋宣纸工艺厂,专门从事书画纸生产和宣纸加工;沈小泉创建了泾县艺然轩书画纸厂,从事皮纸和书画纸的生产。外来投资人全撤了,郑长根便将泾县振兴宣纸工艺厂改

名为泾县汇文堂宣纸工艺厂。

在泾县汇文堂宣纸工艺厂的经营中,郑智源兄弟三人除为厂里服务外,各自还以不同的方式在外承揽业务。郑智源在一次承揽李园宣纸厂700多张书画裱装时,其严谨、细致的工作方式赢得了当年泾县手工造纸界的"红人"——李园宣纸厂厂长张水兵的好感,张水兵当即就开出高薪聘请他。郑智源比较热衷于宣纸加工与书画装裱,而李园宣纸厂只生产生宣,于是认为自己不太适合在那儿,便婉言谢绝了张水兵的邀请,一门心思只做自己喜欢的产品,并热衷向投入门下的青年传授技艺,所以直到泾县凤和堂艺术中心开办的2012年,郑智源身边都有不少徒弟相随。

翟春平,1975年生,泾县凤和堂宣纸加工厂

① 1 调查组成员访谈郑智源
Researchers interviewing Zheng Zhiyuan

② 2 调查组成员与郑智源夫妻交流
Researchers communicating with Zheng Zhiyuan and his wife

3 泾县凤和堂宣纸加工厂营业执照

4 泾县凤和堂文化艺术品中心营业执照

与泾县凤和堂艺术中心的法人代表。由于郑、翟两家父辈曾在村里共事并且交情好,村会计出身的郑智源的父亲在开办企业后常到翟家串门,见翟春平干活利索,便动员翟父将其送去自己厂里做学徒。1994年,翟春平开始随郑智源学习熟宣加工与书画裱装。1995年,郑智源的父亲突然去世,"汇文堂"在没有当家人的情况下,郑家三兄弟经过商议,决定依然保留企业名称,兄弟三人都不使用,用于纪念亡父的功绩。近年来,已转向书画艺术品交易的大哥郑智铭偶尔使用"汇文堂"名称;二哥郑智奇则创办了"春晖堂",从事与宣纸加工有关的业务。

1996年春节,遵照郑父生前的约定,翟春平与郑智源成婚。婚后,以翟春平的名义利用自家空闲的老屋院落开办了2个槽(四尺、六尺槽各1个)生产书画纸,主要是为别人贴牌生产。初办纸厂时,生产管理交由翟父,郑智源夫妇主要在外承接熟宣加工和书画裱装业务。稍有规模时,因丁家桥镇的业务较多,便租用了丁桥油厂的原厂房,加工的产品主要销往日本、韩国、马来西亚、法国等国家。

据翟春平介绍,2001年,因为女儿要上幼儿园,他们不得已放弃了丁家桥的加工场地,搬到县城绿宝街,租赁了一间上下三层的门面房,店面用于展示产品及招揽门客生意,楼上用于熟宣加工和书画裱装。其间,翟家在村里另辟场地建了楼房,翟家老宅专门用作书画纸的生产。新房建成后,翟父也逐步退出了造纸管理的一线,日常管理交给了翟春平,由她负责县城和五星村两地业务。2012年,县城荷花塘城市文化广场北区有两间上下两层门面房抵债给了郑智源家后,夫妻俩便搬至此成立了泾县凤和堂文化艺术品中心。2014年,随着书画艺术品市场的下行,加上两地兼顾比较累,郑智源夫妇就停止了五星村的原纸生产。在经营艺术品中心的过程中,夫妻俩不再招收徒弟,由翟春平负责书画经营,间或有书画家上门,以纸交换书画艺术作品或现金交易。郑智源则在五星村的原厂房根据书画家的需求专门量体裁衣地制作加工纸,在技术上不断精进,琢磨出不少尖端的从未有人涉猎过的产品,也深受书画家喜爱。

在调查组最后一次调查时,郑智源的女儿已经大学毕业,学的就是艺术专业,并在安徽省内一所高校就职。女儿设想,先在外面闯上一段时间,再回来继承父母从事的文化产业。她认为,在外拓展自己视野后,能更好地回家传承父母的衣钵。

三 泾县风和堂宣纸加工厂的代表纸品及其用途与技术分析

3 Representative Paper and Its Uses and Technical Analysis of Fenghetang Xuan Paper Factory in Jingxian County

（一）代表纸品及其用途

在调查中得知，泾县风和堂宣纸加工厂由于不量产，因此所有的产品均走精细化途径。以洒金、洒银系列产品为例，市场上绝大多数为对中低档书画纸的加工，即便以宣纸做的加工，其稳定性也不太好，容易掉金（银）粉，但"风和堂"的洒金、洒银系列均以高品质宣纸加工，不仅稳定性好，而且在书写时墨走在金（银）点时不受影响。"风和堂"的粉蜡笺也有特色，除了色彩、纹饰多样外，最重要的是可以折叠。目前，市场上的粉蜡笺主要存在以下缺陷：一是不上墨或对墨要求特别苛刻；二是折叠时易断裂或卷筒时易造成色彩不稳定，只能戴手套慎重接触纸。"风和堂"粉蜡笺不仅可卷曲、折叠，而且吃墨效果可媲美生宣，对墨的选择没有任何要求（与纸同色除外）。除此之外，一些特定纸张也很受客户的欢迎，如"风和堂"脸谱粉蜡笺等，特别是为四川甘孜州德格藏医院复制的清代中期《大藏经》抄经纸，与原件对比，视觉上没有任何差异，书写效果也能达到《大藏经》抄写要求。

调查组综合调查对象，以及考虑到使用各种纸加工的同质化程度高等原因，在"风和堂"选择了硬黄皮纸、磁青竹纸、碎瓷纹粉蜡笺三种纸。主要原因是：一是用皮纸加工成硬黄纸在市场上少见；二是以竹纸加工成熟纸的少；三是用宣纸制成粉蜡笺，使用时有生宣效果，又呈现出碎瓷纹饰的纸张仅此一家。这三种纸的主要用途有：硬黄皮纸主要用于抄经、书法、绘画等；磁青竹纸主要用于抄经、绘画

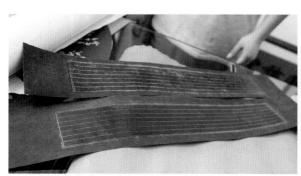

⊙1 抄经纸 Paper for copying scriptures

⊙2 郑智源向调查组成员介绍产品 Zheng Zhiyuan introducing products to researchers

（佛像题材，主要使用朱砂、金等颜料）等；碎瓷纹粉蜡笺主要用于书法创作等。

（二）技术分析

1. 代表纸品一："风和堂"硬黄纸

测试小组对采样自泾县风和堂宣纸加工厂生产的硬黄纸所做的性能分析，主要包括厚度、定量、紧度、抗张力、抗张强度、撕裂度、色度等。按相应要求，每一指标都需重复测量若干次后求平均值，其中定量抽取5个样本进行测试，厚度抽取10个样本进行测试，抗张力（不分横纵）抽取10个样本进行测试，撕裂度（不分横纵）抽取5个样本进行测试，色度抽取8个样本进行测试。对泾县风和堂宣纸加工厂硬黄纸进行测试分析所得到的相关性能参数见表6.14。表中列出了各参数的最大值、最小值及测量若干次所得到的平均值或者计算结果。

表6.14 "风和堂"硬黄纸相关性能参数
Table 6.14　Performance parameters of "Fenghetang" Yinghuang paper

指标	单位	最大值	最小值	平均值	结果
厚度	mm	0.196	0.145	0.161	0.161
定量	g/m^2	—	—	—	92.0
紧度	g/cm^3	—	—	—	0.571
抗张力	N	30.4	21.8	27.8	27.8
抗张强度	kN/m	—	—	—	1.853
撕裂度	mN	753	625	710	710
撕裂指数	$mN \cdot m^2/g$	—	—	—	7.4
色度	%	12.68	6.63	9.25	9.3

○3 "风和堂"硬黄纸透光摄影图
A photo of "Fenghetang" Yinghuang paper seen through the light

由表6.14中的数据可知，"风和堂"硬黄纸最厚约是最薄的1.35倍，经计算，其相对标准偏差为0.105，纸张厚薄较为一致。所测"风和堂"硬黄纸的平均定量为92.0 g/m^2。通过计算可知，该厂硬黄纸紧度为0.571 g/cm^3，抗张强度为1.853 kN/m，抗张强度较小。所测该厂硬黄纸撕裂指数为7.4 $mN \cdot m^2/g$，撕裂度较小。

所测"风和堂"硬黄纸平均色度为9.3%。色度最大值是最小值的1.913倍，相对标准偏差为0.235，色度差异相对较小。

2. 代表纸品二："风和堂"磁青竹纸

测试小组对采样自泾县风和堂宣纸加工厂生产的磁青竹纸所做的性能分析，主要包括厚度、定量、紧度、抗张力、抗张强度、撕裂度、色度、吸水性等。按相应要求，每一指标都需重复测量若干次后求平均值，其中定量抽取5个样本进行测试，厚度抽取8个样本进行测试，抗张力抽取20个样本进行测试，撕裂度抽取8个样本进行测试，色度抽取4个样本进行测试，吸水性抽取10个样本进行测试。对"风和堂"磁青竹纸进行测试分析所得到的相关性能参数见表6.15。表中列出

表6.15 "风和堂"宣纸加工厂磁青竹纸相关性能参数
Table 6.15 Performance parameters of "Fenghetang" Ciqing bamboo paper

指标		单位	最大值	最小值	平均值	结果
厚度		mm	0.172	0.148	0.159	0.159
定量		g/m²	—	—	—	92.8
紧度		g/cm³	—	—	—	0.583
抗张力	纵向	N	43.5	33.2	39.2	39.2
	横向	N	29.9	21.6	26.7	26.7
抗张强度		kN/m	—	—	—	2.2
撕裂度	纵向	mN	447.0	347.5	394	394
	横向	mN	396.4	350.9	378	378
撕裂指数		mN·m²/g	—	—	—	4.1
色度		%	5.3	5.0	5.1	5.1
吸水性	纵向	mm	15	11	12.6	15
	横向	mm	18	16	17.4	4.8

了各参数的最大值、最小值及测量若干次所得到的平均值或者计算结果。

由表6.15中的数据可知，"风和堂"磁青竹纸最厚约是最薄的1.16倍，经计算，其相对标准偏差为0.044，纸张厚薄较为一致。所测"风和堂"磁青竹纸的平均定量为92.8 g/m²。通过计算可知，所测"风和堂"磁青竹纸紧度为0.583 g/cm³。抗张强度为2.2 kN/m，抗张强度较小。所测该厂磁青竹纸撕裂指数为4.1 mN·m²/g，撕裂度较小。

所测"风和堂"磁青竹纸平均色度为5.1%。色度最大值是最小值的1.06倍，相对标准偏差为0.030，色度差异相对较小。所测该厂磁青竹纸吸水性纵横平均值为15 mm，纵横差为4.8 mm。

⊙1

"风和堂"磁青竹纸透光摄影图
A photo of "Fenghetang" Ciqing bamboo paper seen through the light

四 "风和堂"加工纸生产的原料、工艺流程与工具设备

4 Raw Material, Papermaking Techniques and Tools of "Fenghetang" Processed Paper

（一）"风和堂"硬黄皮纸

1. "风和堂"硬黄皮纸生产的原料

（1）主料：原纸（构皮、狼毒、竹浆）。该纸定制生产。

（2）辅料：黄檗、黄连、生山栀、朱砂、高岭土、蜡、水、面粉等。

据郑智源介绍，黄檗、黄连、生山栀、高岭土均在当地药店购买；朱砂在墨厂购买，自行按需研磨；蜡购回后需另行加工。

加工纸的生产用水量很少，对水源要求也不如生产原纸高。泾县风和堂宣纸加工厂使用的水源从水井中随用随抽。据调查组现场测试，泾县风和堂宣纸加工厂所用水pH为6.5，呈弱酸性。

⊙2 黄檗与生山栀 Phellodendron amurense and Gardenia Jasmine
⊙3 黄连 Coptis
⊙4 朱砂 Cinnabar
⊙5 高岭土 Kaolin soil
⊙6 蜡 Wax
⊙7 抽水电器设备 Pumping equipment
⊙8 现场测水 On-spot water test

2. "风和堂"硬黄皮纸生产的工艺流程

据郑智源介绍，以及综合调查组2018年8月5日在风和堂宣纸加工厂的实地调查，归纳"风和堂"硬黄皮纸生产的工艺流程为：

壹	贰	叁	肆	伍	陆	柒
配色	上杆	晾干	刷染	再次晾干	再次刷染	第三次晾干

拾叁	拾贰	拾壹	拾	玖	捌
研光	打蜡	裁边	起纸	上墙	托裱（覆背）

壹　配色　1

把适量的黄檗、黄连分别熬煮，然后按需配比调色。

贰　上杆　2

用自调的糨糊将纸张一侧粘贴在晾纸杆上。

叁 晾干 3

夏天通常晾十几分钟，冬天需晾至第二天。

⊙3

肆 刷染 4

染色前用湿布把裱画桌抹潮，再用糨糊刷蘸取适量染料，从纸没有黏沾棍的一头向另一头刷染。在刷染时，从纸内向纸边均匀刷，时刻注意防止纸张褶皱。整张纸经过刷染后，根据目测对色弱的地方进行补刷，直至整张纸上色均匀。

⊙4

伍 再次晾干 5

刷染结束后，将纸头上的木棍拿起，将湿纸从裱画桌上轻轻揭起，挂在工作间绷好的钢丝绳上。

⊙5 ⊙6

陆 再次刷染 6

将晾干的纸取下来，放在裱画桌上按原刷染程序，将纸背面重新刷染。

⊙7

⊙3 晾干 Dyeing the paper
⊙4 刷染 Dyeing the paper with a brush
⊙5/6 晾纸 Dyeing the paper
⊙7 刷染 Dyeing the paper with a brush

柒 第三次晾干

7　⊙8

刷染结束后，将纸头上的木棍拿起，将湿纸从裱画桌上轻轻揭起，挂在工作间绷好的钢丝绳上。

⊙8

捌 托裱（覆背）

8　⊙9 ⊙10

先将固定在纸上的木杆取下，再用水将裱画桌面打湿，将两面染色后的纸平整地贴在桌上，再往纸上刷一层糨糊（面粉需经去筋、防霉、防蛀等处理），覆上一张宣纸。

⊙9

⊙10

玖 上墙

9　⊙11

将覆背后的纸刷粘到墙上。为便于下揭，在上墙后的纸下部留一小口。

⊙11

拾 起纸

10　⊙12

用起子轻挑小口，等纸从墙上揭下一条边时，将整张揭下。

⊙12

拾壹 裁边
11 ⊙13
将从墙上揭下的纸用美工刀进行裁边规整。

拾贰 打蜡
12 ⊙14
先把蜡涂到刷子上,再将涂到刷子上的蜡均匀地刷到纸上。

拾叁 砑光
13 ⊙15 ⊙16
将打蜡后的纸先用印章石砑光,再用玉石仔细砑光。

⊙13
⊙14
⊙15
⊙16

(二)
"凤和堂"磁青竹纸

1. "凤和堂"磁青竹纸生产的原料

　　(1) 主料:从福建将乐县西山购买的竹纸。

　　(2) 辅料:青花、蓼蓝、高岭土、墨碇、水、面粉等。

⊙17
⊙18

⊙18 调制好的颜料 Prepared pigment
⊙17 蓼蓝 Indigo plant
⊙15/16 砑光 Smoothing the paper
⊙14 打蜡 Waxing
⊙13 裁边 Trimming the paper

2. "风和堂"磁青竹纸生产的工艺流程

据郑智源介绍,以及综合调查组2018年8月5日在风和堂宣纸加工厂的实地调查,归纳"风和堂"磁青竹纸生产的工艺流程为:

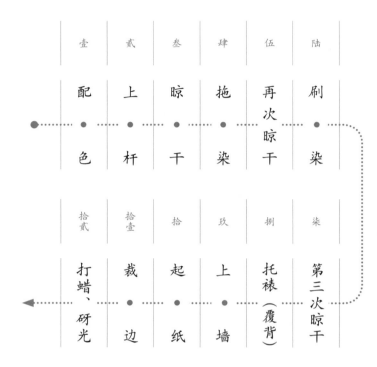

壹 配色 → 贰 上杆 → 叁 晾干 → 肆 拖染 → 伍 再次晾干 → 陆 刷染 → 柒 第三次晾干 → 捌 托裱(覆背) → 玖 上墙 → 拾 起纸 → 拾壹 裁边 → 拾贰 打蜡、研光

壹 配色 1

把适量的蓼蓝进行熬煮,将青花研细后熬煮,高岭土需要捶打得很细,墨碇需要研磨形成墨汁,然后按需配比调色。

贰 上杆 2

用自调的糨糊将纸张一侧粘贴在晾纸杆上。

叁 晾干 3

夏天通常晾十几分钟,冬天需晾至第二天。

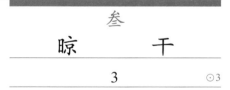

肆 拖染 4

将配好色的染料放入拖纸盆内,再将纸从上杆处浸入拖纸盆的染料中,依次从染料中拖过。

⊙4

伍 再次晾干 5

将拖染过的纸从拖纸盆处直接拿到钢丝绳上挂好。

⊙5

陆 刷染 6

染色前用湿布把裱画桌面抹潮,再用糨糊刷蘸取适量染料,从纸没有黏沾棍的一头向另一头刷染。在刷染时,从纸内向纸边均匀刷,时刻注意防止纸张褶皱。整张纸经过刷染后,根据目测对色弱的地方进行补刷,直至整张纸上色均匀。

⊙6

柒 第三次晾干 7

刷染结束后,将纸头上的木棍拿起,将湿纸从裱画桌上轻轻揭起,挂在工作间绷好的钢丝绳上。

⊙7

捌
托裱（覆背）
8 ⊙8

先用水将裱画桌打湿，将两面染色后的纸平整地贴在桌上，再在纸上刷上一层糨糊（面粉需经去筋、防霉、防蛀等处理），覆上一张竹纸。

⊙8

玖
上　　墙
9 ⊙9

将覆背后的纸刷粘到墙上。为便于下揭，在上墙后的纸下部留一小口。

⊙9

拾
起　　纸
10 ⊙10

用起子轻挑小口，等纸从墙上揭下一条边时，将整张揭下。

拾壹
裁　　边
11 ⊙11

将从墙上揭下的纸用美工刀进行裁边规整。

⊙10

⊙11

拾贰
打蜡、砑光
12 ⊙12 ⊙13

先把蜡涂到刷子上，再将涂蜡后的刷子均匀地刷到纸上。将打蜡后的纸先用印章石砑光，再用玉石仔细砑光。

⊙12

⊙13

⊙ 8 托裱（覆背） Glueing the paper
⊙ 9 上墙 Pasting the paper on the wall
⊙ 10 起纸 Peeling the paper down
⊙ 11 裁边 Trimming the paper
⊙ 12 / 13 打蜡、砑光 Waxing and smoothing the paper

（三）"风和堂"碎瓷纹粉蜡笺

1. "风和堂"碎瓷纹粉蜡笺生产的原料

（1）主料：从泾县当地购买的宣纸。

（2）辅料：蓼蓝、墨碇、水、面粉等。

2. "风和堂"碎瓷纹粉蜡笺生产的工艺流程

据郑智源介绍，以及综合调查组2018年8月5日在风和堂宣纸加工厂的实地调查，归纳"风和堂"碎瓷纹粉蜡笺生产的工艺流程为：

壹 配色 → 贰 上杆 → 叁 晾干 → 肆 刷染 → 伍 再次晾干 → 陆 制花 → 柒 打蜡、砑光

壹 配色 1 ⊙14

把适量的蓼蓝进行熬煮，墨碇需要研磨形成墨汁，然后按需配比调色。

贰 上杆 2 ⊙15

用自调的糨糊将纸张一侧粘贴在晾纸杆上。

叁 晾干 3 ⊙16

夏天通常晾十几分钟，冬天需晾至第二天。

肆 刷染 4 ⊙17

染色前用湿布把裱画桌面抹潮，再用糨糊刷蘸取适量染料，从纸没有黏沾棍的一头向另一头刷染。在刷染时，从纸内向纸边均匀刷，注意防止纸张褶皱。整张纸经过刷染后，根据目测对色弱处进行补刷，至整张纸上色均匀即可。

⊙14 配色 Color scheming
⊙15 上杆 Sticking the paper to the pole
⊙16 晾干 Drying the paper
⊙17 刷染 Dying the paper with a brush

伍 再次晾干 5 ⊙18

刷染结束后，将纸头上的木棍拿起，再将湿纸从裱画桌上轻轻揭起，挂在工作间绷好的钢丝绳上。

陆 制花 6 ⊙19

在纸晾至半干时做成碎瓷形花纹，具体做法属技术机密，郑智源表示目前不便公开。

柒 打蜡、砑光 7 ⊙20 ⊙21

先把蜡涂到刷子上，再将涂到刷子上的蜡均匀地刷到纸上。将打蜡后的纸先用印章石砑光，再用玉石仔细砑光。

⊙18

⊙19

⊙20　⊙21

（四）"风和堂"硬黄皮纸、磁青竹纸及碎瓷纹粉蜡笺生产的工具设备

壹 棕刷 1

用于平整纸张，在当地文房四宝商店购买。实测风和堂宣纸加工厂的棕刷尺寸为长18 cm（带柄），宽15 cm。

贰 小糨糊刷 2

用于刷糨糊，在当地文具商店购买。实测风和堂宣纸加工厂的小糨糊刷尺寸为长16 cm（带柄），宽6 cm。

叁 大糨糊刷 3

用于在纸上大范围刷染或刷糨糊，在当地文具商店购买。实测风和堂宣纸加工厂的大糨糊刷尺寸为长24 cm（带柄），宽20 cm。

⊙22

⊙23

⊙24

肆 糨糊桶 4

用于装糨糊，在当地塑料品商店购买。实测风和堂宣纸加工厂的糨糊桶尺寸为顶部外框直径19 cm，高13 cm。

⊙25

伍 起子 5

用于剥离粘贴在墙上已加工好的纸，由竹子自制而成。实测风和堂宣纸加工厂的起子尺寸为长39 cm，宽0.7 cm。

⊙26

陆 晾纸杆 6

用于晾干、拖染、刷染固定的四方形木棍，自制。实测风和堂宣纸加工厂的晾纸杆尺寸为长91 cm，宽1 cm。

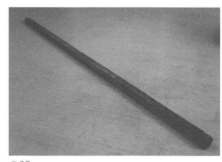
⊙27

柒 挑纸杆 7

在托裱时用于卷生纸，便于将纸完整地覆盖在湿纸上；也可用于挑湿纸，降低纸张的破损率。材料为PVC管，在建材商店购回后，按需裁断。实测风和堂宣纸加工厂的挑纸杆尺寸为长121 cm，宽0.8 cm。

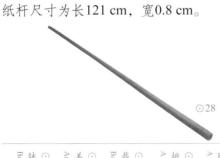
⊙28

捌 裁纸尺 8

由塑料板自制而成。实测风和堂宣纸加工厂的裁纸尺尺寸为长123 cm，宽6 cm，厚0.3 cm。

⊙29

玖 美工刀 9

用于裁纸，在当地文具商店购买。

⊙30

拾 钵、风炉 10

用于染料原材料的蒸煮工具，均在泾县陶窑厂购买。实测风和堂宣纸加工厂的钵（带柄）尺寸为高9 cm，直径18 cm；风炉尺寸为高12 cm，直径19 cm。

⊙31

拾壹 印章石 11

用于第一遍砑光，呈四方形，在当地文房四宝商店购买。实测风和堂宣纸加工厂的印章石尺寸为长20 cm，宽5 cm。

拾贰 玉 石 12

用于第二遍砑光，在玉石店购买。实测风和堂宣纸加工厂的玉石尺寸为长7.5 cm，宽4.2 cm，厚2 cm。

拾叁 洒金筒 13

由塑料筒自制而成，筒壁钻有小孔，便于将筒内金银碎末均匀洒在纸上。实测风和堂宣纸加工厂洒金筒尺寸为A桶高12 cm，直径8 cm；B桶高12 cm，直径6 cm。

⊙32

⊙33

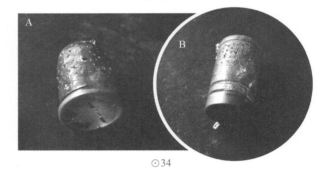
⊙34

五 泾县风和堂宣纸加工厂的市场经营状况

5 Marketing Status of Fenghetang Xuan Paper Factory in Jingxian County

与泾县多数加工纸厂家不同的是，泾县风和堂宣纸加工厂在前期生产原纸和熟宣加工时，因没有自己的品牌，多年来以来料加工和上门加工业务形态为主，尽管摊子铺得不小，员工也有一批，但没有形成品牌与市场的对应关系，抗风险能力低。风和堂文化艺术中心成立后，才注册了自己的商标，将生产范围圈定在技术实力最强的宣纸加工一项，并依托多年来积累的人脉关系，为特定的小众消费者提供性价比高的产品，企业也因提高了产品的附加值而增强了实力。

⊙35 "风和堂"粉蜡笺纸品

访谈时从郑智源夫妇处获取的信息如下：泾县风和堂宣纸加工厂系列产品售价以单张纸来定价，每张纸的售价最低不低于100元，最高可达万元。调查组2018年8月第二次调查并采样分析的三款纸品中，长143 cm、宽76 cm的硬黄纸售价为580元/张；长65 cm、宽33 cm的磁青竹纸售价为100元/张；长130 cm、宽35 cm的碎瓷纹粉蜡笺售价为180元/张。

六 泾县风和堂宣纸加工厂的品牌文化与民俗故事

6
Brand Culture and Stories of Fenghetang Xuan Paper Factory in Jingxian County

交流中郑智源回忆：1996年创办泾县风和堂宣纸加工厂时，正好接待了著名画家杨晓阳，谈到取名，杨晓阳提示可以用王羲之《兰亭集序》中"天朗气清，惠风和畅"的"风和"二字，既示历史悠久，能与传统文化根脉相承，也能优雅地显示出企业风貌。郑智源、翟春平夫妇当年又联想到泾县县城标志性风物——荷花塘，"塘"与"堂"谐音，加上规模较小的文房四宝企业习惯用"堂""斋""轩"等命名，当下便决定以"风和堂"为名。郑智源表示："风和堂"这个名称也得到了业内外人士的认可，有书画家认为翟春平性格风风火火，做事干练又注重和气生财；郑智源性格温和，有一种风清气和的感觉，因此认为郑智源、翟春平夫妇性格相得益彰。经书画家演绎后,"风和堂"也因此引申出更多的含义来。

调查组从郑智源处获悉一件趣事。杨晓阳在给郑智源、翟春平夫妇提议并题写"风和堂"后不久,即接任国家画院院长,画院下属一机构创办了一个文化公司,杨晓阳忘了自己的提议和题名,又为下属机构题写了"风和堂"。此后,两家先后提交了"风和堂"商标注册申请,所幸的是郑智源、翟春平夫妇早于该机构两个月获得"风和堂"的商标注册权。其间,该机构希望郑智源、翟春平夫妇能将"风和堂"商标转让,但郑智源、翟春平夫妇多年来已对"风和堂"倾注了心血,加上国内诸多书画家都曾为"风和堂"题词作画,尽管该机构许以优厚的条件,郑智源、翟春平夫妇还是坚持继续拥有"风和堂"商标的注册与使用权。

⊙1

七 泾县风和堂宣纸加工厂的传承现状与发展思考

7
Current Status of Business Inheritance and Thoughts on Development of Fenghetang Xuan Paper Factory in Jingxian County

访谈中,调查组成员带着疑虑询问郑智源,既然"风和堂"产品销路很好,为什么不招工人或带徒弟扩大生产,而仅仅是夫妻俩独自支撑。郑智源的说法是:因他制作的产品都有个人在传承基础上创新的因素,意图将传统产品向更优化的方向深度挖掘,每一款产品都有一定的核心机密,如果广招人员扩大生产,在现在的行业环境下,对保守技术机密不利。在泾县就有多起这样的实例,一款产品研发成功后尚未投入规模化生产就被泄密了,结果研发成本尚未收回就遇上倾销的竞争对手,导致血本无归。还有一点就是"风和堂"除客户指定用其他原纸加工外,所有的产品都用宣纸加工,而市场上的加工纸大多以中低档书画纸加工,有的还用机制书画纸加工,对一般消费者而言,在外形上难以区分,如果没

⊙1 "风和堂"商标注册证
Trademark registration certificate of "Fenghetang"
⊙2 翟春平与杨晓阳合影
Photo of Zhai Chunping and Yang Xiaoyang

○3

有一定的核心机密做保障,一旦被别家仿冒就无生存空间了。所以,多年来郑智源因不确定招人能否保守技术机密的因素,坚持个人独立完成所有订单,宁愿不接大订单也不多招人。

据郑智源分析,以粉蜡笺为例,有的加工纸企业以机制书画纸制作,每天能生产很多,不仅技术上达不到"风和堂"的要求,而且颜色会在很短时间里退化。如:他一个人平均每天只能加工五六张"风和堂"粉蜡笺,不仅耗时久,而且每张纸仅成本一项就需要投入100多元。在大众因认知与辨别水平不够导致对产品优劣的分辨能力低的环境下,"风和堂"的纸走向市场,特别是网络和社交营销的市场,就无法与低成本、效果差的产品相抗衡了。

谈到传承,郑智源、翟春平夫妇表示:女儿已经接受了高等教育,也表示在不久的将来会回家乡传承,但真正上手也不是一蹴而就的,需要较长时间积累和不断提高动手能力才有希望完整传承。当调查组成员表示若女儿以后不能回来继承怎么办,郑智源、翟春平夫妇倒是表现得较为乐观。夫妇俩表示,遗憾的是政府相关部门将关注的眼光投入到会炒作、会经营的规模大的企业,对他们夫妻俩这样钻研特色产品与坚守经典技艺的人很少有政策上的扶持,也期盼政府部门将保护措施覆盖到"风和堂"这类企业。

○4

○3 使用多年的工作围腰
The work wrap used for years

○4 "风和堂"脸谱粉蜡笺
The facial makeup drawn on "Fenghetang" Fenla paper

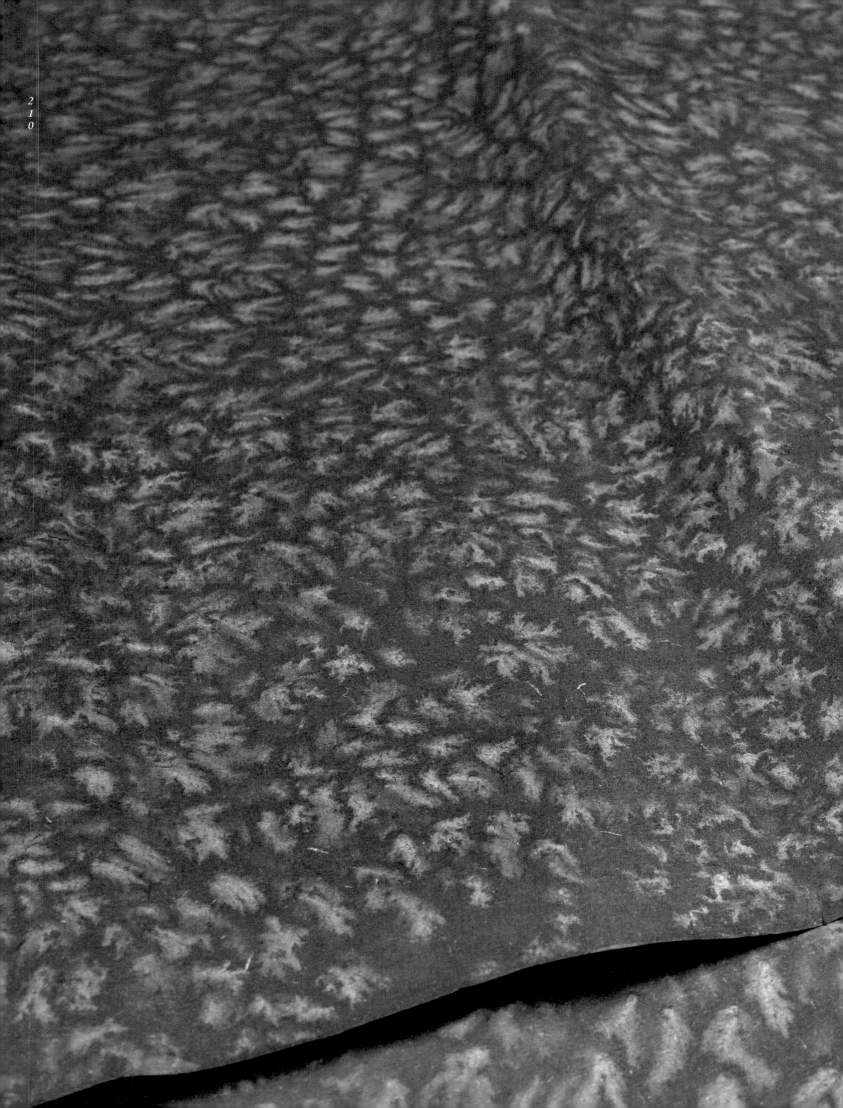

加工纸

泾县凤和堂宣纸加工厂

Processed Paper of Fenghetang Xuan Paper Factory in Jingxian County

加工纸

泾县风和堂宣纸加工厂

Processed Paper of Fenghetang Xuan Paper Factory in Jingxian County

「风和堂」仿古硬黄纸透光摄影图
A photo of "Fenghetang" antique Yinghuang paper seen through the light

泾县风和堂宣纸加工厂加工纸

Processed Paper of Fenghetang Xuan Paper Factory in Jingxian County

"风和堂"仿古磁青纸透光摄影图
A photo of "Fenghetang" antique Ciqing paper seen through the light

第七章
工具

Chapter VII
Tools

第一节
泾县明堂纸帘工艺厂

安徽省
Anhui Province

宣城市
Xuancheng City

泾县
Jingxian County

调查对象
丁家桥镇
泾县明堂纸帘工艺厂
工具

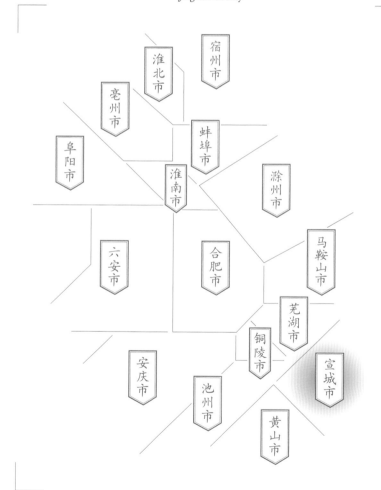

Section 1
Mingtang Papermaking Screen Craft Factory in Jingxian County

Subject
Tools of Mingtang Papermaking Screen Craft Factory in Jingxian County in Dingjiaqiao Town

一

泾县明堂纸帘工艺厂的基础信息与生产环境

1

Basic Information and Production Environment of Mingtang Papermaking Screen Craft Factory in Jingxian County

调查时泾县明堂纸帘工艺厂的厂址在泾县丁家桥镇街道的卫生院对面（而了解到的旧作坊地址在镇辖新渡村），地理坐标为东经118°19′32″，北纬30°38′54″。X072县道从其厂门前经过，交通尚属便利。2015年11月12日和2017年6月3日，调查组两次前往泾县明堂纸帘工艺厂进行实地调查，调查时获知的基础信息如下：泾县明堂纸帘工艺厂负责人为陈明堂，厂以人名命名。共有员工10人，年产手工制作的抄纸竹帘近200张，生产场区占地约500 m²。

⊙1
纸帘作坊内景
Internal view of papermaking screen workshop

二

泾县明堂纸帘工艺厂的历史与传承情况

2

History and Inheritance of Mingtang Papermaking Screen Craft Factory in Jingxian County

关于泾县明堂纸帘工艺厂的技艺发展史，调查中陈明堂是这样描述的：他出生于1968年，从事纸帘制作工艺已有30余年。1985年，年仅17岁的陈明堂到丁家桥镇小岭村随编帘老艺人曹鸿江做学徒，成为小岭首位曹姓造纸世家以外的编帘学徒工。

泾县明堂纸帘工艺厂成立于1988年，生产点最初在丁家桥镇新渡村溪边村民组，是利用家庭住房开辟的作坊，当时只有2～3个员工，主要技术力量是陈明堂及其姐姐、妹妹。2000年，考虑到运输与往来方便，举家迁至较为繁华热闹的丁家桥镇区的街道上，员工迅速增多，生产规模也不断扩大，最多时制帘工人达10多个，尚不包括因受场地限制，领料回家的制帘人员；最高年产大小纸帘200余张。

泾县明堂纸帘工艺厂位置示意图

Location map of Mingtang Papermaking Screen Craft Factory in Jingxian County

路线图
泾县县城 → 泾县明堂纸帘工艺厂
Road map from Jingxian County centre to Mingtang Papermaking Screen Craft Factory in Jingxian County

考察时间
2015年11月 / 2017年6月

Investigation Date
Nov. 2015 / June 2017

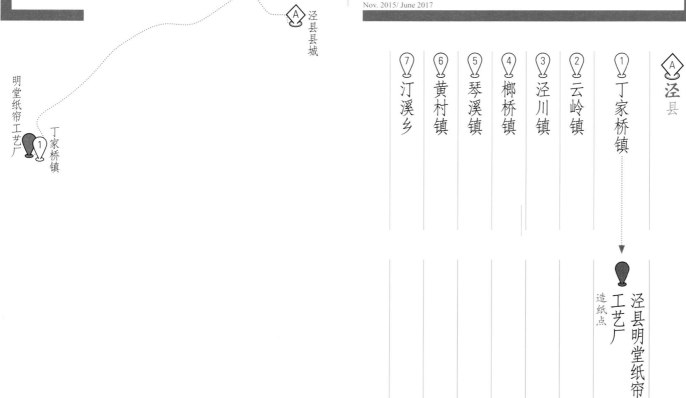

地域名称
A 泾县县城
① 丁家桥镇
② 云岭镇
③ 泾川镇
④ 榔桥镇
⑤ 琴溪镇
⑥ 黄村镇
⑦ 汀溪乡

造纸点名称
① 泾县明堂纸帘工艺厂 造纸点

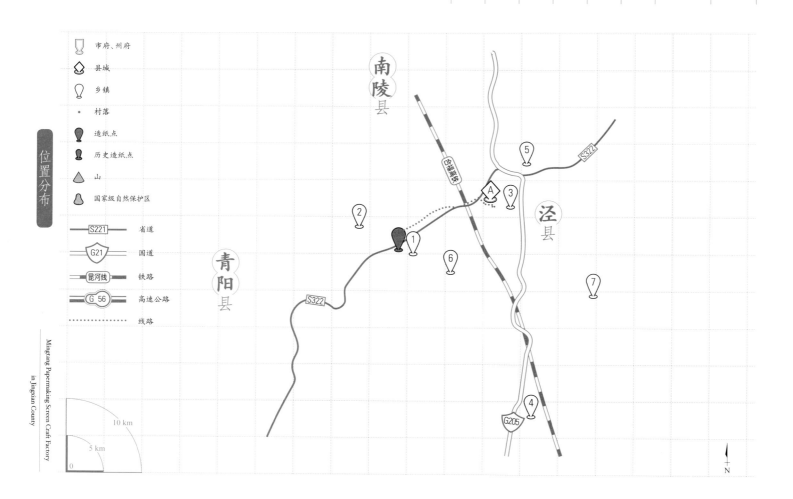

位置分布

图例：
市府、州府
县城
乡镇
· 村落
造纸点
历史造纸点
△ 山
国家级自然保护区
S221 省道
G21 国道
昆河线 铁路
G56 高速公路
······ 线路

据陈明堂介绍，明堂纸帘工艺厂生产的高峰是2012~2013年，那时年产量约为210张（一张"一改二"纸帘平均为2.5 m²），年销售额为30万元左右。陈明堂兄弟姐妹共5人，都已各自成家，其中有3户人家的女性都在明堂纸帘工艺厂工作。姐姐陈明玉1966年出生，现居丁家桥镇必胜村民组。1988年陈明堂出师后将纸帘的编制工艺教给姐姐，之后，陈明玉就一直在明堂纸帘工艺厂编制纸帘。妹妹陈明霞1975年出生，同样是在1988年陈明堂出师后向其传授了纸帘工艺，在明堂纸帘工艺厂工作了15年后，由于颈椎问题不再工作，调查时在家带孩子。妻子黄秋珍1968年出生，1992年与陈明堂结婚后也在厂里工作，主要从事纸帘的后期维修和刺绣工作。

据陈明堂口述，2017年上半年厂里共有工人10名，其中在厂工人5名，包括妻子黄秋珍。另外5名员工从厂里领取原材料在家制作纸帘，然后由陈明堂统一收购。

⊙1 工人在作坊编制纸帘
Workers making the papermaking screen in the workshop

⊙2 纸帘技师陈明堂
Papermaking screen technician Chen Mingtang

三 泾县明堂纸帘工艺厂纸帘生产的原料、工艺流程与工具设备

3 Raw Materials, Processing Techniques and Tools of Mingtang Papermaking Screen Craft Factory in Jingxian County

（一）

纸帘生产的原料

1. 主料

据陈明堂介绍，该纸帘厂制作抄纸竹帘的主料多为苦竹。

苦竹，禾本科，因常被用作伞柄（纸伞和油布伞），故泾县当地又称之为"伞柄竹"。苦竹竿呈圆筒形状，高度可达4 m，下部数节的竹节间距长25～40 cm，直径约15 mm，比一般竹种的竹节明显要长。苦竹4～5月开花，花呈绿色或淡紫色，因所产竹笋味道很苦而难以食用，故名"苦竹"。宣纸制帘采用苦竹，主要是因为苦竹节稀、尺寸长（短者尺余，长者可达2尺），特别适合制作无节的长竹丝。

苦竹主产于中国安徽、江苏、浙江、江西、福建、湖南、湖北、四川、云南、贵州等省份。山地普遍野生，在低山、丘陵、平地均能生长。竹节长，韧性强，是制作箫笛、篮筐、伞柄、支架、旗杆和帐杆的好材料。同时，苦竹嫩叶、嫩苗、根茎等均可入药。

泾县明堂纸帘工艺厂生产纸帘所需的原料，早期使用的是从安徽省芜湖市南陵县购买的苦竹，但是由于南陵的苦竹竹节较短而不利于生产，所以在20世纪90年代中叶改为从安徽省金寨县购买当地人称为管竹（即泾县当地称为苦竹）的竹子，一般都采用3～4年生长期的苦竹，其韧性和长度均达到制帘所需的最佳状态。

据陈明堂介绍，其生产纸帘所需的苦竹来自两个地区：一是产自皖西大别山金寨县的竹子，2015年采购价格为1.7元/kg，不过现在已经减少对

⊙1 作坊早年遗留下来的苦竹
Amarus bamboo left in the early years of the workshop

⊙2 编帘用的苦竹篾丝
Amarus bamboo splits for making the papermaking screen

原竹的采购,因为二次加工耗时耗力,间接增加了成本,故逐渐采购半成品苦竹丝,2015年采购价格为8元/kg。陈明堂表示,随着物流的发展,有需要随时都可以购买到,很方便。二是从浙江温州采购的成品篾丝,采购价格为140元/kg,然后找人加工(价格为90元/kg),后一部分的原料价格相对较高,因此主要以采购半成品竹料为主。

2. 辅料

(1) 锦线。锦线主要是起到固定每根竹篾的作用,使得竹篾与竹篾紧密地结合在一起,不易散开,从而形成一张完整的帘幕。据陈明堂介绍,其采购的锦线主要从温州以及上海购买,

2015年国外进口线达到1 000~1 400元/kg,国产普通线则是60元/kg,价格相差很大,通常0.5 kg线可以编织20张普通四尺宣纸帘。

(2) 生漆,又称"土漆""国漆"或"大漆"。生漆是从漆树上采割的乳白色胶状液体,接触空气数小时后表面干涸硬化而生成漆皮。生漆具有耐腐、耐磨、耐酸、耐溶剂、耐热、隔水和绝缘性好、富有光泽等特性,是手工艺品和民用家具的优质涂料。

据陈明堂介绍,泾县明堂纸帘工艺厂所用的生漆多采购自四川、湖北、陕西等地,其中四川美星工艺厂的生漆采购量占60%以上,2015年采购价格在280~300元/kg,通常0.5 kg只能涂刷2张四尺"一改二"的竹帘。

(二) 纸帘生产的工艺流程

根据调查组对泾县明堂纸帘工艺厂进行的实地调查和访谈,归纳纸帘生产的工艺流程为:

壹 浸泡 → 贰 剖篾 → 叁 削尖 → 肆 抽丝 → 伍 编帘 → 陆 装芒杆 → 柒 上绷架 → 捌 刷漆 → 玖 上图 → 拾 绣花

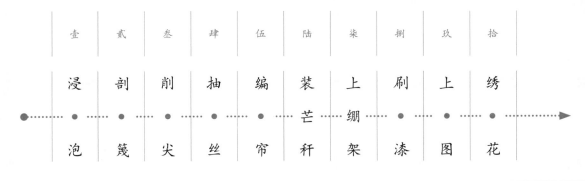

3/4 普通线(左)与进口线(右)
5 制帘用的生漆

工艺流程

壹 浸泡 1

半成品竹子一般在水中浸泡2天左右，受天气和水温影响，夏天浸泡2天即可，冬天气温低于0℃时一般需要浸泡3天。

贰 剖篾 2 ⊙1⊙2

用竹刀把半成品竹片切簧、去芯，剖成与牙签同样粗细的竹条。

⊙1

叁 削尖 3

将剖好的竹篾的一头削尖，将有瑕疵的竹篾剔除。

⊙2

将细竹篾穿过一个特制的圆孔里，另一头用钳子夹住，将整根竹篾从圆孔里穿过，使每根竹丝粗细相同。

肆 抽丝 4 ⊙3⊙4

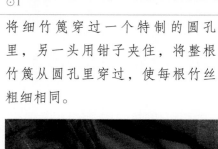
⊙3

⊙4

伍 编帘 5 ⊙5

编织时将竹丝放置在一个木质的工作台架上，在木架的横杆上进行编织，每隔2 cm左右挂有一个竹片制成的梭子，梭子上绕有锦线。编帘时要求竹丝与竹丝间保持一定的间距，线与线之间的间距保持在1.8 cm左右。保持适当间距是为了捞纸过滤水的同时纸浆料不会过多漏掉，

如果间距太大会造成捞出来的纸面留下明显纹路。据陈明堂介绍，他们厂按照泾县以捞制宣纸和书画纸为主的客户要求，一般是1 cm编有11.5根竹丝。编帘需要3~4人配合完成，11个左右的人每小时可以编织完成1张四尺"一改二"的帘子。

⊙5

⊙1 剖竹条
⊙2 剖竹篾
⊙3/4 抽丝
⊙5 正在聚精会神编帘的女工

陆 装芒秆 6

编好的帘子需要在横向的两边各装一根芒秆作为固架，保持竹帘不变形。芒秆一般选用泾县当地河谷溪流边上生长的长草秆，表面光滑坚硬，中间呈絮状，质量较轻。

⊙6

⊙7

柒 上绷架 7

将装好芒秆的帘子两端剪齐，整体固定在一个由4根木棍围成的长方形木床中间，使竹帘完整打开、绷直。

⊙8

捌 刷漆 8

泾县明堂纸帘工艺厂的竹帘需要刷2遍土漆，第一遍是用特制的漆球在竹帘上滚刷，第二遍是用漆把刷两面，让油漆浸入竹丝缝隙。同时，还需要对帘子进行检查，对缝隙大、缺丝、断丝的地方加以修补。刷完的竹帘需要在室内阴干，不可在太阳下暴晒，刷漆时需要保持一定的温度和湿度。

⊙9

⊙10

⊙6 芒草秆 Chinese silvergrass straws
⊙7 装芒秆 Loading Chinese silvergrass straws
⊙8 绷架上的竹帘 Bamboo screen on the stretcher
⊙9 用滚球刷漆 Painting with a rolling ball
⊙10 用漆把刷漆 Painting with a brush

⊙11

玖　上　图　9　⊙11

从20世纪末开始，在泾县宣纸行业流行生产特制纪念纸和名家订制产品纸，不仅在宣纸制作的原料配比、加工工艺上会有所改变，同时还会在宣纸上面印有暗纹，暗纹的技术就是靠纸帘上绣有的帘纹来决定的。一般在绣纹之前需要根据客户的需求给竹帘绘制图案，俗称上图。上图时一般在纸帘下面放上客户提供的图案，用工笔涂料在竹帘上勾勒出大致图案形状。

⊙12

拾　绣　花　10　⊙12

根据上图工序绘画出来的图形，用针线沿着上图勾勒的形状一针一线地绣出图案。

（三）
纸帘生产的工具设备

壹 剖竹刀 1

把竹片剖成一根根竹条的铁质工具。

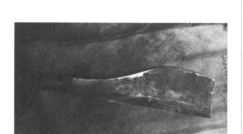
⊙13

贰 剖篾台 2

将剖好的竹条再进一步剖成竹篾的工作台，主要由一组两个刀片组成，固定在一个宽木凳上。

⊙14

叁 抽丝台 3

将竹篾定型、塑造成一定细度竹丝的工具，主要由木凳和若干铁片组成，在铁片上打有不同大小的圆孔，圆孔的大小决定着竹丝的粗细。

⊙15

肆 卡 尺 4

购置的测量工具，用以测量抽出来的竹丝直径的尺子。

⊙16

伍 漆 把 5

刷土漆时用的工具，主要使漆水能均匀地渗透到帘缝中，由材质较硬的猪鬃毛制成。

⊙17

四 泾县明堂纸帘工艺厂的市场经营状况

4 Marketing Status of Mingtang Papermaking Screen Craft Factory in Jingxian County

泾县明堂纸帘工艺厂的客户主要为泾县当地捞制宣纸、书画纸的厂家。随着新的消费用途的出现，泾县明堂纸帘工艺厂也开发出不同用途的工艺竹帘，主要用于窗户帘幕和茶几用垫等方面。据陈明堂介绍，工艺品竹帘一般都销往外地，泾县本地并未流行，以订单制作为主，销量占比并不大，制作工艺没有捞纸竹帘功能性要求高，只要外观相对美观即可。工艺帘虽然只是初试水，但是价格却比纸帘稍贵，达到650～900元/m^2，2013～2015年，每年工艺帘销量约占该厂年销售量的10%。

纸帘的制作主要用于手工造纸行业，所以纸帘的销售受到手工造纸行业的影响很大。据陈明堂口述，2013年开始行业景气度下行，2015年整个宣纸和书画纸行业受到国内消费大环境的影响，泾县当地企业减产或停产的较多，纸帘的市场销售较2014年减少了20%～30%。

为了拓展多元化市场，保持小企业可持续发展的生存空间，陈明堂通过造纸行业朋友的介绍，泾县明堂纸帘工艺厂近几年也开始接到省内其他地区以及省外的部分订单，如山东、浙江已有预约订货。

⊙1

调查组成员和陈明堂观看纸帘纹
Researchers and Chen Mingtang watching the pattern of the papermaking screen

五 泾县明堂纸帘工艺厂的民俗故事

5 Stories of Mingtang Papermaking Screen Craft Factory in Jingxian County

据陈明堂介绍,他本人是从1985年开始跟随师傅曹虹江学习纸帘生产工艺的。曹虹江1937年出生,祖籍小岭,家里世代制造纸帘,但由于年代久远加上师傅已经去世,陈明堂表示师傅的祖上已无从考证,但是当时拜师学艺时的一些事情还记忆犹新:1985年经父辈的朋友张道林介绍向曹虹江拜师学艺,拜师前需要在镇里和家里举行拜师宴2次,请师傅吃饭喝酒,以表拜师诚意。在3年学徒期间,没有工资并需要向师傅交纳自己的伙食费(出粮食),每个月有2~3天的休息时间,但是不能离开镇子太远。春节前腊月二十四放假,方便学徒回家送灶王爷和过年,最晚正月初八必须回到师傅家学习和工作。

访谈中陈明堂表示,在制作竹帘这一行业内拥有的传统习俗并不多,从他做学徒起,知道的传统习俗也就只有一个:在制帘师傅吃饭的时候,离开的工作位置,小孩和大人都不可以随便去坐,去坐的话就意味着在和制帘师傅"抢饭碗"。

⊙2
已经废弃的生产纸帘的旧时工具
Abandoned tools for producing the papermaking screen

六 泾县明堂纸帘工艺厂的传承现状与发展思考

6
Current Status of Business Inheritance and Thoughts on Development of Mingtang Papermaking Screen Craft Factory in Jingxian County

1. 若干传承发展问题

访谈时陈明堂表示：当前泾县纸帘生产技艺和传统手工造纸面临着同样的挑战，问题主要有5个方面：一是由于习艺难度大，见效慢，年轻人不愿学，现有的从业者大多50岁左右，已经出现很典型的后继无人的传承困境。二是由于苦竹、漆树使用面较为狭窄，农民培育少，在泾县及附近地区出现了原料选购难的问题。三是因漆帘工序使用的土漆对有些人会造成较严重的过敏，业内有人对其怀有恐惧心理，也是使从业者越来越少的一个影响因素。四是随着机械化制纸技术和设备在泾县的普及，地方上部分手工书画纸厂转产半机械和机械书画纸，不需要用到纸帘，造成宣纸和书画纸捞纸帘的需求面萎缩。五是加工周期长，价格与耗时不匹配，经济效益不理想，年轻人不愿入行。

2. 保护和发展泾县竹帘业态的若干思考

（1）按照纸帘行业传统的惯例，泾县纸帘

① 访谈陈明堂
Interviewing Chen Mingtang

② 新旧生产工具对比（61号为20年前的生产工具）
Comparison of old and new production tools (No. 61 is the production tool 20 years ago)

主要用于捞制宣纸、书画纸，用途较为单一，而且较少以外地客户为拓展对象，市场空间和销售人群面窄。针对这一制约当代发展的问题，陈明堂认为可以适度采取推出产品新用途的方式来拓宽销路，比如制作窗帘、凉席、竹编制类艺术品等。这一点在泾县明堂纸帘工艺厂已有初步实践，不仅开拓了新的市场，更关键的是最大限度地保留了传统制作工艺。

（2）纸帘制作是宣纸制作技艺的附属行业，除了中国宣纸股份有限公司聘用了专门的制帘工人外，其余泾县宣纸、书画纸生产企业所需纸帘均依靠外购。丁家桥镇是宣纸、书画纸生产的集聚区，历史上衍生了若干诸如陈明堂这样的家庭作坊式纸帘制作户，当地政府可以考虑整合各制帘作坊资源，做强一个大的品牌制帘厂，以此提高纸帘的附加值，使制帘艺人从中得到实惠，也可使制帘技艺得以更有活力地传承。

⊙3
历史遗物（苦竹原料）
Historical relics (amarus bamboo as raw material)

第二节
泾县全勇纸帘工艺厂

安徽省
Anhui Province

宣城市
Xuancheng City

泾县
Jingxian County

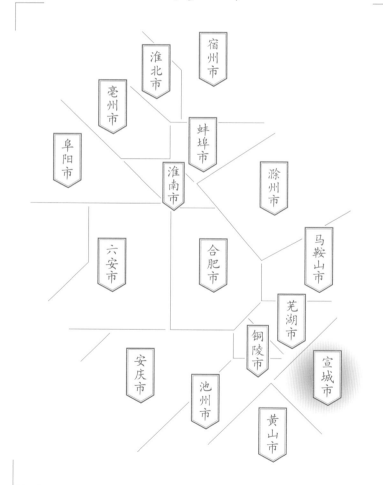

调查对象
丁家桥镇
泾县全勇纸帘工艺厂
工具

Section 2
Quanyong Papermaking Screen Craft Factory in Jingxian County

Subject
Tools of Quanyong Papermaking Screen Craft Factory in Jingxian County in Dingjiaqiao Town

一

泾县全勇纸帘工艺厂的基础信息与生产环境

1

Basic Information and Production Environment of Quanyong Papermaking Screen Craft Factory in Jingxian County

泾县全勇纸帘工艺厂坐落于泾县丁家桥镇的工业园区内,地理坐标为东经118°19′35″,北纬30°39′23″。紧邻S322省道。2015年11月9日,调查组前往纸帘生产现场实地调查获知的信息如下:全勇纸帘工艺厂负责人为肖全勇,工厂工人有11个左右,年销售金额在50万~60万元,年生产约200张纸帘。主要业务为各种规格的手工捞制宣纸和书画纸竹帘制作,同时承接宣纸暗纹、水印、图案定制帘的编制。

⊙1

二

泾县全勇纸帘工艺厂的历史与传承情况

2

History and Inheritance of Quanyong Papermaking Screen Craft Factory in Jingxian County

访谈中肖全勇介绍的传承信息如下:他学习纸帘制作技艺是师从小岭村纸帘编制的老艺人曹康明。曹康明,1955年生,作为土生土长的小岭曹氏造纸世家的传人,世代生活在小岭村,调查时据曹康明回忆,家族有记载的传承可追溯到上三代曾祖父,从那时起就开始了制作纸帘的历史。曹康明曾祖父曹守品、祖父曹俊生、父亲曹云西都是以制作纸帘为营生,到曹康明是第四代,按照保守的推算,至少已有120年的明确传承。

⊙1 销售店门面 Sales shop facade
⊙2 制纸帘人曹康明 Papermaking screen maker Cao Kangming

泾县全勇纸帘工艺厂位置示意图

Location map of Quanyong Papermaking Screen Craft Factory in Jingxian County

路线图
泾县县城
↓
泾县全勇纸帘工艺厂

Road map from Jingxian County centre to Quanyong Papermaking Screen Craft Factory in Jingxian County

考察时间
2015年11月

Investigation Date
Nov. 2015

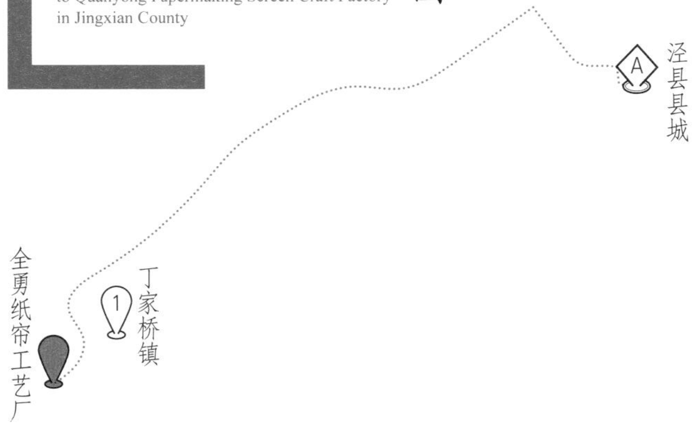

地域名称

A 泾县
① 丁家桥镇
② 云岭镇
③ 泾川镇
④ 榔桥镇
⑤ 琴溪镇
⑥ 黄村镇
⑦ 汀溪乡

造纸点名称

泾县全勇纸帘工艺厂 造纸点

位置分布

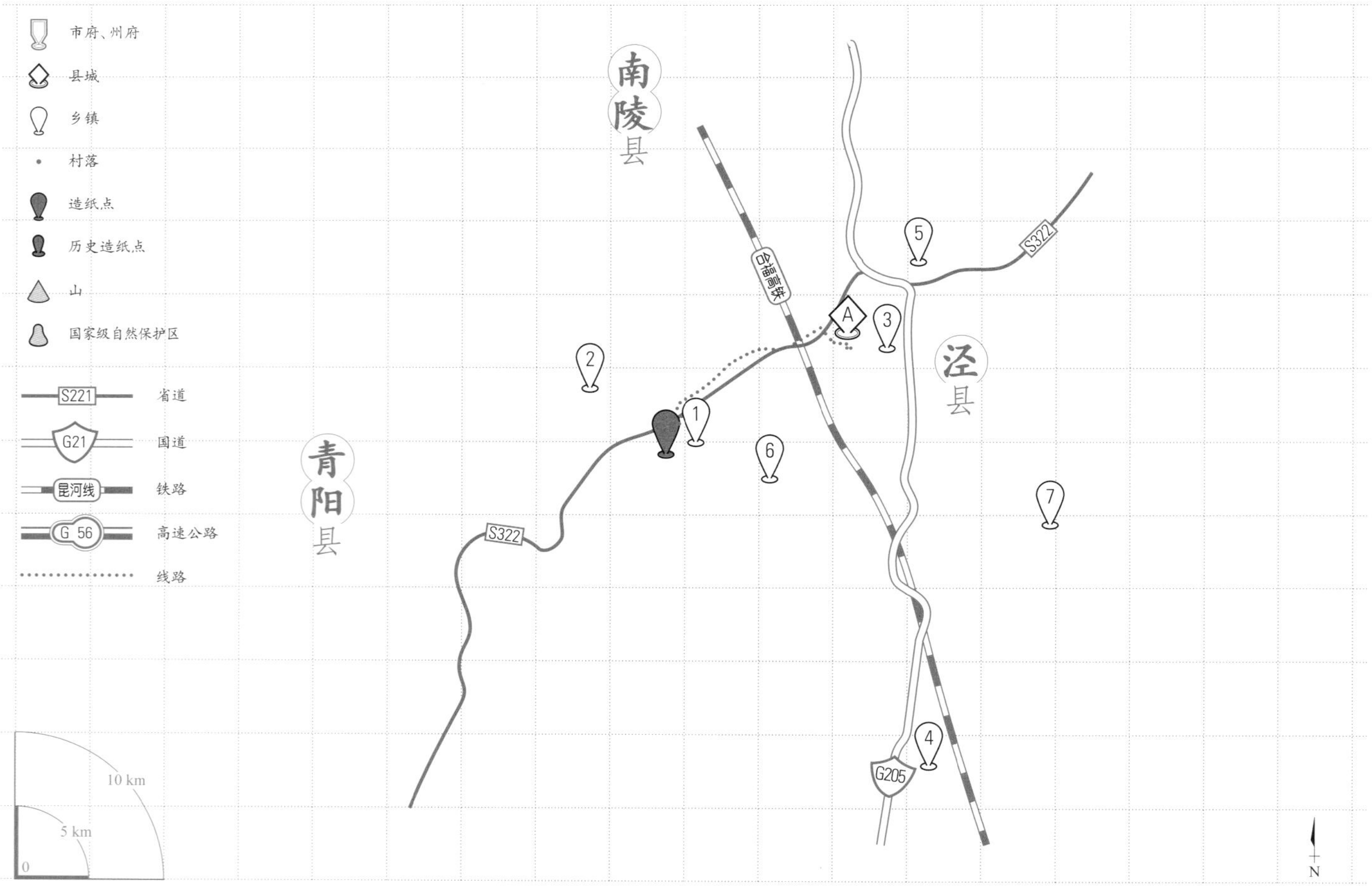

图例：
- 市府、州府
- 县城
- 乡镇
- 村落
- 造纸点
- 历史造纸点
- 山
- 国家级自然保护区
- S221 省道
- G21 国道
- 昆河线 铁路
- G56 高速公路
- 线路

表7.1 泾县全勇纸帘加工技艺传承情况
Table 7.1 Inheritance of Quanyong Papermaking Screen processing techniques in Jingxian County

第一代 曾祖父	第二代 祖父	第三代 父亲	第四代 师父	第五代 徒弟
曹守品	曹俊生	曹云西	曹康明	肖全勇

⊙1
半自动化纸帘机器
Semi-automatic papermaking screen machine

曹康明8岁开始从父祖辈处接触纸帘制作，20岁开始掌握纸帘制作的全过程工艺。调查时，他的下一代子女都在外工作，他有两个女儿，大女儿结婚后生活在江苏省南京市高淳区，二女儿生活在湖南省，目前二人都不再从事跟造纸相关的工作。他的妻子叫童丫姑，1959年生，目前与丈夫曹康明一起从事纸帘制作。子女中均没有从事竹帘制作工作的。他颇为无奈地表示，编纸帘的匠人在以前是被别人看不起的，是别人都不愿做的又累又脏的苦活。现在，只剩下夫妻二人留在家中，平时接点纸帘制作的活儿。

肖全勇从1985年在师傅曹康明那里学习纸帘制作技艺，直到现在一直从事纸帘加工，不断地进行学习与实践探索。肖全勇，1969年生，目前主要是自己和妻子在纸帘厂进行打理，夫妻两人一方面制作纸帘，同时也负责纸帘的市场销售等工作，目前独立经营全勇纸帘工艺厂，同时与师傅曹康明一直保持紧密合作。其纸帘厂面临着两个现实问题：一是随着市场萎缩和人力经济的冲击，人力成本不断加大，有纸帘加工技术的工人很难找到。二是机械化和半机械化生产设备降低了纸帘加工的经济成本，提高了生产效率。为了突破工厂可持续生产和纸帘技艺革新的瓶颈，1992年肖全勇去浙江省衢州市一家纸帘加工厂进行实地考察学习，萌发了引入半自动化的纸帘加工方法的想法，试图在一些简单、机械化的编制环节用机械的方法代替人力。与此同时，面临机械化的局限性有两个方面：一是增加了原料成本。二是纸帘精细程度不如手工制作，报废率比手工制作高。但对于乡镇造纸厂而言，他们更倾向于选择半机械的纸帘，主要原因是价格便宜，制作中低端纸品对纸帘的要求不是很高。这对于全勇纸帘工艺厂而言利润更高，有利于纸帘技艺的正常经营与技艺的传承。但同时纸帘厂也存在一些高端手工定制的纸帘。

三 泾县全勇纸帘工艺厂纸帘生产的原料、工艺流程与工具设备

3
Raw Materials, Processing Techniques and Tools of Quanyong Papermaking Screen Craft Factory in Jingxian County

（一）纸帘生产的原料

1. 主料

据曹康明介绍，小岭当地纸帘生产的主料为苦竹。

⊙1

据肖全勇介绍，全勇纸帘工艺厂制帘所需的苦竹，以3~4年生长期的竹子为最佳，其直径可达3~4 cm，越粗的苦竹越好。外购的苦竹有两种：一是产自皖西大别山安徽省金寨县的管竹，这种竹子一般较细长，适合做笛子等管弦乐器，2015年，来自金寨县含有水分的竹子采购价格为90元/50 kg，晒干后只剩下20 kg重。二是从浙江省温州市采购的竹子，有时也会砍伐泾县小岭村当地产的苦竹来用。

2. 辅料

（1）锦线。锦线主要是起到固定每根竹篾的作用，使得竹篾与竹篾紧密地结合在一起，不易散开，从而形成一张完整的帘幕。

（2）生漆，又称"土漆""国漆"或"大

⊙2

⊙1 小岭村外生长的苦竹
Amarus bamboo growing outside Xiaoling Village
⊙2 晒干后的苦竹竿
Amarus bamboo sticks after drying

漆"。生漆是从漆树上采割的乳白色胶状液体，一旦接触空气后就转为褐色，数小时后表面干涸硬化而生成漆皮。生漆是手工艺品和民用家具的优质涂料。

泾县全勇纸帘工艺厂所用的生漆多采购自四川、广西、陕西等地，2015年采购价格为250元/kg，这里面也包含了采购所需的交通成本费用。

⊙3

⊙4

（二）纸帘生产的工艺流程

根据调查组对全勇纸帘工艺厂进行的实地调查和对肖全勇等人的访谈，归纳纸帘生产的工艺流程为：

壹 切片 → 贰 剖篾 → 叁 浸泡 → 肆 抽丝 → 伍 晾晒 → 陆 编帘 → 柒 装芒杆 → 捌 剪齐 → 玖 刷漆

壹 切片 1

将竹子按节断开，再用剖竹刀劈成一根根竹条。

⊙5

贰 剖篾 2

用剖竹刀把竹片切篾、去芯，剖成像牙签一样粗细的竹条。

⊙6

3 编帘使用的锦线
Thread for making the papermaking screen

4 陶盆中待用于漆帘的生漆
Raw lacquer to be used for painting the papermaking screen in pottery pots

5 切片后的竹条
Bamboo canes after cutting

6 肖全勇展示剖篾
Xiao Quanyong showing how to split the bamboo

叁 浸泡 3

将劈好的干燥竹片在水中浸泡3天左右，待抽丝前拿出来，让其软化，方便抽丝。

肆 抽丝 4

将细竹篾穿过一个特制的圆孔里，另一头用钳子夹住，将整根竹篾从圆孔里穿过，使得每根竹丝变得粗细相同。

伍 晾晒 5

将抽好的竹丝晾晒干，主要是为了防止竹丝潮湿，编好帘后水分蒸发变形造成帘纹缝隙加大，从而影响捞出纸的质量。

陆 编帘 6

编织时，将竹丝放置在一个木质的工作台架上，在木架的横杆上进行编织，每隔2 cm左右挂有一个竹片制成的梭子，梭子上绕锦线。编帘时要求竹丝与竹丝间保持一定的间距，帘纹的缝隙根据使用线的粗细决定，要求最大不超过2 cm。原则上2个技术工人3～4天可完成1 m²的帘子。

柒 装芒秆 7

芒秆用于固定编好的帘子的两端，保持竹帘不变形。泾县全勇纸帘工艺厂一般选用泾县当地河谷溪流边上生长的长草秆，表面光滑坚硬，中间呈絮状，质量较轻。

捌 剪齐 8

将装好芒秆的帘子两端剪齐，整体固定在绷床中间，使得竹帘完整打开。

⊙7 在水中浸泡的竹丝 Bamboo silk soaked in water
⊙8 抽丝 Threading

玖　刷　漆

刷漆时需要保持一定的温度和湿度，确保油漆充分覆盖住竹丝。同时，还需要对帘子进行检查，对缝隙大、缺丝、断丝的地方加以修补。刷完的纸帘需要在室内阴干，存放10天左右方可交给用户使用。

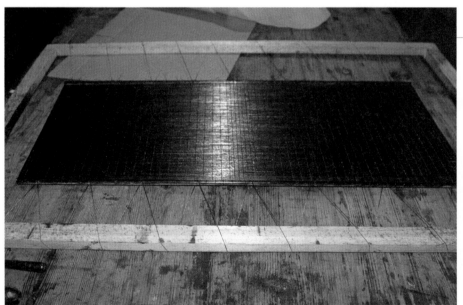

⊙ 9
编帘
Making the papermaking screen

⊙ 10
绷床上刷过漆的纸帘
Painted bamboo screen on the stretcher

（三）纸帘生产的工具设备

壹 剖竹刀 1

把竹片剖成一根根竹条的铁质工具。

⊙1

贰 绷床 2

刷漆前把纸帘充分绷开的长方形木架，由4根木棍组成。

⊙2

叁 抽丝台 3

将竹篾定型、塑造成一定粗细规格的工具，主要由木凳和若干铁片组成，在铁片上钻有不同大小的圆孔，圆孔的大小决定着竹丝的粗细。

⊙3

肆 卡尺 4

购置的测量工具，用以测量抽出来的竹丝直径的尺子。

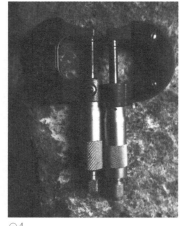

⊙4

四 泾县全勇纸帘工艺厂的市场经营状况

4
Marketing Status of Quanyong Papermaking Screen Craft Factory in Jingxian County

调查时，泾县全勇纸帘工艺厂的主要客户来自泾县当地的手工宣纸与书画纸生产厂商，这部分约占销售总量的70%，其余部分的竹帘产品转向生活化的工艺品市场，主要用作一些窗户挂帘、窗帘以及茶几用垫等。纸帘制作作为宣纸行业的附属产业，传统的渠道销售受到手工造纸行业业态变化的影响很大，据肖全勇介绍，2015年泾县宣纸行业受到国内环境影响，很多造纸厂家减产停产，纸帘的市场销售较前一年也减少很多。

⊙5

2015年泾县全勇纸帘工艺厂年销售额50万～60万元，年销售纸帘200张左右。纸帘厂的员工不固定，存在两种形式：一种是固定员工形式，即在工厂制作纸帘，大约有8～11个固定员工；另一种是外包形式（打包前期工艺销售给散工，原料自己出，回收后进行后续的工艺加工），主要是给周围村子里带孩子不方便外出工作的家庭妇女，做一些散工养家糊口。整个过程中，纸帘厂与固定员工一直保持密切合作关系，而对散工则根据人力情况和接单情况进行不定期的联系。

⊙5 调查成员与肖全勇边走边聊
A researcher and Xiao Quanyong talking while walking

五 泾县全勇纸帘工艺厂的民俗故事

5 Stories of Quanyong Papermaking Screen Craft Factory in Jingxian County

在制帘师傅吃饭的时候，离开的工作位置，小孩和大人都不可以随便去坐，去坐的话就意味着在和制帘师傅"抢饭碗"，所以一般制帘师傅很忌讳，现在都还一直保留这样的习俗。

⊙1 在曹康明家中进行的访谈
An interview in Cao Kangming's house

六 泾县全勇纸帘工艺厂的传承现状与发展思考

6 Current Status of Business Inheritance and Thoughts on Development of Quanyong Papermaking Screen Craft Factory in Jingxian County

调查时，泾县纸帘生产技艺和传统手工造纸面临着同样的代际断层压力，技艺失传的问题已经相当突出，新生代村民外出打工比较普遍，愿意从事纸帘生产制作的少之又少。加之编制纸帘工作辛苦，利润低，生产周期长，一年2人最多能生产竹帘200 m²，耗时耗力回报低。调查中了解到的信息是，曹康明这些传统制帘人的下一代几乎无人愿意继承父祖辈从事的这一行业。

纸帘制作是宣纸业的附属行业，泾县本地的零散制作主要分布在丁家桥镇，以现存业态的经济效益和消费空间而言，已很难集中形成规模效应；与此同时，单一纸帘产品的销售渠道仅仅作为手工宣纸（书画纸）抄纸的工具，在手工造纸业态自身萎缩明显的背景下，其需求的萎缩趋势更加突出。和作为人类非物质文化遗产的宣纸比起来，编制纸帘这一附属小产业得到的保护、扶持和重视程度要低得多。同时，像中国宣纸股份有限公司这样有代表性的企业为了打造面向宣纸文化旅游和名家高端定制纸，自身又建设了专门的纸帘编制部门，因此作为民间家庭作坊式的泾县传统纸帘生产厂坊的销售量就更加雪上加霜了，基本上处于艰难维持状态。

纸帘之所以在机械化能力高度发达的今天仍然未被机器代替，其中重要的原因之一是手工造纸行业独特工艺要求的支撑，另一重要因素是手工制作纸帘的技艺目前仍有机器无法取代的价值点，包括高端纸的工艺诉求、材料成型方式诉求、审美诉求等。因此，如何让作为附属的纸帘作坊能与主流手工造纸厂家形成紧密的嵌入式生产业态，以及进一步拓展新的消费形式与产品，均属像泾县全勇纸帘工艺厂这样的传统小微企业所需要思考的。

⊙2

⊙3

⊙4

⊙2 泾县全勇纸帘工艺厂区外围景观
External view of Quanyong Papermaking Screen Factory in Jingxian County
⊙3 采访肖全勇
Interviewing Xiao Quanyong
⊙4 泾县全勇纸帘工艺厂周围环境
The surroundings of Quanyong Papermaking Screen Factory in Jingxian County

第三节
泾县后山大剪刀作坊

安徽省
Anhui Province

宣城市
Xuancheng City

泾县
Jingxian County

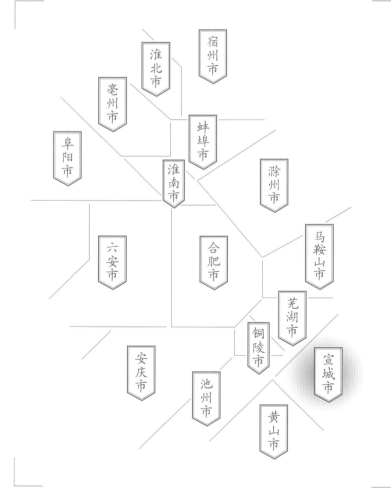

调查对象
丁家桥镇
泾县后山大剪刀作坊
工具

Section 3
Houshan Shears Workshop
in Jingxian County

Subject
Tools of Houshan Shears Workshop
in Jingxian County
in Dingjiaqiao Town

一

泾县后山大剪刀作坊的基础信息与生产环境

1
Basic Information and Production Environment of Houshan Shears Workshop in Jingxian County

在泾县手工造纸体系里，宣纸制作经捞、晒成纸后，再经逐张检验、裁边规整纸张的工序被称为剪纸，剪纸所用的剪刀为特制的，超过一般民用剪刀多倍规格，被称为"大剪刀"或"宣纸剪"。根据泾县的宣纸制作技艺传统分工，这种大剪刀附属于宣纸制作本身的技艺，属于传承宣纸制作技艺的旁支，一直在丁家桥镇后山一带聚集，因而当地习称"后山大剪刀"。

丁家桥镇后山行政村地理坐标为东经118°17′38″，北纬30°38′50″。2015年11月13日，调查组前往后山村青弋江河畔北侧的张力伟作坊进行田野调查。该作坊是一家世代制剪的家庭作坊，家中3个成员均会制作大剪刀，按制作进度，一人一天即可完成一把大剪刀的制作。调查当天恰逢张力伟作坊停产，只对其生产现场、制作工具进行了拍摄，对其家族的制作历史、工艺等方面进行了调查。随后，调查组又到位于青弋江南岸的后山村俞村村民组的俞宋桃剪刀作坊进行了补充调查。

⊙1

⊙2

⊙1
张力伟家庭剪刀作坊外景
External view of Zhang Liwei Family Shears Workshop
⊙2
俞宋桃在点火起炉
Yu Songtao making a fire

路线图
泾县县城
↓
泾县后山大剪刀作坊

Road map from Jingxian County centre to Houshan Shears Workshop in Jingxian County

Location map of Houshan Shears Workshop in Jingxian County

泾县后山大剪刀作坊位置示意图

考察时间
2015年11月

Investigation Date
Nov. 2015

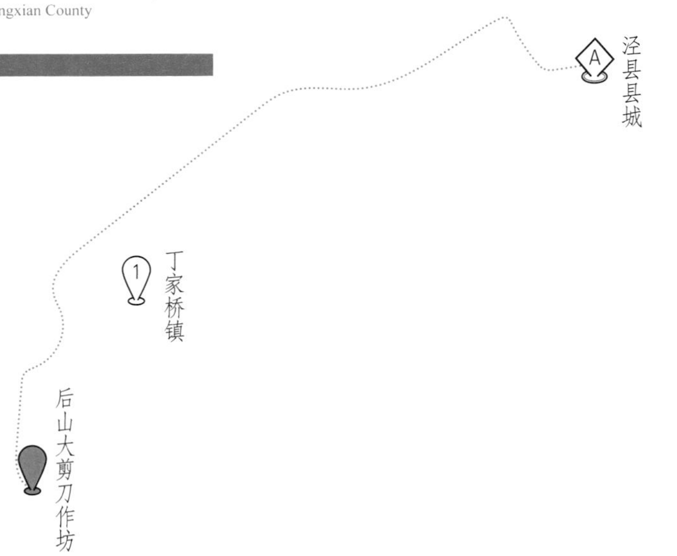

地域名称

A 泾县
① 丁家桥镇
② 云岭镇
③ 泾川镇
④ 榔桥镇
⑤ 琴溪镇
⑥ 黄村镇
⑦ 汀溪乡

造纸点名称

泾县后山大剪刀作坊 生产点

位置分布

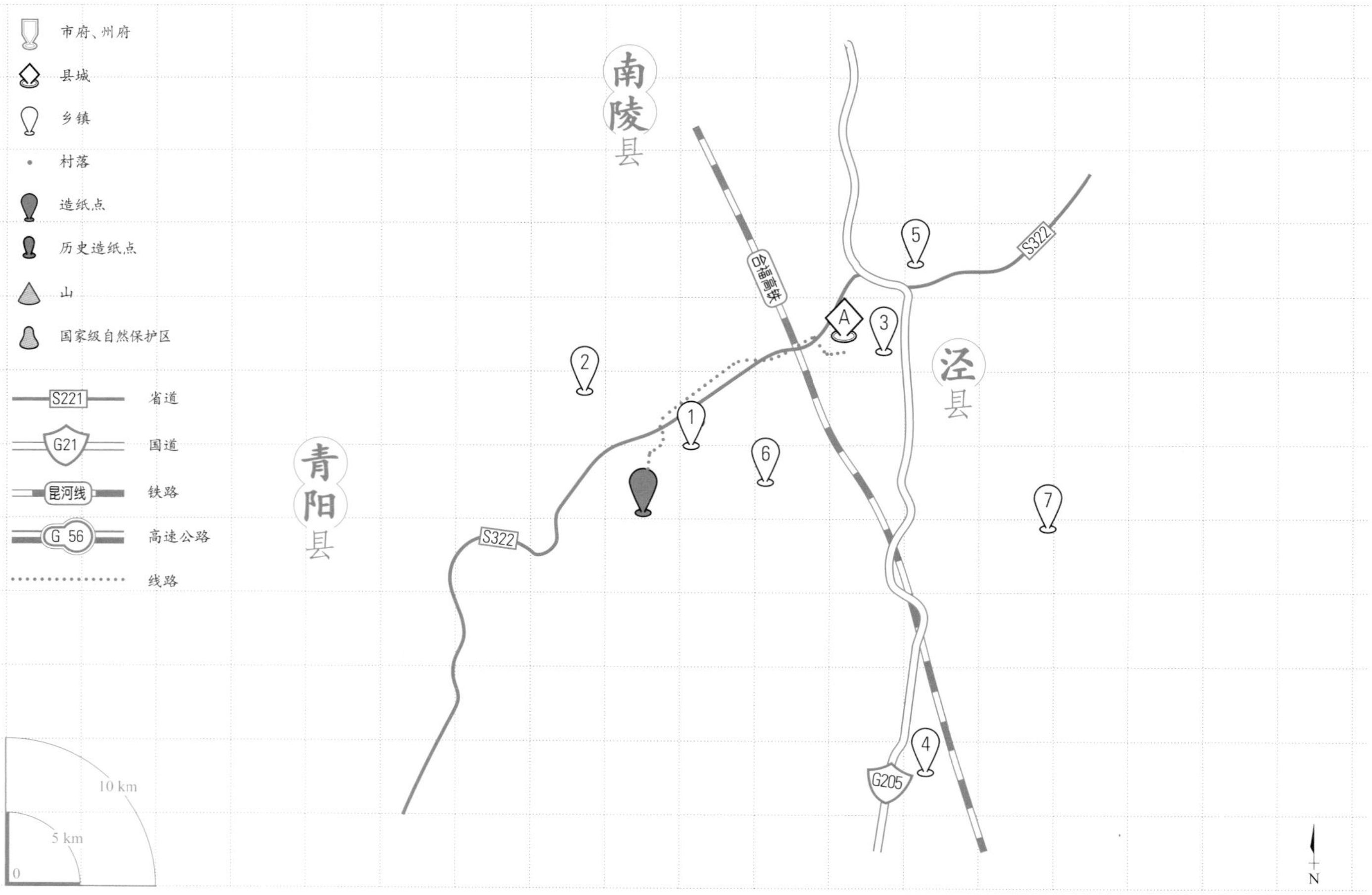

- 市府、州府
- 县城
- 乡镇
- 村落
- 造纸点
- 历史造纸点
- 山
- 国家级自然保护区

- S221 省道
- G21 国道
- 皖赣线 铁路
- G56 高速公路
- ········ 线路

南陵县
泾县
青阳县

S322　G205

二 泾县后山大剪刀作坊的历史与传承情况

2 History and Inheritance of Houshan Shears Workshop in Jingxian County

关于后山大剪刀家庭作坊群落的历史与传承，调查组成员对后山大剪刀制作师傅张力伟进行了较深入的访谈，据张力伟回忆，他自己家中至少有3代人从事制剪工艺（张力伟已不记得祖父名字及从业情况），其父张时昌，1949年出生，1963年前后，14岁的张时昌开始跟随父亲学习制剪，时因人民公社"合作化"体系，且因后山打制剪刀的人多，故就在后山张氏祠堂组建了后山剪刀组，专门打制剪刀，其中包括布剪、裁缝剪、桑剪、铜剪、焊剪、镴剪、角剪及宣纸剪等多种用途的剪刀。1962年，张时昌进入丁桥"下门"剪刀社，1980年张时昌离开剪刀社，开始在家起炉子打剪刀。

据张力伟介绍，现在宣纸行业使用的大剪刀就是张时昌在旧宣纸剪刀的基础上改进创制的，这种大剪刀全身黑灰色，刀身长25～26 cm，手柄长10 cm，刀身宽10 cm，刀背厚0.4～0.5 cm，每把剪刀的重量为0.8～0.85 kg。

张力伟，1983年出生，2000年初中毕业后，先后在北京、上海打工，2005年，开始跟随父亲学习制剪，2007年时已可以独立打造出合格的后山大剪刀。

俞宋桃，1966年生于丁家桥镇后山村俞村组，1979年，因家贫弃学，到丁桥剪刀社随姨父张尚志做3年学徒。1982年6月，俞宋桃按照剪刀社规定的学徒时间出师后，又按照社里规定谢师一年。从1982年6月开始，直到1993年10月，俞宋桃一直在丁桥剪刀社（后改成丁桥剪刀厂）打制剪刀。1993年10月，丁桥剪刀厂倒闭后，俞宋桃回家务农，以打制和经营剪刀为副业。2012年，俞宋桃申请提交"后山"牌剪刀商标，是泾县制作剪刀行业中唯一注册商标的人。2013年，在俞宋桃的积极主张和努力下，"后山剪刀制作技艺"成功进入安徽省第四批非物质文化遗产代表作名录；2015年，俞宋桃被公布为安徽省第五批非物质文化遗产代表性传承人。

⊙1 大剪刀 Shears

⊙2 张力伟 Zhang Liwei

三 泾县后山大剪刀作坊生产的原料、工艺流程与工具设备

3 Raw Materials, Processing Techniques and Tools of Houshan Shears Workshop in Jingxian County

(一) 后山大剪刀生产的原料

1. 主料

后山大剪刀必不可缺的重要材料是铁板，据张力伟介绍，后山大剪刀制作的铁板是在泾县的钢铁市场购买的。在传统农耕社会时期，没有现成的铁板购买，只能自行配置。

2. 辅料

（1）煤。煤是制作大剪刀重要的辅助原料，通过烧煤提高燃点软化铁板，便于打造。据张力伟介绍，制作大剪刀使用的是在泾县购买的安徽优质淮煤，燃烧后的淮煤能量高，能超过6 000卡，泾县当地所产的煤达不到打造剪刀的要求。2015年调查时购买价为1 000元/t。

（2）柴。木柴主要用作点炉子时的引火柴，以松针等易燃的茅柴为主。

⊙1 打制剪刀的铁板 / Iron plate for making shears

(二) 后山大剪刀生产的工艺流程

据张力伟、俞宋桃介绍，归纳后山大剪刀生产的工艺流程为：

壹	贰	叁	肆	伍	陆	柒
裁料	出胚	雕弯	镶钢（压钢）	打头片	打手柄	打眼

捌	玖	拾	拾壹	拾贰	拾叁	拾肆	拾伍	拾陆	拾柒	拾捌
砂轮打光	开口	锉片	淬火	敲口整形	磨口	制销子	铣眼	钉铰	试剪	上油

工艺流程

壹　裁料

1　⊙2

将市场上购买来的铁块按剪刀尺寸裁成条状，每条0.5 kg左右。此道工序也称备料。

贰　出胚

2　⊙3

将裁好的细铁反复锻打，除去杂质，制成剪刀的雏形。

叁　雕弯

3　⊙4

将剪刀头片和尾部用小铁锤锻打成一定的弧度。

肆　镶钢（压钢）

4　⊙5

将裁好的钢片通过锻打使之与原刀坯粘连。因两种材料粘连部位有空隙，锻打时有大量的火星溅出。

⊙2 已裁切的铁条 Iron bars after cutting
⊙3 锻打 Forging
⊙4 雕弯 Forging with a small hammer into a certain arc
⊙5 镶钢（压钢）Inlaid steel

伍 打头片 5

将剪刀的刀身部分锻打成型。

陆 打手柄 6

将剪刀的手柄部分锻打成型。

柒 打眼 7

在剪刀刀身上冲眼。

捌 砂轮打光 8

用砂轮磨打刃口。在没有砂轮前，用钢锉锉。

玖 开口 9

使用锉刀、铲刀制成剪刀刀口。

拾 锉片 10

用锉刀将刀片锉平，将刀口内口锉平整。

拾壹 淬火

将半成品剪刀上炉膛加热至一定火候，使用山泉水淬火，以此冷轧定型。

⊙12
淬火
Quenching

拾贰 敲口整形 12

将淬火后变形处用小锤整形。

⊙13

⊙14

拾叁 磨口 13

使用粗、细刀石将剪刀刀口磨锋利，要求外口钢铁分明、内口光滑平整。

拾肆 制销子 14

使用锤、钻等将细铁制成剪刀销子。

⊙15

⊙13 整形 Shaping
⊙14 磨口 Sharpening the blade of shears
⊙15 制销子 Making shears pin

拾伍
铣　眼

15　　　⊙16

用圆挫将剪刀刀柄眼反复摩擦，再将刀眼口磨光、磨圆。

⊙16

拾陆
钉　铰

16　　　⊙17

将左右两片刀身用销子钉铰成一体。

拾柒
试　剪

17　　　⊙18

将制作好的后山大剪刀试验剪纸，以此检验有无问题。

拾捌
上　油

18　　　⊙19

给剪刀内口、销子等处涂上油，并用手将油涂抹均匀，以防生锈。

⊙17

⊙18

⊙19

（三）

后山大剪刀生产的工具设备

壹 铁钳 1

锻打、淬火时夹住剪刀刀胚的工具。

⊙1

贰 铁锤 2

锻打的主要工具之一。

⊙2

叁 水盆 3

磨刀时盛水的工具。

⊙3

肆 砂轮 4

锉磨工具。

⊙4

伍 铲子 5

添加燃煤的工具。

⊙5

陆 电子秤 6

把握剪刀重量的计量工具。

⊙6

四 泾县后山大剪刀作坊的市场经营状况

4 Marketing Status of Houshan Shears Workshop in Jingxian County

20世纪60～80年代集体合作制的剪刀社解体后，后山大剪刀作坊一直为家庭式作坊，最多年产400多把大剪刀。调查时的2015年前后，由于整个宣纸市场不太景气，后山大剪刀作坊实际上处于停产阶段。以张力伟家庭作坊的生产状况看，2015年生产了200把大剪刀，市场售价为300元/把，不包括材料设备成本和人工投入，全部销售额为6万元。作坊经营状况处于明显下滑阶段，而且宣纸行业的需求疲软，售价也疲软，来自材料和人工成本上升的压力越来越大。张力伟表示：经济效益不理想，正考虑转型改行。

五 泾县后山大剪刀作坊的品牌文化与民俗故事

5 Brand Culture and Stories of Houshan Shears Workshop in Jingxian County

1. "天下第一剪"的由来

2001年5月21日，时任中共中央总书记江泽民参观中国宣纸集团公司时，在生产车间看到了这种尺寸特别大的专用剪刀，十分感兴趣地拿起剪刀体验，赞许道："工欲善其事，必先利其器。如此锋利的大剪刀，真不愧是天下第一剪。"此后，后山村大剪刀的"天下第一剪"美誉就传遍全国，成为宣纸行业里津津乐道的文化轶事。

2. "叮咚踢踏后山张家"

后山村张氏为制剪世家，历史悠久，在当地有"穿靴带顶茂林吴家，开仓卖稻云岭陈家，冲担打杵小岭曹家，叮咚踢踏后山张家"的乡间古谚。"叮咚踢踏后山张家"是对后山张氏制作大剪刀代表性工作场景的形象描摹。

3. 被迫抵押的金牌

清朝道光年间，后山村制剪工人张三荣生产的"后山剪刀"闻名全国，当时的后山村主要打制的是民用剪刀，并非宣纸专用大剪刀，属于地方特色物产。当地民间传说，在清朝某年间，后山剪刀被送到京城参展并获得金奖，领奖回来的路上，各州府鸣炮庆祝，张氏一路办酒席酬谢，结果到达浙江临安时，所带的盘缠所剩无几，张氏不得已将金牌抵押给驿站。回到后山后，因没有足够的资金赎回金牌，就这样，一块彰显后山剪刀的御赐金牌就再也没出现在人们眼前。

六 泾县后山大剪刀作坊的传承现状与发展思考

6
Current Status of Business Inheritance and Thoughts on Development of Houshan Shears Workshop in Jingxian County

后山大剪刀依附于宣纸技艺，具有一定的独特性和专有性，蕴含着丰富的宣纸文化基因。由于制作技术难度大，合格的从业者需要长时间经验积累，工作辛苦，而且总体收入偏低，年轻人多不愿学，已经是后继严重乏人。2015年调查时，制剪技艺传承者年龄普遍较高，平均年龄已达50岁以上，而且从事这一行业的制剪工人在整个丁桥镇的后山区域只剩下不到10人，逐年减少趋势明显。

由于后山大剪刀依附于宣纸技艺而生存，营销渠道目前比较单一，在使用时，又不属于易耗品，更新时间很长，需求量因此更显小，尽管一直没有被机械替代，但始终处于价值和销量比不佳的状态。当前，在宣纸市场不景气的环境下，宣纸、书画纸产量锐减，后山大剪刀的制作时常被迫中断，制剪刀户难以凭制作剪刀技艺养家糊口，如果没有新的市场拓展和国家非物质文化遗产体系的扶持，已处于前景堪忧的境遇中。

⊙1

⊙2

⊙1 俞宋桃的作坊
Yu Songtao's mill

⊙2 张力伟作坊里的火炉
The stove in Zhang Liwei's mill

Appendices

Introduction to Handmade Paper in Anhui Province

1 History of Handmade Paper in Anhui Province

1.1 Features of Craft Culture in Anhui Province

In the early Chinese cultural and historical map, Anhui Province located at the bordering part of cultures from all directions, connecting Wuyue in the southeast, neighboring Central Plains in the north, and meeting Chu State in the west. It was named "the head of Wu and the end of Chu" in the Spring and Autumn Period (770 B.C.~ 476 B.C.). Yangtze River, Huaihe River and Xin'an River flow eastward to the sea, and the Grand Canal span from the north to the south, which contribute to the convenient waterways leading to all directions. Hence, the area enjoys typical characteristics of cultural diversity and convergence.

Since ancient times, Anhui has developed unique culture of science, technology and craftwork. During the Spring and Autumn Period (770 B.C.~ 476 B.C.), Sun Shu'ao, the prime minister of the Chu State, started the Quebei Water Conservancy Project in Shouchun, capital of Chu State (now Shouxian County in Anhui), which was acclaimed as one of the four greatest ancient Water Conservancy Projects in China, now known as the famous Anfengtang Reservoir. Quebei has a history of 2 600 years, which is the earliest large-scale Water Conservancy Project ever recorded in China. A series of descriptions of the project has been documented from the Han Dynasty.

Liu An, the King of Huainan in the Western Han Dynasty (202 B.C.~ 8 A.D.), edited a treatise *Huainanzi* with his followers, which recorded a wealth of classical techniques and novel ideas of science and technology. The legendary *tofu* is believed to be invented by Liu An and his followers, and spread to the world. Until now, grand *Tofu* Festival has been held in the site of Huainan. While the legendary physician Hua Tuo, who was born in Bozhou, northern Anhui, passed down *Wuqinxi* (Exercise of the Five Animals), *Baduanjin* exercise, *Mafeisan* (a kind of anaesthetic) and many other marvelous stories to the later generations.

Portrait of Hua Tuo

During the Three Kingdoms (220~280) and the Western and Eastern Jin (221~265) Period, Anhui was famous for the water conservancy construction of the farmland. The development of the well-known low-lying paddy fields brought large-area connected network of polders, which could ensure the stability of farming despite drought or flood. The agricultural productivity was improved so significantly that grains produced in Jianghuai area was shipped into the ten big granaries in the capital of the Song Dynasty. This practice was honored as "The ancient good method could benefit the future."

During Sui (581~618) and Tang (618~907) Dynasties, the canal throughout Anhui delivered the goods from South China to Luoyang and Chang'an. Through the canal, the products and tributes in South China were shipped constantly from Yangtze River basin to Huai River and Yellow River basins, so as to circulate all over China, including Porcelain from Changsha kiln in Hunan, Miseci Porcelain from Yue ware in Zhejiang, tribute paper from Xin'an and Xuancheng, and tea and silk, etc.

During Song (960~1279) and Yuan (1271~1368) Dynasties, Wang Zhen, magistrate of Jingde County, Anhui Province, wrote a large-scale agricultural treatise *Nong Shu*, and printed it using the wooden movable type printing system invented by Bi Sheng, a commoner in the Song Dynasty. This became the first large-scale practice of movable type printing in the world. *Printing by Setting Movable Types*, composed by Wang Zhen himself, attached to *Nong Shu*, described the experiences and techniques of wooden movable type printing.

During Ming (1368~1644) and Qing (1636~1912) Dynasties, Prince Zhu Zaiyu innovatively described the twelve-tone temperament, which started a new era for the modern music temperament. Many books, which symbolized the highest level of science and technology and craftwork in ancient China, were spread at home and abroad with profound influences, such as *Little Notes on the Nature of Things* by Fang Yizhi, *Chinese and Western Mathematics* and *Research on Ancient and Modern Calendar* by Mei Wending, *Collection of Algorithm* by Cheng Dawei, *Collections of Lacquer Painting* by Huang

Relic of country road in Jianghuai watershed area

Cheng, *Engraving on Ink Mould* by Cheng Junfang, *Lists of Chinese Ink* by Fang Yulu, and *Ten Bamboo Studio Manual of Painting and Calligraphy* by Hu Zhengyan, etc.

Ten Bamboo Studio Manual of Painting and Calligraphy by Hu Zhengyan

These outstanding masters and their achievements, which were the scientific, technological and humanistic accomplishments made by Anhui people, or people who traveled here officially and made achievements in Anhui, are all wide-circulated classics of Chinese craft culture, benefiting the whole world. Three features can be summarized as: high-level originality, rapid spread and popularity, and far-reaching impact. These agree well with the following facts, such as bordering Yangtze River and Huai River, the convenient waterways, migration of the northern population to the south, delivery of the southern goods to the north, the blending and innovation of culture of Central Plains and South Areas of Yangtze River.

As the distinctive pattern of Anhui craft culture, the handmade papermaking practice of Anhui, represented by Xuan paper, mulberry bark paper ("Han Paper" as its representative brand) and paper made in ancient Huizhou ("Chengxintang" as its representative brand), naturally reflected the blending and innovative cultural features mentioned above.

Making 3-zhang-3-chi large-sized Xuan paper of Hongxing (red star) brand

A photo of mulberry bark paper made in Yuexi County

1.2 Origin and Development of Handmade Papermaking Industry in Anhui Province

Due to the transformation of states, vicissitudes of dynasties, combination and splitting up of administrative divisions, frequent replacement of affiliation relationship, the range of Anhui has been varying constantly. According to the ancient and current documents in the present region of Anhui, the descriptions of local chronicles in the past dynasties and other literature, together with the data from field investigation of our team, we can roughly draw a profile of the handmade paper development in Anhui from the Jin Dynasty (265~420) and divide it into the following five main historical stages.

1.2.1 The Origin and Initial Evolution of Anhui Handmade Paper Practice

Anhui was known as "a land flowing with milk and honey" in the lower Yangtze region since ancient times. Demarcated by Yangtze River, the vast northern plains and Jianghuai hilly terrain were not only granaries in the past dynasties, but also the strategic fields for many battles. While the pretty southern mountainous area in Anhui was not only engaged in farming, but also presented many precious artistic works due to the elegant landscape and gathering of talents, such as the famous four treasures of the study, four kinds of carvings of Huizhou, Hui-style architecture and Wuhu iron picture. And the handmade paper is inevitably one of the representative varieties. As for the exact time of the origin of the handmade paper of Anhui, there is no consesus based on the researches and records up till the deadline of the survey (August 2017).

According to the local handicraft documents of mulberry bark paper in Qianshan and Yuexi Counties acquired in this field work, and the folk oral history of "Han Paper" (big Han paper, middle Han paper and small Han paper), papermaking practice could date back to the Eastern Han Dynasty (202~220). In the investigation, the two national mulberry papermaking skill inheritors, Liu Tongyan and Wang Bailin, and Wang Chun, a local mulberry tree paper researcher from Cultural Center of Yuexi County all claimed that the mulberry bark paper in these areas originated in the Han Dynasty based on ancestors' statements, but they couldn't find any reliable proofs from local literature. Located in the hinterland of Dabie Mountains, this area was originally the core region of ancient Wan State. Its local culture developed before the Han Dynasty, and it belonged to the area where the descendants of a celebrity named Boyi lived in. However, this area belongs to typical mountainous area. Even now, the land and water transportation of the papermaking villages is quite inconvenient. If it is true that the papermaking technique of the Han Dynasty was introduced into the ancient Wan State shortly after its invention, that was an indeed very surprising cultural communication phenomenon considering the transportation status.

A papermaking village in Guanzhuang Town of Qianshan County

A papermaking village in Maojianshan Town of Yuexi County

Although there was no conclusive data from history, we assume it is likely that the handmade paper of Anhui started from Wei and Jin Dynasties, based on the facts that as the former site of the Three Kingdoms, Anhui developed prosperous agricultural economy in Sui and Tang Dynasties, cultivated convenient waterways, and enjoyed the frequent communications of officials and businessmen from different places. The specific clues are as follows:

Firstly, the capital of Eastern Han Dynasty was settled in Luoyang, Henan Province. At that time, Luoyang was the place where high-level papermaking prospered, and "Caihou Paper" made by Cai Lun got popular quickly; while Xuchang, the capital of Wei State, was close to Luoyang. The northern plain of Anhui is neighboring Luoyang and Xuchang; furthermore, the leading Cao family and the intellectual elites such as Ji Kang were from Bozhou City of Northern Anhui. Due to the above geographical factors, including circulation of culture of Central Plains, requirements of local construction, it is quite possible that advanced technology and culture of handmade papermaking spread to North Anhui. For example, at that time, Bozhou, hometown of Cao Cao, possessed the large troop road and Cao family tombs.

It's undoubtful that paper made in Central Plains was circulated and used, and there is the possibility that the above-mentioned factors contributed to the introduction of papermaking skills to North Anhui. Hemp was used as the material for papermaking in the early stage, while the northern plain yielded hemp and mulberry in abundance, therefore, the area enjoyed the advantage of raw materials.

Secondly, the Rebellion of Eight Princes in the Western Jin Dynasty (291~306) caused a large-scale migration of northern intelligents to the southern area of Yangtze River. Emperor Yuan of the Jin Dynasty, Sima Rui, moved the capital to Nanjing. Accordingly, the techniques and craftsmen, representing elite culture and advanced productive force of Central Plains, also spread to the south area thereupon. Several northern aristocratic families moved to the ancient Xin'an, which was later called Huizhou area. The clear historical source about the process could be found in *Records about Xin'an* by Luo Yuan, a Chin-Shih in the Song Dynasty, *Records of Great Clans in Xin'an* and *Records of Great Families in Xin'an* in the Ming Dynasty. Nanjing borders on South Anhui, and the southward movement of political and cultural center of Central Plains brought the local demands for paper. In addition, the southern area of Yangtze River provided various and sufficient papermaking materials. All these brought great opportunities and impacts to the widespread dissemination and development of papermaking technique. At that time, the regions in south of Changjiang River, such as Jiangsu, Zhejiang and Anhui, etc. were the earliest benefited regions.

The special geographical location and abundant papermaking materials of Anhui, together with the co-existence of the Eastern Jin Dynasty and Southern Dynasties, offered spacious room in developing the handmade paper. The reasons can be summarized as follows. Firstly, the supply of paper from northern area was blocked, while the consumption demand in the neighboring area increased rapidly. Secondly, there is an abundant supply of raw materials in Southern Anhui. The attempt to use different kinds of raw and auxiliary materials prompted the development of technology. Thirdly, the convenient waterways accelerated the spread of advanced skills. The technical circulation and exchange in the agricultural cultivation period promoted the development of handmade paper. Fourthly, the unique location advantage and geographical conditions of Southern Anhui attracted various cultures and people who mastered advanced skills, which resulted in rapid progress of farming civilization and craftsmanship. During Tang Dynasty and Five Dynasties period, Hui paper, Chi paper and Xuan paper were chosen as tribute and enjoyed the national reputation. These high-level papermaking system couldn't have been achieved overnight, which probably have emerged and developed into the preliminary form at this stage.

Due to lack of the documents about handmade paper of Anhui, at present it is difficult to describe the details of handmade paper at its preliminary stage, which calls for further literature study and archaeological discoveries.

1.2.2 High-quality and Diversity of Anhui Handmade Paper During the Tang and the Five Dynasties Period

During the Sui and Tang Dynasties, the handmade paper represented by Shezhou (covering the Huizhou area after the Song Dynasty), Xuanzhou, and Chizhou in the southern area of Anhui, had shown high quality and diversity, which made them the brilliant figures among Chinese medieval culture of handmade paper. According to the recordings in *Volume 2 Chapter 6 of Yuanhe Maps and Records of Prefectures and Counties* by Li Jifu, *A Comprehensive Guide* by Du You in the Tang Dynasty, and *The New Book of Tang: Geographical Chronicles* by Song Qi and Ouyang Xiu in the Song Dynasty, there were 11 states offering the tribute paper in the Tang Dynasty, including Xuanzhou, Shezhou and Chizhou. In the year of 743, Jiangxi, Sichuan, South Anhui and East Zhejiang offered paper as tribute. "The most prominent clan in Xuancheng, Xuanzhou offers tribute of local products, including silver, brassware, silk, cloth, red carpet, rabbit fur felt, bamboo map, paper, brush pen, *Rhizoma Dioscoreae*, *Coptis chinensis Franch*, and plant pigments."

A photo of *The New Book of Tang: Geographical Chronicles*

According to the records in *Volume 150 of the Old Book of Tang*, Wei Jian, the governor of Shaanxi, offered tribute on behalf of Xiao Jiong in 3rd of Tianbao Year (744), and "Xuancheng prefecture carried Kongqingshi (a kind of magic stone), paper, brush pen and *Coptis chinensis Franch* by ship." *Volume 28 of the Six Legal Documents of the Tang Dynasty* recorded "paper used in government produced in Xuanzhou and Quzhou." The recordings explained that it was because of the superb quality that the handmade paper produced in Xuancheng prefecture and Shezhou in the Tang Dynasty was chosen as the tribute to imperial court routinely.

The famous painting theorist Zhang Yanyuan in the Tang Dynasty, who served as the inspector of Shuzhou in Anhui, wrote in *Famous Paintings in the Past Dynasties*: "The eastern area of Yangtze River had good environment, and local people were good at skills. The collectors stored hundreds of pieces of Xuan paper, polished by wax for copying. The ancient people preferred to copy the paintings by paper, most of which had never lost the flavour."

A photo of *Famous Paintings in the Past Dynasties*

Because the paper for copying calligraphy and painting called for quality of being thin, transparent and dense, it was necessary to polish it by wax. Through the process of smoothing, pulping, filling the powder, waxing and soaking the paper in alum, etc. to make the copies sustain the flavour. The detailed description of Zhang Yanyuan reflected the high quality of the processed Xuan paper at that time.

The Annals of Ningguo Prefecture during the Reign of Emperor Jiaqing compiled in the Qing Dynasty recorded: " Paper was made in Xuancheng, Ningguo, Jing County

A photo of *The Annals of Ningguo Prefecture during the Reign of Emperor Jiaqing*

and Taiping, so it was named Xuan paper." Thus, Xuan Paper was named after its origin. Meanwhile, the recording reflected that Xuan paper was distributed extensively—almost every county in this prefecture produced this kind of paper.

From the Tang Dynasty and the Five Dynasties Period to the Two Song Dynasties, the flourishing of papermaking practice in Anhui under the support of most counties of South Anhui, brought convenience for the application of paper in many ways. In addition to the regular usage such as calligraphy, painting, umbrella, paper fan, printing and paper money (e.g. jiaozi, a paper currency in Northern Song Dynasty), paper was applied creatively to certain special fields—for example, being used to make paper armour and paper clothes for war.

Based on the literature in existence, paper armour originally appeared in the middle and late Tang Dynasty. According to the recordings in Volume 113 of the New Book of Tang by Song Qi and Ouyang Xiu: "During the reign of Emperor Xuanzong of Tang, Xu Shang was sent by Emperor Xuanzong to patrol the border regions. He was then made the military governor of Hezhong. While he was in military service, there was a group of Tujue remnants who crossed the Yellow River and alleged allegiance to him. He settled them down in Kuang Town of Shandong and organized them into an elite fighting force of 1 000 men, who used paper armours, which were firm enough to withstand the powerful arrows." According to Volume 132 of History of Jin compiled by Tuotuo of the Yuan Dynasty, "Xiao Huaizhong was put to death and his clans were abused and exterminated severely for his failure to capture Sa Ba, the rebel. Wanyan Xiang, an officer, replied: I was in the army. Though Hu Tu and Ze commanded a powerful troop of 13 000 soldiers, the enemies were equipped with paper armours. Most of them were obliged to follow and should have been easy to deal with, but when Hu Tu showed signs of cowdice and hesitation, the rebels just escaped."

From the Southern Tang Dynasty to Northern Song Dynasty, it was popular to make paper armour for military use in Huainan area. According to Volume 293 of Comprehensive Mirror to Aid in Government by Sima Guang: "In the Southern Tang Dynasty, harassed by the troops of later Zhou (Emperor Shizong), the local people in Huainan area got together in the mountain, built the fortress, held the agricultural machines as their arms, made paper as their armours. People called them the Baijia (white armor) Army. Troops of Later Zhou were defeated many times in the attacks." It reflected the firm paper armour could play a solid protective role in the war.

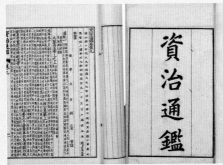

A photo of *Comprehensive Mirror to Aid in Government*

According to Volume 197 of the History of Song by Tuotuo (1314 ~1355) of the Yuan Dynasty: "In April of the first year, the emperor gave an imperial decree to troops of Jiangnan and Huainan, to make 30 000 pieces of paper armours for the archers defending Shaanxi." This recording illustrated that the papermaking practice in Anhui was so prosperous that the paper armours were made to furnish the Shaanxi garrison troops. According to Volume 22 of Songwen Jian by Lv Zuqian in the Southern Song Dynasty: "I had witnessed the sufferings of troops in Southern Song Dynasty when they went on expedition to Liao State. Their inflicted wounds were exposed in freezing coldness, while their paper armours would be grubby when raining. They mashed the shoots of plants and cooked the rotten fish for food. The civilians suffered and perished." This document reflected the terrible environment in wartime; meanwhile, the defects of paper armour in rainy days were fully exposed.

A photo of *The History of Song* by Tuotuo in the Yuan Dynasty

The famous historian in Southern Song Dynasty, Li Tao (1115~1184) recorded in his Volume 132 of A Sequel to Comprehensive Mirror to Aid in Government: "The enemies stopped their harassment in awe. When I was the controller general of Jiangning Prefecture, we used many books of accounts to make the paper armour. When Cao Bin conquered the Southern Changjiang River area and Hezhou, the soldiers were rewarded with pork meat." This showed paper armour was once made in Jiangning Prefecture during the period. Jiangning Prefecture in the Song Dynasty belonged to the jurisdiction of Nanjing, Jiangsu Province, covering modern areas of Jiangning, Jurong, Lishui, as well as dangtu, Wuhu, Fanchang, Guangde, Tongling and Qingyang currently belonging to Anhui Province.

According to *Volume 30 of Major Events from the First Year of Jianyan Reign* by Li Xinchuan: "When the troops from the Jin State invaded Anji County, the magistrate Zeng Chao assembled militia to stone-made walls to guard the pass. They all abandoned paper armours and bamboo spears to escape, which were burned by the invaders after entering into the county." And volume 57 of the same book also recorded: "The executive secretariat said: militia in southern and eastern counties band together to arrange the paper armours in private. According to *Edicts in Xining*, people who made five pieces of paper armours privately would be hanged as punishment." The former referred to the fact that the local armed forces in Anji County equipped themselves with paper armours and bamboo spears, while the latter story showed that people who made paper armours privately would be severely punished to maintain the political stability. Anji County currently belongs to Zhejiang Province, bordering on Guangde County of Anhui Province.

A photo of *Major Events from the First Year of Jianyan Reign*

The famous militarist Zheng Ruozeng (1503~1570) of the Ming Dynasty wrote in his famous book *Jiangnan Jinglue*: "Spear, sword, axe, iron helmet, armour, paper armour, crossbow and bow and arrow were made by bowyers in Taicang State, under the supervision of an officer from the Weapon office." The paper armour was directly classified into weapon category.

From the records above, we concluded that high-quality paper armour in the Tang and Song Dynaties was applied in the military field commonly, and the technique for manufacturing paper armour was quite superb. At that time, Huainan in Anhui Province

A photo of *Jiangnan Jinglue* (old version)

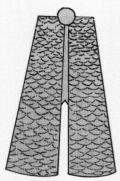
Paper Armour depicted in *A Corpus of Military Equipments written* by Mao Yuanyi

and Southern Anhui area were the important production places of paper armour.

As for the manufacturing method of paper armour, Zhang Binglun and Wu Xiaoxi in their *History of Science and Technology in Anhui* depicted that paper armor making was elaborated on in Mao Yuanyi's 240-volume *A Corpus of Military Equipments*, which was believed to be an encyclopedia of ancient military science. The specific producing procedures include: soft paper and silk cloth were piled up alternately to 1 to 3 inches and then nailed tightly, which was the important defensive equipment in the Song Dynasty for both infantry and sailors for it can "resist knife and arrow attack".

The manufacturing method was specified as well. According to *Volume 7 of Collection of Cuiwei's Military Actions Conquering East and West* by Hua Yue of the Song Dynasty: "The manufacturing method for paper armour was different according to different usage. The infantry preferred longer, the cavalry shorter, crossbowmen wider, and pikemen narrower. Otherwise its usage will be limited, e.g. the fat soldiers will find it too tight." For different purposes of service, the forms were different. At that time, it was superior to use paper armour than iron armour in the southern areas, because "The soldiers couldn't bear heavy load when marching in the south. Furthermore, the armour was prone to get rusty and decayed in the damp weather and became useless." This was one of the main reasons for the popularity of paper armour in the south.

In addition to paper armour, other feature paper in Anhui history during Sui, Tang and the Five Dynasties were also documented in literature.

From middle and late Tang Dynasty to the early Northern Song Dynasty, diverse "processed paper" emerged in China, such as the famous "Xuetao paper" produced in Chengdu of Sichuan Province and "Ten-color paper" produced by Xie Jingchu in Zhejiang Province. During the Five Dynasties period and the Southern Tang Dynasty, "Chengxintang paper", mainly produced in Shezhou, the southern area of Yangtze River, became famous. Chengxintang was originally the name of a place for banquet when Li Bian took the office as military governor in the Southern Tang Dynasty. When it came to the reign of Li Yu, last emperor of the Southern Tang Dynasty, who was adroit at poetry, painting and calligraphy, regarded the paper produced in Shezhou of Anhui as a treasure. He spared a room called Chengxintang to store the paper in the imperial palace in Jinling, and established a bureau to supervise the making of this kind of paper, hence the paper was named as "Chengxintang paper".

However, Chengxintang paper was the tribute paper for the emperor exclusively. Even the imperial princes and court ministers didn't have the chance to appreciate it, not to mention the common intellectuals. According to *Volume 32 of The Histories of the Ten States*: "Li Tinggui was good at making inkstick, who migrated to Shezhou from Yishui with his father. Originally, he was surnamed 'Xi' and awarded the emperor's family name 'Li'. Although his younger brother Tingzhang and his son Wen both carried on his practice, he remained the best of all. Chengxintang paper, dragon-tail inkstone, and Tinggui inkstick were regarded as 'The Three Treasures of Chinese Study' in south area of Yangtze River. When a nobleman mistakenly lost a pill of Tinggui inkstick into the pool, he didn't take it thinking that it had been ruined by the water. One month later, when he ordered his servant who's good at swimming to get a lost gold vessel in the pool, he got the lost inkstick again. To his surprise, the inkstick was new as before. That's why the magic Tinggui inkstick was viewed as a treasure." In Luoyuan's *Volume 10 of Records about Xin'an*, he wrote: "In the past, Emperor Li was intoxicated with calligraphy and brush painting, and he favored the top three stationeries—Chengxintang paper, Tinggui inkstick and dragon-tail inkstone." Litterateur, calligrapher and painter Li Rihua(1565~1635) of the Ming Dynasty mentioned in his *Volume 4 of Notes from Liuyan Studio*: "When Song Xiegong served in Hui District, the relatives of his concubine offered four treasures of Xin'an as gifts: Chengxintang paper, Wang Boli brush pen, Li Tinggui inkstick and Yangdouling inkstone."

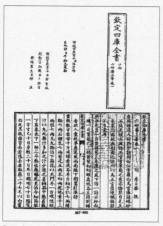
Scanning records about *Notes from Liuyan Studio*

Volume 38 of Reading Records on Tianzhong Mountain by Chen Yaowen of the Ming Dynasty recorded: " Poets of the Tang Dynasty used Koryo tribute paper; Japan produced pine bark paper, and the ancient Nanhai county produced *Aquilaria sinensis* bark paper, which was white with tiny veins; Youtai paper used Herba Sphagni as the making material, which was also called Celi paper. Xue Daoheng in his poem described in spring, papermakers used Herba Sphagni to make paper, which was used for calligraphy and painting. Jipi bark paper was produced in Fusang (ancient name for Japan). Nowadays, only mulberry bark paper, rattan paper, Yuezhong bamboo paper and mulberry bark paper was produced in China. Hui paper was known as Chengxintang paper in the Southern Tang Dynasty. Both being dyed paper, Wu paper was superior to Shu paper for not being heavy and thick." The article pointed out directly Chengxintang paper was the processed Hui paper, and the historical status of Chengxintang paper also highlighted the position of handmade paper made in Anhui region in China during the Sui and Tang Dynasties.

1.2.3 Anhui Handmade Paper Being the Representative High-quality Paper During the Song and Yuan Dynasties

After the perishment of the Southern Tang Dynasty, Chengxintang paper broke the royal monopoly situation, and fell into the hands of intellectuals in the Northern Song Dynasty. It was viewed as treasure soon, and eulogized by poets. The historian Liu Chang in Northern Song Dynasty (1019~1068) once obtained hundreds of pieces of Chengxintang paper from the palace. He couldn't help composing a poem: "Chengxintang paper was sold at the price of 100 gold coins, which was stored in the Chengxin chamber in the palace and never appeared in the folks. However, I got a few pieces from the palace luckily." Afterwards, Liu Chang gave Ouyang Xiu ten pieces of the paper as present, and Ouyang Xiu composed a poem as a reply: "Though you gained Chengxintang paper, no one dared to write on them. I didn't know where you got the paper, which featured pure, stiff, with smooth surface and clean texture." Then he regifted Mei Yaochen two pieces of the paper, and Mei indited: "Ouyang Xiu once gave me two pieces of the paper and I was reluctant to use it for its value. The paper was suitable for him to spread his outstanding talents." It is clear that Chengxintang paper fell into the hands of scholars in the Northern Song Dynasty and was looked upon as treasure. Even Ouyang Xiu, the great essayist and statesman, didn't dare to write rashly. No doubt that it is related to its rarity due to monopoly by Emperor Li and the war, but more importantly is decided by the outstanding quality.

In addition to imitation of Chengxintang paper, there were many other types of paper in South Anhui in the Song Dynasty. According to *A Sequel to Comprehensive Mirror to Aid in Government* written by Li Tao in the Song Dynasty: "In June, the seventh year of Xining, the court distributed making pattern of Xuan paper to Hangzhou, demanded 50 000 pieces production per year. No matter how important the official document was, Xuan paper shouldn't be used in government." This shows at that time there were certain regulations on the usage of Xuan paper. On one hand, the court urged to increase the production; on the other hand, it was stipulated that Xuan paper shouldn't be used randomly for official documents. However, Xuan paper mentioned here was the paper from Xuanzhou made in the Song Dynasty, which could be different from the Xuan paper made of *Pteroceltis tatarinowii* Maxim. bark and straw in the following generations.

A photo of *A Sequel to Comprehensive Mirror to Aid in Government*

According to *Records about Xin'an* (1173) by Luo Yuan: "There were seven types of tribute paper in Xin'an area in the Song Dynasty: Changyang, Xiangyang, Dachao, Jingyun, Sanchao, Jinglian and Xiaochao, which were called seven colors. And 1 448 632 pieces of paper were offered as tribute to the court every year." "Paper enjoyed various texture, among which Maiguang, Baihua, Bingyi and Ningshuang were classified as high-quality paper. And the famous Longxu paper was produced in Longxu area, now the jurisdiction of Shexian and Jixi." This book also pointed out the reason why Xin'an area produced high-quality paper: "The clear water contributed to the smoothness, whiteness and stiffness of the paper" and "Biao paper, Maiguang, Bingyi and Baihua paper was tribute paper in the Song Dynasty."

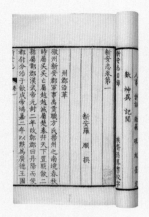

A photo of *Records about Xin'an* (old version)

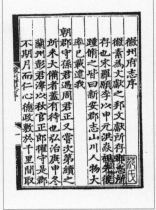

Scanning records about *The Annals of Huizhou Prefecture* during the reign of Emperor Hongzhi

A photo of *Four Treasures of the Scholar's Study* (old version)

The Annals of Huizhou Prefecture during the reign of Emperor Hongzhi mentioned the paper in the Song Dynasty: "There were several types of paper, e.g. Jinzha, Dianzha, Yuban, Guanyin, Jinglian and Tangzha, which were all produced in Yurui, Hemu and Liang'an villages in Xiuning County." The tribute paper recorded was similar to that written in *Records about Xin'an*. As for Yuban paper, *Fuxuan Yelu*, written by Chenyou in the Southern Song Dynasty, recorded it in detail: "Yuban paper in Xin'an area, which featured fine texture, but less resistance to shear stress. The paper became smooth and stiff after pulping. If stored correctly, it wouldn't breed worms for a long time." Thus, there were many patterns of paper in Xin'an area in the Song Dynasty, and certain types of paper had reached the "tribute" level.

It was noticable that the large-size paper was produced in Yixian County and Shexian County in the Song Dynasty, which was recorded in *Four Treasures of the Scholar's Study* by Su Yijian: "Ningshuang and Chengxin paper was produced in Yixian County and Shexian County. The length of the paper could reach 50 chi." The common paper screen and papermaking container couldn't satisfy the production of such large paper and the operation also demanded highly-skilled technique and cooperation among people, in order to ensure the uniformity of the thickness of the paper. The making procedure was cited as follows: "People in the county arranged the mulberry bark for several days, then soaked the material in the long boat. Dozens of people made the paper, then steamed the material on big stove repeatedly. Thus, the thickness of the paper during the process was almost the same." The wisdom of papermakers in Yixian County and Shexian County could be observed. Therefore, the researchers Zhang Binlun and Wu Xiaoxi presented that the long-specification paper in the Song Dynasty was almost produced in these two counties in South Anhui. "2-*zhang* Xuan" and "6-*zhang* Xuan", the special types of Xuan paper during the Ming and Qing Dynasties, should be the heritage of the large-size paper in the Song Dynasty.

In the Song Dynasty, people in Huizhou area used paper to make clothes. According to *Volume 4 of Four Treasures of the Scholar's Study* by Su Yijian: "The dwellers in mountainous areas often wear paper clothes instead of silk clothes, following the Buddhism beliefs. The disadvantages of silk clothes lie in the fact that it prevents the inner energy to go out, and 10 years will suffice to make the clothes-wearer become unhealthy. The procedures of making paper clothes are as follows: Pile 100 pieces of paper as a stack and boil them with walnut and frankincense in it. Or put them on the steamer and frankincense is spilled continuously during steaming process. Roll the paper on a shaft tightly, stroke rolls along the shaft to the other end, and then the paper stack would be pushed to the end in case it would form crease due to the pushing. People in Yixian County and Shexian County produced the paper cloth, the size of which was as large as the gate. The scholar-bureaucrats would wear the cloak made by paper to keep out the cold when going out." The information reflected Yixian County and Shexian County produced the gate-sized large cloth at that time. Then, this garment-making method spread to Japan and the origin of the paper clothes adopted in the foreign realm of art and literature could be traced back to the traditional making method in South Anhui in the Song Dynasty.

As for Chi paper produced in Chizhou District, Anhui Province, Wang Anshi, the then prime minister imposed a poem elaborating on the making process and usage of the paper in his *On Giving Chi Paper as Present*, in Volume 11 of *Linchuan Gentleman Anthology*.

In addition to the imitation of "Chengxintang paper" in the Tang Dynasty, there were many other types of paper in Xuanzhou prefecture: "Jinbang, Huaxin, Luwang, Bailu and Juanlian" in Jing County, "Biyun Chunshu paper, three-colored paper with dragon and phoenix, Yinjin Tuanhua" and "Jinhua paper"; "Chi paper" in Chizhou, and famous "Xibai paper" in Wuwei; and "Longxu paper" was still produced in Jixi, which was well known from the Tang Dynasty.

Scanning records about *Linchuan Gentleman Anthology*

Yuan Dynasty was an important period for the development of Xuan papermaking practice as well. The innovation of material (*Pteroceltis tatarinowii* Maxim. bark and straw mixed in certain proportion) and processing technique by the Caos in Xiaoling of Jingxian provided high-quality paper for the development of Chinese ink wash painting. "The Four Masters of the Yuan Dynasty" represented by Wang Meng, Huang Gongwang, Wu Zhen and Ni Zan, broke the material and technique constraints of the traditional royal silk and processed painting paper, spurred landscape rendered in an impressionistic manner with coarse brushstrokes and wet ink washes technique. Xuan paper, which was good at water-absorbing and color-separation, provided broad room for revealing performance and imagination of ink wash painting. While the prosperity of paper-based freehand style painting and its becoming the mainstream in the Ming and Qing Dynasties, also promoted the sustainable development of Xuan paper practice. A new era for the painting and calligraphy paper came into being formally in the Yuan Dynasty.

Based on reliable literature, Jingxian County, which was world-famous for Xuan paper, had entered orderly-inherited family production state at least since the late Song and early Yuan Dynasties. According to the preface to *Genealogy of the Caos* compiled in the early Republic of China by the Caos family in Xiaoling village, Jingxian County: "In the late Song Dynasty, the whole country was in chaos caused by the war and the public left their hometown to avoid the war. Cao Dasan migrated to Xiaoling from Qiuchuan to avoid the war. He noticed that there were few farmland to plough in the mountainous area, so he resumed his papermaking practice to support his family." It showed that the Caos had lived by papermaking in Xiaoling village in late Song and early Yuan Dynasties. The family's practice also witnessed the origin of Xuan papermaking practice.

Photos of *Genealogy of the Caos* printed by Jishantang in the eleventh (1873) year of Tongzhi Reign in the Qing Dynasty

According to a noteworthy book *Shulin Qinghua* by Ye Dehui: "Zhang Yanyuan in the Tang Dynasty wrote in *Famous Paintings in Past Dynasties* that the collectors liked to arrange hundreds of pieces of Xuan paper, polished by wax for copying. Meanwhile the Zhuge family in Xuancheng were also skilled in papermaking." This manifested that Zhuge family who were the representative of ink brush making had also been engaged in early Xuan papermaking practice. But we didn't know the basis for Ye's inference.

Throughout the Song Dynasty, no matter the unified Northern Song Dynasty, or the Southern Song Dynasty established by royal family members who migrated to the South, the number of papermaking places and paper types in Anhui exceeded the past generations greatly, which played so significant a role in the domestic papermaking industry, that Hui paper and Chi paper were sold to the areas thousands of miles away, such as Sichuan Province. *Paper Manual* by Fei Zhu in the Yuan Dynasty recorded: "Shu paper weighed heavily, so a person could only load 500 pieces a time. Chuan paper was expensive because it was unavailable to the remote areas. However, people in Chengdu preferred the lightness of Hui paper, Chi paper and Bamboo paper, which were sold several times the price of Chuan paper when they are available in Chengdu. When officer Fan served on his post for two years, he used Shu paper exclusively and saved a lot of money. That is why he blamed those government officials who used Hui paper and Chi paper at high cost." Hui paper and Chi paper were sold with a high price to the main papermaking place of Sichuan Province, which was due to the exquisite skills of paper makers, the excellent materials and water in South Anhui.

Although the social productive forces were destroyed greatly in the Yuan dynasty, there was still abundant production of paper in Huizhou district. According to *History of Anhui* by Li Zegang: In the Yuan Dynasty, the tribute paper included Fubei paper, Xingtai paper, Lianfangsi paper, and so on. The number reached 220 000 pieces per year. In addition, there were millions of paper orders from the government.

As for papermaking method in Huizhou during the Yuan Dynasty, *The Annals of Huizhou Prefecture: Products* recorded: "The process of manufacturing handmade paper can be generalized into the following steps: separate the useful fibre from the rest of raw materials (60% out of the raw material), wash in streams, beat and ferment to make the material white. Then boil the material and ferment again, and pick out the impurities. Beat again and put the material in a cloth bag to make pulp. Pour the pulp in the papermaking container, and add in papermaking mucilage, which was called Huashui functioned as adhesive. Scoop the pulp to make paper with a screen followed by pressing and drying to get the final paper." The recording not only detailed the process of papermaking, but also mentioned the usage of juice from the branches of Yangtao Kiwifruit as papermaking mucilage. The significant improvement of the papermaking quality in the Song and Yuan Dynasties could be attributed to the widely usage of papermaking mucilage.

1.2.4 Expansion and Innovation of Jingxian Xuan Paper During the Ming and Qing Dynasties

During the Ming and Qing Dynasties, especially in the Ming Dynasty, the printing practice and printing techniques in Anhui had developed at top speed. Official printing and private block-printing almost prevailed the whole province. Huizhou was a printing-intensive area, which possessed block printing, assorted block printing, embossed designs printing, wooden movable type, clay type and tin-casting printing. The large scale and excellent presswork were unprecedented in the history of Anhui.

During the reign of Emperor Wanli in the Ming Dynasty (1573~1620), Huizhou leaped onto one of the printing centers in the whole country as a result of economic development, cultural prosperity, availability of ideal printing materials and the high-quality timber, ink stick and paper. Additionally, there are other factors contributing to the achievement. One is the unparalleled engravers of woodblock printing in Huizhou. According to statistics, during the Ming and Qing Dynasties, the number of engravers in Huizhou reached up to more than 400 people, whose names could be verified, and called "Xin'an engravers". "In the Ming Dynasty, block printing was popular in Hangzhou, which was all engraved by craftsmen in Shexian County with exquisite performance." The other reason is the abundant funds. The merchants of Huizhou had arrived at their period of great prosperity since the reign of Emperor Chenghua in the Ming Dynasty (1465~1488). The outstanding merchants came forth in large numbers, among whom there were intellectuals and scholars. Typing and storing books had become a trend. The representatives consisted of Wu Mianxue (from Shexian County, Huizhou, 1368~1644), Wang Tingne (from Xiuning County, Huizhou, 1573~1619) and Hu Zhengyan (from Xiuning County, Huizhou, 1570~1671). Wang Tingne compiled *Collections in Huancuitang*, and Hu Zhengyan published *Yincun Xuanlan Chuji (Collection of Seals)* and *Seal Carving*. They made a lot of contributions to the development for future generations.

At that time, there were several famous private book-engraving mills in Xuancheng, Jingxian, Ningguo, Fanchang, Guichi, Lu'an, Tongcheng, Taihe, Hezhou and Fengyang, etc. As for wooden movable type, clay type and tin-casting printing in the Yuan Dynasty, Wang Zhen, the magistrate for Jingde County, created more than 30 000 wooden movable types, and printed *Book of Agriculture* compiled by himself. During the Qing Dynasty, Cheng Weiyuan in Huizhou used the wooden movable types to print *Dream of the Red Chamber* (i.e. the Chengjia edition and the Chengyi edition). Chao Yiduan in Lu'an used the wooden movable types to print the multi-volume series *Chinese Classics as Arranged by Subject Matter*. And Zhai Jinsheng in Jingxian invented clay blank based on the clay movable types of Bi Sheng, and printed *Xiuyetang Collection* and *Genealogy of the Zhais from Shuidong, Jingchuan*. In 1760 under the reign of Emperor Qianlong, Cheng Dun in Shexian reprinted *Brick Inscriptions of the Qin and Han Dynasties* with tin casting, which was the only surviving product by tin-casting printing in China currently. The peak of movable types printing could illustrate the abundant supply of raw materials for papermaking industry in Southern Anhui in the Ming and Qing Dynasties. Moreover, the superior quality of the local paper led to emergence of diverse usages.

In the Ming Dynasty, the production of Xuan paper in Anhui had entered an important developing stage. The technique had become higher with more varieties of paper. At the same time, there appeared a lot of poems in praise for Xuan paper. Shen Defu in the Ming Dynasty wrote in *Feifu Yulue* (a book about calligraphy): "Paper made in Jingxian could be used for calligraphy and painting after being pasted on the wall for a long time, but no longer absorbed ink." Wen Zhenheng in the Ming Dynasty recorded in *Treatise on Superfluous Things*: "In the Ming Dynasty, paper such as Lianqi, Guanyin, Zouben and Bangzhi were all of inferior quality ... Sajin paper in Wuzhong and Tan paper in Songjiang couldn't be stored for a long time. Liansi paper produced in Jingxian was the best."

Fang Yizhi wrote in *Little Notes on the Nature of Things*: "Lianqi paper was produced in Jiangxi in the reign of Emperor Yongle. Paper for official documents offered to the Emperor was produced in Yanshan, while notice paper was produced in Changshan of Zhejiang and Yingshan of Luzhou. The Suxin paper was produced in the 5th year of Xuande Emperor Period, including varieties of paper decorated with golden dots, five-color golden paper, and Ciqing wax paper. As for the mulberry paper, we recommended the one produced in Xingguo and Jingxian, which is white and suitable for calligraphy." The literature showed that at that time the intellectuals held that Xuan paper produced in Jingxian was of superb quality.

Wu Jingxu in the Ming Dynasty recorded in *Lidai Shihua (Notes on Poets and Poetry)*: "Xuan paper featured great tensile strength, smooth surface, pure and clean texture and great resistance to crease. It could be used as tribute paper, and some paper was made from cotton. Usually of the size of official notice paper, and 3 to 4 pieces could be made at one time." This proved the variety of Xuan paper, including not only high-class tribute paper, but also the ordinary Mian paper and the special variety paper. Zha Shenxing in the Qing Dynasty mentioned in *Renhai Ji*: "Tribute paper, Mian paper, Suxin paper produced in the 5th year of Emperor Xuande Period, and white paper, paper decorated with golden dots, five-color powdered paper, five-color golden paper, five-color large-size paper, Ciqing Paper. Among those, Chenqingkuan paper ranked first."

This kind of comparative analysis manifested the high quality of Xuan paper. According to another version from the academic research circle, Xuan paper in the Ming Dynasty was also called "Xuande paper" following the title of the emperor's reign, enjoying a fame together with "Xuande Furnace" and "Xuande kiln". This showed in the Ming Dynasty no matter the production of Mian (using uncooked material) or that of Tribute paper (Processed Xuan paper), both reached a high level, far ahead of those produced in previous generations, which became more outstanding than other paper types in different parts of the country.

Hui style woodcut in the Ming Dynasty: *Fengnan Illustration*

The remains of Wooden Movable Words in Southern Anhui in the Qing Dynasty

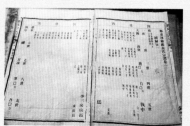

A photo of *Renhai Ji* (old version) written by Zha Shenxing in the Qing Dynasty

Jin Nong in the Qing Dynasty recorded "6-zhang (the size of paper) Xuan paper in the reign of Emperor Xuande" in his *Dongxin Huazhu Tiji (Inscription of Bamboo Painting)*. *Records about Jingxian* in the 18th year during the reign of Emperor Qianlong in the Qing Dynasty recorded: "After 1718, the imperial storehouse dispatched people to purchase papers. Among which, the largest was called Luwang, 1.6 zhang in height, claiming a tradition from Prince Lu in the Ming Dynasty." This verified the production of large-size Xuan paper in the reign of Emperor Xuande. The imperial storehouse during the reign of Emperor Kangxi followed the standard of the Ming Dynasty in paper purchase, which was called "Luhuang paper" or "Luwang paper".

With the rapid development, Xuan paper entered its prosperous era during the reign of Emperor Kangxi and Qianlong. A successful candidate in the highest imperial examinations, Chu Zaiwen during the reign of Emperor Kangxi composed *Ode to Ribbing-patterned Xuan Paper* when traveling officially in Jingxian. It is a very valuable documentation on early Xuan paper in Jingxian County.

This literature elaborated on the details of Xuan papermaking in Jingxian about 300 years ago. From the personal perspective of Chu Zaiwen in his field investigation, Jingxian during the reign of Emperor Kangxi had become a large top-class papermaking base.

In addition to the prosperity of Xuan paper, Hui paper and Chi paper had already enjoyed a fame in the Song Dynasty. Until the Ming and Qing Dynasties, the famous products of Hui paper still presented the orderly-inherited favorable situation.
Liu Ji (?~1128), a famous general in the Song Dynasty, expressed his appreciation for Hui paper in *Songshan Ji*: "Recently I got Hui paper with good quality so I bought several hundred pieces of them."

Yushan Minsheng Ji, co-edited by Gu Ying (1310~1369), a writer in the Yuan Dynasty and his friends, recorded: "I took the honey plum and the Hui paper to Yushan, on which my friends composed poems to express their feelings." Gu Ying also mentioned in *Caotang Yaji (Elegant Meetings in Thatched Cottage)*: "I gave my friends Hui paper as present." In all the literature above mentioned Hui paper was regarded as a gift, which played the role of strengthening communications among friends.

In *Fangyu Shenglan: On facilities*, Zhu Mu (?~1255), listed Zeng Wenqing bamboo paper (Shaoxing), Bai Juyi *The Song of the*

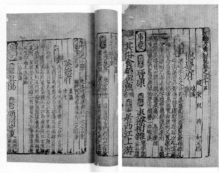

Photo of *Fangyu Shenglan* (old version) written by Zhu Mu in Southern Song Dynasty

Scanning records about *Songshan Ji* written by Liu Ji in the Song Dynasty (included in *The Complete Library in the Four Branches of Literature*)

Pipa Player (Jiangzhou), Bai Juyi Zitai brush (Ningguo), Bai Juyi red-yarn blanket (Ningguo), Wang Jiefu Xiechi paper (Chizhou), and Huang Luzhi Qingjiang paper (Chizhou). Sang Shichang of the Southern Song Dynasty wrote in his *Lanting Kao*: "Jian Zhai (the nickname of Chen Yuyi, a poet in the Song Dynasty) practiced calligraphy delightedly on the Chi paper." These documents mentioned the early application of Chi paper respectively. In addition to Hui paper and Chi paper, there were plenty of accounts on the handmade paper in other counties of Southern Anhui in the Ming and Qing Dynasties. For example, *The Annals of Ningguo County* recorded: "Ningguo was the mountainous area, which produced abundant papermaking materials. At first there was no papermaking practice there. During the reign of Emperor Guangxu in the Qing Dynasty, the papermaking mill was first built imitating Jiangxi Province practice, and then developed gradually. Now there were more than 40 papermaking mills throughout the area, with the products of Biaoxin, Wangao and Qiangu paper, etc. The paper made was sold to Wuhu, Xuancheng, Gaochun and Dongba, nearly 20 000 *dan* recorded every year. Occasionally many producers stopped the business because of the low market demands and price. Now they make some improvements in order to survive."

Bao Mingfa wrote in *Wenjishan Ji*: "Mt. Wenji in Ningguo County produced high quality bast paper, with the abundant mulberry tree bark being the material for papermaking. The paper featured fine texture, paralleling to the tribute paper in Xuancheng and Jingxian."

The Annals of Nanling County by Xu Naichang, the famous scholar and bibliophile in Nanling, recorded: "Keng paper was produced in Shannan Town. The large one was called bast paper, and the small one bark paper. Paper to make umbrellas was produced as well. *Pteroceltis tatarinowii* Maxim. bark could be used to make paper."

In 1905, an article *The Reform of Anhui Paper* published in *Jiangxi Official Report* mentioned: "In Kaiyuan Town, Guichi County of Anhui, some Liu intended to reform the papermaking practice by adopting the Japanese method, to save the lossmakers. First they built 100 rooms and then hired foreign craftsmen to start the practice upon completion."

According to the revised edition of *Volume 6 of The Annals of Fanchang County*: Products in the reign of Emperor Daoguang: "Paper was produced in Fanchang of various types, including Luwang, Bailu, Huaxin, Juanlian, Liansi, Gongdan, Xueshu, paper to make umbrella (made from bark), Qianzhang paper, Huo paper, Xiabao paper and Gaoyi paper."

Similar local chronicles and literature were quite considerable, which showed the wide distribution and prosperity of handmade paper production in Southern Anhui.

A small papermaking village in Shexian County: Liuhe Village

Su Yijian (958~996) in the Northern Song Dynasty recorded in *Volume 4 of Four Treasures of the Scholar's Study:* "Muzhi, the grandson of Lei Kongzhang, still keeps the letters from Zhanghua, which was written on paper made of mulberry bark." The fact that Zhanghua was born in 232 A.D. showed the production history of mulberry bark paper could date back to 1 700 years ago. *The Annals of Anqing Prefecture*: *Products* in the 33rd year of Emperor Jiajing Period in the Ming Dynasty recorded: "Qianshan produced tea, paint, and paper." The entry on papermaking and printing in *History of Anqing Area* wrote: "In Yuexi, Qianshan, Taihu and Guichi, local Hanpi paper, Pu paper, Baima paper and painting and calligraphy paper was made. In 1949, there were 191 papermaking containers in Yuexi, which yielded 71 tons of paper."

1.2.5 Development of Anhui Handmade Paper from Late Qing Dynasty to the Contemporary Era

During the late Qing Dynasty and the Republic of China period, brand of Jingxian Xuan paper began to sparkle, featured by winning prizes frequently in various international exhibitions. For example: In 1908, Xuan paper won the first place in Shanghai Commodity Exhibition Contest. In 1910, "Bailu" brand of Xuan paper won the "Top Award" in the Southeast Asia Industries Association. In 1910, "Hongji" brand of Xuan paper from Jingxian won the "Superb Award" in the Southeast Asia Industries Association. In 1915, "Taoji" brand of Xuan paper won the gold medal in Panama-Pacific International Exposition. In 1926, "Wangliuji" and "Caoxingtai" brands of Xuan paper won the gold medal in the Philadelphia Exposition, etc.

Xuan paper won the prizes frequently in the international exhibitions, which made it an important bridge between Chinese and foreign cultures. For example, in 1930s, Mr. Lu Xun sent a letter to Mr. Xidi (Zheng Zhenduo), saying: "I prefer Xuan paper, which was durable and soft, suitable for thick books." Moreover, Mr. Lu Xun had once gave Xuan paper to Piskarev, a Soviet wood carver, as a gift. Later on, he received Soviet woodblock prints from the artist in return. This Soviet wood carver made a comment on Xuan paper: "As for printing the woodblocks, Xuan paper from China is the best choice, which was unparalleled in the world. It's wet, gentle, enduring, able to absorb ink and smooth. All these features made it the most suitable paper for hand copying and wood carving."

With Xuan paper gaining a high reputation, the investigation and research on it had moved forward. Hu Yunyu mentioned in *Series of Puxuezhai Books*: *on Paper*: "Material came first for papermaking." The status of raw material in papermaking was first proposed. In 1936, China Paper Research Institute launched a survey of Xuan papermaking practice. In the same year, Wei Zhaoqi published *A Survey of Xuan Papermaking Industry*, with an account of the distribution of raw materials and papermaking containers, the local production situation, the equipment in paper mills, making method, types of Xuan paper, the cost and marketing of the paper, etc. In the following year, Zhang Yonghui published *A Review of Anhui Xuan Papermaking Industry*. In addition to the introduction to the raw and supplementary materials for producing Xuan paper, such as *Pteroceltis tatarinowii* Maxim. bark, straw, mucilage, adhesives, lime, alkali, and bleacher. The article also depicted procedures and manufacturing methods: bast producing procedure, straw producing procedure, papermaking, pressing, baking and inspection. The article mentioned specifically that "The specially-made paper materials were supplied for all Xuan papermakers. We'd better maintain the traditional handmade papermaking methods, except for the improvements, thus lowering the price." This method had been widely employed in the Xuan papermaking practice.

At the same time, reports on Xuan paper had appeared abroad due to its good reputation. In late 19th century, Japanese Printing Bureau, Ministry of Papermaking sent an official to China, and after investigation, he published *Chinese Papermaking Industry*, which included the production of Xuan paper. In 1883, A Japanese wrote *Investigation Diary of Papermaking in the Qing Dynasty*, which also recorded producing technique of Xuan paper. In the early 20th century, Japanese Uchiyama Asamon wrote an article *The Production of Xuan Paper* in 1906, which was published in *Japanese Industrial and Chemical Magazine* (volume 9, No.98). The reports all attempt to collect techno-economic intelligence of Xuan paper.

According to the records of *Xuan Paper of Anhui* by Wei Zhaoqi: "In the early Republic of China, Wuhu was taken as the foothold and paper was sold to the domestic markets in Shanghai, Beijing, Jiangsu, Zhejiang, and Hankou." The business flourished then.

With the Anti-Japanese War breaking out in 1937, all these crucial cities for commercial transition and marketing became enemy-occupied areas, which seriously disturbed the trade channel of Xuan paper. The price of Xuan paper varied, which even changed from day to day in some places. In addition to the impact of machine-made paper, the marketing of the Xuan paper, sold with the form of stacks, got into serious crisis. The production of Xuan paper in Jingxian county found it difficult to continue. On the eve of the establishment of People's Republic of China in 1949, the industry of Xuan paper almost stopped and craftsmen became destitute and homeless and they had to find other ways for survival.

According to statistics from Li Dexuan's investigation of the Xuan paper industry in the 26th year of the Republic of China (1937): Xiaoling had only "23 households" at the time. In addition to the heavy taxes imposed on Xuan paper manufacturers during the Republic of China, Xuan paper manufacturers faced the impact of mechanical papermaking. "Capital cannot be concentrated, manufacturers failed to switch to mechanical papermaking to retain workers, reduce costs and increase production efficiency. These are the main reasons why the industry cannot improve. Besides, neighboring Japan succeeded in imitating, and the former sales outlets established by foreigners and in Northeast Province were robbed, the business gradually declined." According to the memories of the locals. due to the economic collapse, inflation and falling price of the bank bill during the period of the Republic of China before the founding of PRC, worsened by decreased production fields and lack of food, the paper factories were closed one after another and the production of Xuan paper suffered disastrous decline. As a result, the manufacturing equipment was damaged, the workshops collapsed, and the raw materials were abandoned. Only few shabby papermaking containers were left and discontinued on the eve of the founding of People's Republic of China in 1949. Xuan paper workers could only survive by chopping firewood.

In the same period, other kinds of paper developed with the advancement of society.

In 1932, *Domestic Products Research Magazine* published an article entitled *Business News: Industrial Information: Papermaking Industry of Wuyuan in Anhui Province*. It said: "China was the birthplace of papermaking practice, whose handmade paper spread all over the country. Huizhou and Ningguo Prefectures hold the largest production base of paper at home. In addition to the famous Xuan paper of Jingxian County, the bast paper produced in Wuyuan (belonging to Huizhou District at that time) was also known in Hubei and Jiangxi Provinces."

In 1943, Li Yinwu mentioned in *How to Develop the Papermaking Industry in Western Anhui*, published in *Politics in Anhui*: "Handmade paper was the special technique in Western Anhui. The main production regions include Qianshan, Yuexi, Lihuang (now Jinzhai County), Shucheng, Lu'an and Huoshan. Among them Qianshan, Yuexi, Lihuang (now Jinzhai County) and Shucheng boast top production, followed by Lu'an and Huoshan while less in other counties. As for the paper types, before the Anti-Japanese War, the main products include Bast paper, Huajian paper, Fangmaobian paper, Biaoxin paper and Shao paper, which were not only offered to the local areas, but also sold to Shanghai, Anqing and Southern Henan Province, etc. After the War, most papermakers stopped their production, till the capital of the province transferred to the Lihuang County. Paper types also adapted to social needs, and paper for publication, Fangmaobian paper and envelope paper was made accordingly to meet social needs. According to our survey, the papermaking status of the counties in Western Anhui was shown in the following table."

After the foundation of the People's Republic of China, Xuan paper workers who switched jobs before returned gradually with the help of government and associations, and Xuan paper production resumed under the leadership of the original proprietors. After going through joint operation, public-private joint management and state-run management, China Xuan Paper Group Company (former Jingxian Xuan Paper Factory of Anhui Province) kept decades of monopoly from 1950s to 1960s, and remained its dominating status till the 21st century while other papermakers also boomed during the period. From 2015 to 2016 when our team investigated in Jingxian County, this county emerged nearly 20 Xuan paper producing enterprises. In Huangshan District (former Taiping County) of Huangshan City, Baitian'e Xuan Paper Factory reopened which was migrated from jingxian County. More than 300 paper (mainly painting and calligraphy paper and processed paper) producing enterprises were opened. There were up to 1 000 shops selling paper and "Scholars' Four Jewels" were opened in major cities across the country, mostly run by persons who owned a papermaking mill back in Jingxian County, which was called "Jingxian County phenomenon".

As for bast paper, in 2016, it was still in production in Tanfan Village, located in Guanzhuang Town, Qianshan County, Anqing City, and Banshe Village, located in Maojianshan Town, Yuexi County. Mulberry bark paper was still in production by Jiang Zushu family in Mianxi Village, Shendu Town, Shexian County, and Sanxin Paper Factory in Xiaojiao Village, Haiyang Town, Xiuning County. Other types of paper could scarcely be found in other parts of Anhui.

Researchers visiting the former Guyi Xuan Paper Factory in Jingxian County

Researchers visiting the relics of former Hushankeng Xuan Paper Factory in Jingxian County

Researchers interviewing Xingjie Mulberry Bark Paper Factory in Qianshan County

Table 1.1 Papermaking Status of Various Counties in Western Anhui in 1943

Category	Paper mill		Private makers		Products	Annual output	Note
	number	number of containers	number	number of containers			
Qianshan	3	12	430		4-chi paper, local newspaper, simulated Xuan paper, painting and calligraphy paper	1, 326, 000	unknown
Yuexi			220		4-chi paper, local newspaper, simulated Xuan paper, painting and calligraphy paper	660, 000	unknown
Lihuang	6	31	133	150	Local newspaper, Zhenjiang paper, Jialian paper, imitated Xuan paper, etc.	543, 000	unknown
Shucheng	5	19	50		Local newspaper, bast paper, Huajian paper	207, 000	unknown
Lu'an	4	19	20		Local newspaper, imitated Xuan paper, slogan paper	117, 000	unknown
Huoshan	4	13	11	11	Local newspaper, bast paper, Huajian paper, etc.	72, 000	
Total						2, 925, 000	

Researchers investigating Jinsi Mulberry Bark Paper Facotry in Yuexi County

Researchers investigating Mulberry Bark Paper Mill in Mianxi Village of Shexian County (1)

Researchers investigating Mulberry Bark Paper Mill in Mianxi Village of Shexian County (2)

2 Current Production Status of Handmade Paper in Anhui Province

2.1 The Modern Production Pattern of Handmade Paper in Anhui Province: Coexistence of Recovery and Extinction

Handmade paper in Anhui has developed fast since the Tang Dynasty as a result of rich raw material supplies, the accumulation of techniques and the support of consumption. Various raw materials were employed in practice. Handmade paper not only served as the recording carrier of culture and art, but also spread to ordinary families used for letter and document, calligraphy and painting, contract and certificate, sacrificing tool and household paper.

Before the foundation of People's Republic of China, handmade paper production in Anhui included: Xuan paper for painting and calligraphy, with *Pteroceltis tatarinowii* Maxim. bark and straw grown in the sands as the raw material, mainly used for calligraphy and painting; paper made for household use, folk cultural rites employing paper mulberry bark as raw material; and bamboo paper and rough straw paper, with bamboo and grass as the raw material, used in folk custom, cultural rites and daily living. The production area sites were scattered over the rural areas in the counties of Anhui, especially in the mountainous regions of Southern and Western Anhui. Xuan paper and mulberry bark paper were not only sold all over the country, but also exported overseas. In order to spy upon the mystery of Xuan paper from the late Qing Dynasty, some industrial espionage went deep into the papermaking areas, investigated openly or secretly, and formed documents, which became the early investigating and researching literature for these paper.

In the 1930s and 1940s before the foundation of the People's Republic of China, due to political and economic instability caused by the war, transportation and inflation factors, the export-oriented paper represented by Xuan paper stopped production in large scale. However, the handmade paper production for common people wasn't subject to great influence. After 1949, with the concern of the New Government and the support of relevant organizations, Jingxian Xuan paper production recovered quickly. In the 1950s, after going through the institutional transition from private to public-private partnerships to state-owned enterprises, Jingxian Xuan paper production based mainly on Wuxi and Xiaoling, and the operation system moved forward from private mills to the state-owned factories. After the "Reform and Opening up" in 1978, Xuan paper industry in Jingxian County evolved fast. Private mills and family workshops in Jingxian County also sprung up like mushrooms. As a result, Jingxian County was recognized as "Hometown of Chinese Xuan paper" by China Association of Four Treasures of the Study. In 2002, Jingxian County was authorized as "the Original Birthplace of Xuan paper" (later on, it had been changed as "the Geographical Benchmark of Xuan paper", issued by China's General Administration for Quality Supervision, Inspection and Quarantine in 2005). In 2006, producing technique of Xuan paper was announced as the first batch of representative works of national intangible cultural heritage. In 2009, Xuan paper was included in United Nations Human Intangible Cultural Heritage List of Masterpiece.

Mulberry paper was once well-known paper in ancient Anhui. Parts of mulberry paper production in Anhui region experienced a period of collective ownership after 1949, then shrank gradually due to multiple reasons, which resulted in the extinction of production in some regions. Until the 21st century, the development of Chinese calligraphy and painting, and the efforts to intensify the restoration of ancient documents by relevant departments, stimulated new needs for handmade paper made by different raw materials. The production of mulberry paper revived again. Hence, the producing techniques of handmade paper in Qianshan and Yuexi in Anhui Province were included in national and provincial intangible cultural heritage list of masterpiece in 2008.

Certificate of National intangible heritage inheritance project plague held by Liu Tongyan, a papermaker in Qianshan County

Certificate of National Intangible Cultural Heritage Inheritor by Wang Bolin in Yuexi County

Bamboo paper making in Anhui was prosperous in agrarian society, which could be traced in Shexian County, Jingxian County, Ningguo County, Jingde County and Nanling County. While historical literature reveal that bamboo paper was produced in all the regions with bamboo. In Jinzhai, Huoshan and Shucheng in Western Anhui, old bamboo was used as raw material, different than the fresh bamboo used in Southern Anhui.

Bamboo in Jingxian County

Bamboo paper in Anhui was basically used for sacrificial purposes or as tissue paper. Literature and field investigation also fail to find any trace of high-end use of bamboo paper. Though it's possible that bamboo paper could be used for documents, with the development of society and papermaking skills, together with the emergence of machine-made tissue paper, Huoshao paper and writing paper, the traditional function of bamboo paper had rapidly been weakened, losing the major consumer market. What's more, because of the high production cost and undistinguishable appearance, and usage limited to sacrificial purposes and tissue paper, it was difficult to reverse the trend that handmade bamboo paper was replaced by machine-made one and many papermaking mills of bamboo paper in Anhui Province had quickly vanished. By the time of our field investigation in August 2017, the investigation team failed to find any handmade bamboo paper making site in Anhui Province. Current distribution of handmade paper in Anhui Province is shown in table 1.2.

Abandoned bamboo paper steaming kiln in Longma Village of Jinzhai County

2.2 The Distinctive Feature of Anhui Handmade Xuan Paper Industry in Anhui Province: Evolving Complexity

Xuan paper got a high reputation as painting and calligraphy paper among all the handmade papers in Anhui. In addition to a number of high-level exhibitions and competition honors in the late Qing Dynasty, it almost won all the honors that traditional industrial products could possibly achieve in the contemporary era. Xuan paper not only became the model for handmade paper producers, but also the target of fake products.

At the same time, as the origin and main producing area for Xuan paper, Jingxian County witnessed the establishment of Xuan paper mills with various operation systems, sizes and varieties, which sprang up like mushrooms in 1980s and 1990s, because of the advantage brought by "Reform and Opening-up Policy" in 1978. By the end of 2015, there were more than 400 Xuan paper and painting and calligraphy paper manufacturers in Jingxian County (There were 19~20 Xuan paper manufacturers, among which 16 were acknowledged as original producers. There were actually about 7 Xuan paper manufacturers when investigated in 2016. The rest were painting and calligraphy paper and processed paper manufacturers). These manufacturers yielded 1 000 tons of Xuan paper and over 10 000 tons of painting and calligraphy paper annually with the annual sales revenue of 1 billion yuan RMB approximately. Therefore, Jingxian County became the largest production area of handmade paper in China and even the world, and also the most complex area of handmade paper industry.

Xuan paper and painting and calligraphy paper enterprises in Jingxian County had strong awareness to establish the brand in ferocious competition in such a small area. By the end of 2015, Hongxing (Red Star) Xuan paper trademark produced by China Xuan Paper Co., Ltd. in Jingxian County was recognized as the famous trademark of China. The trademarks of 10 enterprises including "Wangliuji" were awarded famous trademark of Anhui Province. Xuan Paper industry in Jingxian County, led by China Xuan Paper Co., Ltd. developed various commemorating Xuan paper, which entered gradually into the high-end collection field. Xuan paper was used in printing stamps, making the inner page of the award-winning certificate of Olympic Games, paper specifically for hard-pen calligraphy and the print and copy paper that is being experimentally improved. All these trials contribute to wider application fields.

Shanghai World Expo Commemorative Xuan Paper, by China Xuan Paper Co., Ltd.

From the very beginning, Xuan paper industry of Jingxian County expanded their markets in the large and medium-sized cities of the country. In the late period of the Qing Dynasty, paper merchants from Jingxian County carved their niche in major commercial cities like Shanghai, Peking, Wuhan, Hangzhou, Tianjin and other places. Since 1949, marketing model had been transformed into government procurement and marketing, and almost all private merchants vanished then. In early 1980s, there was only one shop selling four treasures of the study in Jingxian County. By the end of 2015, there were more than 100 four treasures of the study shops, centering on Xuan paper in the local area of Jingxian County. These shops were mainly distributed in Lvbao business street, Lotus pond district and Xuan paper district, etc. Therefore, Jingxian became the county boasting great number of four treasures of the study shops.

Winter view of the lotus pond located at the central area of Jingxian County

The prosperity of the local Xuan paper business had driven the emerging of the tertiary industry in the small county which was featured by poor transportation in history. The industry chain, with the Xuan paper as its core, came into being along with the rise of Xuan paper industry. With the diversification of the recruitment system and labor market, the economy centered on Xuan paper industry in Jingxian County was booming. Local service industry including tourism, catering, accommodation and real estate all flourished accordingly.

However, the prosperity and opening-up had also led to a serious of new problems and challenges, i.e. increasing cost of salary, raw materials, products and tools year by year. What's more, prosperous markets and trade made young second or third generation of papermakers unwilling to inherit or engage in handmade paper material processing and papermaking practice. They felt more interested in selling paper online or through social networks like WeChat which could bring considerable pressure for sustainable development to Xuan paper industry.

2.3 Folklores, Customs and Taboos Related to Jingxian Xuan Paper Industry

2.3.1 The Folklores about Xuan Papermaking Skills

1. The Legend of Kong Dan Inventing Xuan Paper

According to a folklore, tradition has it that

in the last years of Eastern Han dynasty, Kong Dan, the apprentice of inventor of paper Cai Lun, wanted to make high-quality paper to draw the portrait of his deceased master, but failed. One day, he accidentally found a *Pteroceltis tatarinowii* Maxim. tree falling alongside a stream, the bark of which blanched by the soaking and washing of the water. He came up with a new idea, and attempted to use the bark to make paper. After repeated tests, he finally made a kind of new paper—white soft Xuan paper.

Based on the legend, in the beginning phase of papermaking practice, the skills were not perfect, so the workers had to scoop one piece of paper and dried it one time, which was labor and time consuming. One day, an old yet healthy man with a crutch came to the papermaking shed and said: "Why do you youngsters look so worried?" Kong Dan found the old man familiar, whose appearance showed some extraordinary background, but could not figure out where they had met before. So Kong Dan replied truthfully: "Master craftsman, we are worrying about the papermaking." The old man then continued: "You can tell me your difficulties, I have some knowledge about papermaking." Kong Dan replied: "To tell you the truth, the wet paper can't be overlapped, or it will be inseparable. We had to scoop one and dry one at a time, which is inefficient. How to solve the problem? Do you have any suggestions?" On hearing this, the old man laughed and said: "Piece of cake!" Then, he used his crutch to stir the pulp, and said: "Ok, you can try again." Kong Dan and his fellow workers made a stack of wet paper, pressed, and then the pile of paper could be split apart easily.

It was said that the old man who had never been heard of since the successful test was the spirit of grand master Cai Lun, who showed up to offer some help. Legend had it that "single-layer Xuan paper" was actually in honor of Kong Dan and the local papermaker in Jingxian County called it "Dan Xuan". For example, 4-*chi* single-layer Xuan paper could be also called 4-*chi* "Dan".

2. The Legend of Luwang Large-size Paper

It was said that over a hundred years ago, Jingxian County produced a kind of large-size paper, called Luwang (Luhuang), commonly known as 6-*zhang* Xuan paper. It was the king of paper due to its large size, and produced only once every summer. Once, the workers were asked to make such paper due to special needs. Though hard efforts were needed, in order to show respect for the boss of the paper mill, and earn more money, they managed to make such paper against tradition.

One day, an old man with white hair and beard leaning on a wooden crutch appeared suddenly in the mill, glared and disappeared instantly. The workers thought the spirit of grand master Cai Lun appeared, then they set off firecrackers and offered sacrifices with eaglewood and offerings. After that, they dared not make the large-size Xuan paper against tradition. In 1964, when Jingxian Xuan Paper Factory was making 6-*zhang* Xuan paper, an anxious veteran worker burned joss incense and kowtowed to Cai Lun secretly, but got banned as feudal superstition.

2.3.2 Cai Lun Convention and Cai Lun Temple

1. Function of Cai Lun Convention and Worshipping Cai Lun

In Xuan papermaking industry in Jingxian County, the spirit of Cai Lun Convention influenced the whole industry, which can be reflected in three aspects: Firstly, every year, on the 16th day of the third month of Chinese lunar calendar (the birthday of Cai Lun legendarily), the whole industry halted production. The proprietors led the papermaking workers, carrying offerings, joss candles and incense to worship Cai Lun in Cai Lun Temple in Xuwan, Xiaoling Village. Secondly, every year, on the 18th day of the ninth month of Chinese lunar calendar (the anniversary of Cai Lun's death legendarily), they also halted production. The proprietors would feast all the laborers and asked for their opinions on the papermaking mills, the salary, the food, etc. And most papermaking workers, who practiced papermaking near Xuwan, burned incense and worshipped Cai Lun in Cai Lun Temple. Thirdly, the proprietors gathered spontaneously, discussing the issues proposed by the papermakers, such as labor and salary, material purchase and product prices, etc. Xuan Paper Trade Union was founded after the Republic of China, and the above-mentioned activities were convened by the head of the union. In the middle and late 20th century, due to the impact of the Great Cultural Revolution, activities worshiping Cai Lun and Cai Lun Convention were temporarily interrupted. In the early 21st century, from 2016 to 2017, the Dingqiao Township Government of Jingxian County and the Caos papermaking family in Xiaoling revived the activities of worshiping Cai Lun and the family ancestor Cao Dasan, and erected a new statue of Cao Dasan.

Portrait of Cao Dasan in *Genealogy of the Caos* (old version)

Cai Lun Convention unites all the Xuan papermaking practitioners so that they form a relationship of counterbalancing and promoting each other. In an interview in 2016, the research members were informed that Cai Lun Convention seemed to fail to form direct constraint to the employees. For example: the proprietors should worship Cai Lun within the fixed time, praying for the blessings of the ancestor of papermakers in the New Year. While the workers only voluntarily worshipped before the dinner on New Year's Eve every year, putting the worshipping tablets of Cai Lun, god of earth and ancestors, etc.

According to the literature research and interviews, the following typical customs in worshiping Cai Lun by the traditional Xuan paper industry are also included:

(1) Generally, when the papermakers encountered technical problems, what first came into their minds was to worship Cai Lun. For example, after practicing papermaking for a period of time, the workers may encounter a certain ceiling in improving his proficiency and skills, similar to the voice change period of boys in their puberty. To overcome the difficulty, they naturally thought of worshipping Cai Lun which was supposed to be personal and should be done secretly, usually in the evening, at home or alone somewhere. They burned joss paper varing in amount, and incense while pleading for the blessing from the papermaking ancestors, and after the secret ceremony they would kowtow to show their gratitude.

(2) On the occasion of making special varieties of Xuan paper, the proprietor would lead the workers to place the incense burner table, burn joss incense, kowtow and burn joss paper, which was a grand event. Usually the proprietor didn't appear at all, while the overseer was in charge of the production details as they were told by the proprietor, e.g. what kind of paper to produce, and how much to produce. Only when ascertaining the production of a new variety, the proprietor

would show up to worship Cai Lun. During the late Qing Dynasty and the Republic of China, Xuan paper industry was affected by the war, so no new variety of Xuan paper was innovated, and there were almost no large-scale sacrificial ceremonies by the workshops because at that time almost no new type of paper was developed as the consequence of the wars. But the custom was kept, thus forming the conscious routine in the industry. In 1957, when Jingxian Xuan Paper Factory resumed the production of 6-*zhang* Xuan paper (Luwang large-size paper), some workers burned joss paper to worship Cai Lun at home spontaneously due to the difficulty in making such large size paper, though such rites were banned at that special time.

2. Cai Lun Temple

According to the experts of Xuan paper and local history intellectuals, Cai Lun Temple in Jingxian County was built in the Ming Dynasty, which was located in the east of Doushi Nunnery on Shentan mountain, Xuwan group of Xiaoling Village. The temple was of brick-wood structure, with hall and attached house, covering an area of about 300 m^2. The statue of Cai Lun was consecrated in the hall and the temple was surrounded by bamboo. There were high pine trees and *Pteroceltis tatarinowii* Maxim. trees around the stone steps of the gate, the stream gurgling and the cloud lingering. When spring came, dozens of peony in the east of the temple bloomed and the fragrance spread over the valley. Cai Lun Temple was rebuilt in autumn of 1935, and the stone monument "Rebuilding Records on Cai Lun Temple" was set up. The descendants of the Caos who were engaged in Xuan papermaking, would halt production on the 16th day of the third month of Chinese lunar calendar (the birthday of Cai Lun legendarily) every year, and the proprietor would lead the papermaking workers, carrying offerings, joss candles and incense to worship Cai Lun. Cai Lun Temple was damaged as the "four olds" in "the Great Cultural Revolution". Only the inscription on the monument "Rebuilding Records on Cai Lun Temple" survived, elaborating on the rebuilding process and significance of the event.

2.3.3 Work Songs

There were some work songs during Xuan paper production, which not only helped to spur strength, but also bring uniform action. The typical one was "Counting Sticks", or "Rowing Sticks", which was one of the inevitable procedures for traditional pulp making of Xuan paper. After the pulp making procedure was undertaken by beating machines in 1958, this procedure gradually faded out of people's sights. This procedure involves blending the bast pulp and straw pulp after beating, pestling and washing, and making the cooked materials. "Counting Sticks" entails the cooperation of four to five people, whose movements should be consistent and uniform. In order to coordinate the movements, the head worker sang the work song.

Work song: Stick Song

2.3.4 Language Varieties Leading to Different Papermaking Expressions

The local dialect used in Jingxian County was Jingxian dialect, which belonged to Xuanzhou Wu dialects as a branch of Chinese. The relationship can be depicted as: Sino-Tibetan family→ Chinese→ Wu dialect branch→ Jingxian dialect. Because the area borders Wu dialect, Hui dialect and Jianghuai Mandarin, there are great differences between eastern and western Jingxian dialects. The county citizens called Jingxian dialect spoken in eastern region as "Dongxiang dialect", the western "Xixiang dialect". Xixiang dialect integrated many morphemes and phonetics of Hui dialect. It was a little bit difficult for the two kinds of dialects to communicate with each other.

The Relics of Cai Lun Temple (Only One Temple Left)

Traditionally, Xuan papermaking skills were mainly inherited in Xixiang, Jingxian County, while rarely practiced in Dongxiang. The inheritance of Xuan papermaking skill in ancient times nearly had no written records, even in local records and township literature of all ages. Around the foundation of People's Republic of China in 1949, some sporadic records gradually appeared. In terms of expression, the differences between Dongxiang and Xixiang dialects resulted in the differences of expressions concerning the processing equipment of Xuan paper. For example, the papermaking screen was called "Lianchuang" in Dongxiang, while "Liancao" in Xixiang. When the tip part was thick in papermaking, Xixiang called it "zuoshao", while Dongxiang called it "tuoshao" with different tones. These differences also enriched the culture connotation of Xuan papermaking skill.

2.3.5 Traditional culture of respecting and cherishing paper

Respecting and cherishing paper shows the tradition of respecting knowledge and its carrier in ancient Chinese local culture. It was once a very typical custom in Jingxian County based on our investigation. For instance, paper which had been written on should all be burnt instead of being thrown away. Some places set up "Cherishing Paper Furnace", which was about two metres high and had three layers. Where there was "Cherishing Paper Furnace", the written paper should all be sent into the furnace and burnt down. According to the interview of the local elders by our team, by 1950s, "Cherishing Paper Furnace" of various shapes and sizes was set up in Maoling, Nanrong and Chengguan districts in Jingxian County. However, all of them were destroyed in "the Great Cultural Revolution" period.

2.3.6 Customs Concerning Papermaking

Jingxian County, known as "land of civilization and intellectuals", advocates simple local culture. Except for traditional weddings, funerals and seasonal customs, the place shelters some paper-specific customs such as *Guohui*, Spring Reception and Year-end Reception, etc.

1. Spring Reception

Spring Reception was basically Kickoff Meeting, a unique activity in Xuan paper industry. After the Xuan paper production operation became state-owned in the middle of 1950s, this custom disappeared. Interestingly, this kind of custom was rarely recorded in the local chronicles or literature. The interviews of our survey recovered certain information:

(1) The Form of Spring Reception
The time for holding Spring Reception was decided by each Xuan paper proprietor. Usually it fell on the first day of papermaking after the Spring Festival, the first day of papermaking practice. On this day, the owners held a feast to entertain all the papermaking workers and one person would be invited from other households who were not involved in papermaking practice. Before the feast, the owners should lead the workers to worship Cai Lun. Different forms of worshipping were adopted. The fancy way was to set the incense burner table in the hall, and led the workers to burn the incense, and kowtow. After the ceremony, they would burn the joss paper and set off the firecrackers before the feast. While for regular owners, they would just set off the firecrackers directly to start the feast, after burning joss paper outdoors.

(2) The Purpose of Spring Reception
The main purpose of Spring Reception was to announce the start of the new year, wishing the business flourishing, with the efforts of all workers, with the help of all village fellows and the protection of Cai Lun. Each family would send one member to the Reception so the practice would get help and blessings.

In the past, Xuan paper practice faced two challenges: fire and water. Papermakers took water from the valley stream. While the water from the mountain stream was apt to dry up in drought season. In such case, the papermakers managed to get water from the remote mountains using split bamboo. The workers chopped the bamboo in half, got rid of the bamboo joints, then connected the bamboo sections one after another to lead water into the papermaking mills. Firewood was the necessary fuel in old times when transportation was poor, with no electricity or coal in the past. Firewood was the fuel for drying paper, which was consumed thousands *jin* a day. Therefore, the owners was mostly fearful of the subversion by the village fellows, who would set fire to burn down the firewood or raw materials, or take away some sections of the bamboo for getting water. It's always too late before the owners of the papermaking mills found out and the quality of paper made had already been destroyed. As a result, the worried owners would take advantage of Spring Reception to win over the neighbors' support for the new year.

Bamboo equipment for transporting water

2. Year-end Reception

The annual "Year-end Reception" is also one of the local Xuan papermaking customs. In the middle of the 1950s, the custom disappeared when the papermaking operation became state-owned. Similar to "Spring Reception", the investigation team could not find any written descriptions in the local documents, which were recovered mostly through our interviews.

(1) The Form of Year-end Reception
Year-end Reception was a ceremony held by the owners at the end of the year before the annual leave, usually no later than the 23rd day of the twelfth month of Chinese lunar calendar. The so-called tradition "No matter the long-term hired hands or casual laborers, they all should cease production so as to enjoy the holiday no later than the 23rd day of the twelfth month of Chinese lunar calendar." On this day, the owners treated all the workers a feast. During the feast, the owners and the workers communicated and summarized their accomplishments of the whole year. The owners may also tell the workers the working schedule for the next year and the start date after Spring Festival, and the workers could speak freely.

Worshipping Cai Lun was the fancy way to start the feast, and the way of worshipping was similar to that of Spring Reception. Other owners just went directly to the business: the feast itself. After the feast, the owners or the workers may have personal conversations with some workers about their work contract. For example, if an owners wanted to fire a worker, he would say to him in private: "I may have to cut production next year, so I may not need so many hands." The worker naturally knew what the boss meant and would take the initiative to resign. Another situation was that the workers themselves did not want to continue. They said to the boss, "My family had a relative and I had to work for him in the coming year." If the owner wanted to keep the worker he would say: "Please let me know if I didn't do so well last year. I promise I will improve next year." Or say, "New Year's coming and bring some meat as the gift home." If the worker wanted to stay, he would tell the boss directly if there were space for improvement or accept the gift from the boss and explain his willingness to stay for the next year.

(2) Purpose of Year-end Reception
Year-end Reception actually was the conclusion of a year, illustrating a harmonious relationship between the owners of the papermaking mills and the employees. In the interview, some old workers memorized that after finishing the ceremony of Year-end Reception, some poor families in the village would pay a visit to the owners, and said "Boss, I want to sell you some fire woods after the Spring Festival" or "Boss, I want to sell you some bark after the Spring Festival." Then the owners would take out two pieces of silver coins wrapped in red envelopes as a gift to the visitor. Otherwise, the papermaking mills might suffer a fire, which was unquestionably related to those family members of the visitor whose real purpose was an intimidation, while two silver coins would suffice to guarantee the visitor's family to celebrate the Spring Festival and the saftety of the papermaking mill. After the festival, the visitor's family would bring some firewoods or bark, symbolically, or in case they didn't offer, the owners would not mention either and just let it go.

3. Idioms in the Hometown of Paper

Through the investigation, the local idioms can mainly be classified as follows:

Firstly, idioms about the papermaking practice. In the hometown of paper, the Xuan paper mill was called "the paper shed", and the workers making Xuan paper were called "Penghuazi (the shed beggars)".

Secondly, idioms about papermaking skills, such as: "First time pouring water along the side, second into the center; First time pouring water loudly, second gently." "Loosening the tip, tightening the head." "People who lift the screen should be flexible, while those holding the screen should keep steady", etc.

Thirdly, idioms about the papermaking life, such as: "The paper cutter dressed more formally than the scooper; people who dried the paper dressed least formally." "A real man should not make paper, for he couldn't afford to have a wife or support a family."

3 Preservation, Inheritance and Current Research of Handmade Paper in Anhui Province

3.1 Characteristics of the Existing Resources of Handmade Paper Industry in Anhui Province

Compared to other provinces in China, Anhui handmade paper industry outshines but is still far from being optimistic. The investigation team of *Chinese Handmade Paper Library: Anhui* conducted the field work and literature survey many times from 2013 to 2017. From the data acquired, the successful handmade paper industry represented by Xuan paper and painting and calligraphy paper in Jingxian County to some extent diluted the challenges from the unbalanced distribution and rapid extinction of papermaking practice in many areas. The main characteristics were extracted as follows:

3.1.1 Handmade Paper Practice Outshines All—Top Output being the Typical Characteristic

In the agrarian society, the handmade paper practice in Anhui region was widely distributed. During the late Qing Dynasty and the Republic of China, there were handmade paper production information recorded in most county annals of Anhui area. Liu Renqing, an expert of papermaking from Beijing Technology and Business University, started to study Anhui handmade paper represented by Xuan paper in 1960s, who had conducted thorough survey of Anhui handmade paper production on the eve of "The Great Cultural Revolution". While the information acquired by our team from 2015 to 2017 was: apart from Jingxian County where papermaking practice assemble, handmade paper production could only be found in Huangshan District (the former Taiping County), Qianshan County, Yuexi County, Chaohu City (county-level city), Xiuning County and Shexian County. In each place, there was almost only one single mill in operation, and the assembled form of production was no longer seen.

In all the handmade paper production areas, the intensive Xuan paper and painting and calligraphy paper production status, represented by Jingxian County, not only

Dyed joss paper in a bast paper mill of Shexian County

almost overwhelmed the industries and brands in the whole Anhui region, but also became the national benchmark of papermaking for artistic purposes. During the peak period, the handmade paper produced in Jingxian County occupied half of the market nationwide. Till the beginning of 21st century, the output had shrunk to around 30 000 tons in mainland China (excluding Taiwan Province). While the annual output of Jingxian County still reached more than 10 000 tons, remaining the largest handmade paper production region in the whole country. This was a typical dominant phenomenon in Anhui handmade paper industry.

3.1.2 Diversity, Richness and Complexity of the Industry

In all kinds of handmade papers in Anhui region, paper made from bark was the representative one, such as Xuan paper in Jingxian County, mulberry bark paper in Qianshan and Yuexi, and mulberry bark paper in Xiuning, Shexian of Huangshan. However, the actual largest output was the painting and calligraphy paper, with *Juncus effusus* (board) as the dominant material. Field investigation of 16 provinces in China reveals that the varieties of Xuan paper and painting and calligraphy paper in Jingxian County were rich and complicated, which is mainly reflected in the following aspects:

1. Abundant Varieties of Xuan Paper

Xuan papermaking skills were passed down in Jingxian County for generations. Because of the different ratio of two kinds of raw materials: *Pteroceltis tatarinowii* Maxim. bark and the straw grown in the sands, together with various size, thickness and vein patterns, the varieties of Xuan paper became abundant.

In terms of the traditional classification of Xuan paper, there were four mature ways based on raw material, size, vein pattern and thickness. However, these four classification ways were not invariable. Before 1950s, paper could be categorized as straw paper, bark paper and hemp paper, according to raw materials; as single-layer paper, double-layer paper, and triple-layer paper, according to thickness; as 4-chi paper, 5-chi paper, 6-chi paper, 7-chi paper, 8-chi paper, 1-zhang-2-chi paper, Duanshan paper, Changshan paper, etc. according to size; as single-thread paper, double-thread paper etc. according to vein pattern. After the 1950s, the varieties of hemp paper, Duanshan paper, etc. gradually faded out of the market and the classification methods were adjusted. For example, the paper products could be categorized as straw paper, bark paper, and superb bark paper, according to raw materials; as single-layer paper, double-layer paper, and triple-layer paper, according to thickness; as 4-chi paper, 5-chi paper, 6-chi paper, 8-chi paper, 1-zhang-2-chi paper, 1-zhang-6-chi paper, 1-zhang-8-chi paper, 2-zhang paper and 3-zhang-3-chi paper, and various other specifications as well, according to size; as single-thread paper, double-thread paper, ribbing-patterned Xuan paper etc. according to vein pattern. The four classification indicators jointly differentiate hundreds of varieties of Xuan paper nowadays, and made it hard for average consumers to distinguish clearly.

2. Mingled Hope and Fear for Xuan Paper and Painting and Calligraphy Paper

In 2002, after obtaining the protection as the unique origin of Xuan paper (geographical benchmark brand), Xuan paper and painting and calligraphy paper mills in Jingxian County developed rapidly. But overall planning and regulation among enterprises, enterprises and brands, enterprises and technicians were lacking. Through several rounds of field work and literature research by our team from 2013 to 2017, and the interviews of many experienced practitioners and researchers from handmade paper industry in Jingxian County, the following four aspects about current situation and development were highlighted:

First was the high-degree regional gathering and rapid increase of the number of production enterprises. By the end of 2015, there were more than 400 Xuan paper, painting and calligraphy paper and processed paper enterprises in Jingxian County (in which there were 16 authorized Xuan paper enterprises, around 7 Xuan paper enterprises in practice when surveyed in 2015), with annual output of 950 tons of Xuan paper, annual sales income of 300 million yuan; and annual output of 10 000 tons of painting and calligraphy paper, annual sales income of 400 million yuan. As for the processing industry of Xuan paper and painting and calligraphy paper, the annual sales income was 100 million yuan. In 2015, the annual sales income of the handmade paper industry of Jingxian County was about 800 million yuan.

Second, the industry lacked inheritors though in prosperity. The handmade paper production was a hard, dirty and tiring work, featured high labour intensity, long-period skill learning, high technical requirements and short career life. The papermaking workers are prone to suffer from occupational diseases such as herniated disk, lumbar muscle degeneration, arthritis and skin disease, etc. Therefore, it's hard both to keep the skillful workers and to attract the youth to engage in the papermaking practice. With failed attempts in many ways, the prosperous industry could not amend the crisis of lacking skilled workers.

Third, the supply of high-quality raw materials for Xuan paper couldn't be guaranteed. Due to the overwhelming substitution of high-yield variety of hybrid rice, plant of traditional local long-stalk rice grown in the sands almost disappeared. It was quite uneasy to get the long-stalk rice straw now. Although the *Pteroceltis tatarinowii* Maxim. bark material base was gradually built up, the processing methods hadn't changed fundamentally. Together with the shortage of labours and high costs of materials, the producing cost of finished products of Xuan paper remains high.

Finally, the industry self-discipline mechanism failed to be constructed. The first aspect was the independent operation of Xuan paper and painting and calligraphy paper production enterprises. The coordination for the purpose of mutual development among enterprises was not achieved. The second aspect was the challenge from fake Xuan paper and brands. Painting and calligraphy paper was sometimes sold as Xuan paper, labelled as national intangible cultural heritage and human intangible cultural heritage. The third aspect was the emergence of OEM (Original Equipment Manufacturer) marketing. The cheap painting and calligraphy paper was purchased from Jiajing and Meishan, Sichuan Province, then sold outside as Xuan paper from Jingxian County, which had disturbed the market order to some extent and affected the normal operation of Xuan paper brand in Jingxian County.

3. Abundance and Chaos of Processed Paper

The ancient data revealed that if written on the paper directly, the ink and watercolor would blur. The reason was that, no matter what kinds of material or skills were used to make the handmade paper, space among the paper fibers just disperse the water and ink. Therefore, people in Jin and Tang Dynasties generally adopted processing technology to reduce the gap between fibers, so as to avoid the blurring of ink.

At first, people used smooth stone to rub the surface of the paper, so as to tighten the space, which was called calendering in ancient times. Modern people call it "polishing", as an initial processing procedure of papermaking. Later on, people used wooden mallet to beat wet paper repeatedly so as to press the space tightly among the paper fibers. The more effective measure was sizing. The initial material for sizing was starch, then evolved to gelatine, which made the paper more water-repellent.

There are more than 60 varieties of processed Xuan paper products currently, e.g. Jade-white Xuan paper, Super-thin Xuan paper, Snow-white Xuan paper, Impregnated Xuan paper, paper decorated with golden spots or silver spots, etc. The processing techniques include waxing, calendering, sizing (desulfonylation), spraying gold foil, spraying silver foil, painting and printing, etc. which involve the different procedures to improve the performance of the unprocessed paper.

During our survey, the above mentioned processing methods were very popular in

Dyed processed paper

Processing workshop of Gold-sprinkled Xuan Paper in "Yixuange" of Jingxian County

Anhui Province, especially in Jingxian County. In Jingxian County, all sizes of processed paper enterprises emerged in an endless stream. According to the statistics of survey by our team at the beginning of 2016, there were hundreds of processing enterprises. Their operating characteristics were that, enterprises of a certain scale usually had both the factory names and brand names, as well as the independent ability of research and development. While the small ones just imitated those processed paper skills dominating the market.

After being processed, the features of original paper had been completely or partially concealed. Therefore, in Jingxian County where processed paper production gathers, imperfect self-discipline mechanism and the domination of low-end processed paper products in the market lead to a new phenomenon of reprocessing machine-made painting and calligraphy papers or lowest-end handmade purchased from Jiajiang, Hongya and Meishan in Sichuan, and then sell the paper as Xuan paper, which resulted in more chaos of processed paper market.

Despite the above mentioned chaos, the richness of processed paper products from Jingxian County could still be described as being prosperous. According to the information acquired in the interviews by our team, from 2015 to 2017 Jingxian County boasts a series of influential papermaking enterprises in China, and the representative ones include Yixuange, Huixuantang, Yiyingxuan, Qinmotang, Fenghetang, Hanyunxuan, Yuwenxuan and Yunlongxuan.

Based on information obtained in our survey, among the processed paper production enterprises in Anhui Province, some are famous for their high-end products, such as Duoyingxuan (Calligraphy and Painting Supplies Co., Ltd.) in Huanglu Town of Chaohu City (county-level city); and some are famous for their large output, such as Yixuange in Jingxian County. As representatives of exellence, they not only created their own independent brands, grasped the unique processing techniques, but also enjoyed high reputation in the industry and art circle as the most representative processed paper production enterprises in the country during our survey. Among them, Duoyingxuan in Chaohu City was the exclusive protection unit in the second batch of national intangible cultural heritage protection project under the entry of "paper processing skill", and it was enlisted among the national masterpiece in 2008. According to our field work in March and August, 2016, various kinds of processed paper were produced by Duoyingxuan, such as Fenla paper, silk Xuan paper, Nijin paper, water print paper and paper decorated with gold and silver spots. Fenla paper was the one worthy particular focus. As a kind of rare processed paper, it appeared in China as early as the Tang Dynasty. The way to make the paper in Duoyingxuan is: with Xuan paper as the raw material, process the material through several complex procedures of dyeing, applying powder, waxing and calendering, gold and silver plating, decorating with golden or silver designs. Fenla paper combined hydrophilic powder and hydrophobic wax, exquisite and smooth, which was the high-quality treasure for practicing painting and calligraphy.

The inheritance protection of "Xuan Paper Products Processing Technique" was undertaken by Yixuange Xuan Paper Craft Co., Ltd. in Jingxian County, which was enlisted into Anhui provincial-level intangible cultural heritage list of masterpiece in 2014.

Powder wax paper decorated with authentic gold in "Duoyingxuan"

A researcher experiencing papermaking process in "Yiyingxuan"

4. Growth and Decline of Bast Paper and Bamboo Paper Production

In Anhui region the handmade paper traditionally gave priority to bast paper, while bamboo paper was also popular. With the radical change of the consumer market, and the development of handmade bamboo paper industry in Zhejiang and Sichuan, bamboo paper production in Anhui Province declined rapidly. The information obtained in the survey from 2015 to 2017 showed that: In the mountain area such as Jinzhai County in the hinterland of Dabie Mountains and Shexian County, small amount of low-end usage bamboo paper practice still existed, but lingering on closing. There was active production in Gufeng Village, Changqiao Town, Jingxian County twenty years ago; while during our survey in 2015, only the ruins of the past mill left. And there was no more information about bamboo paper still in production in other counties.

Interviewing old papermakers in Qingfeng Village of Shexian County

Mulberry bark paper in Western Anhui was traditional handmade paper with a long history, whose representative gathering producing sites include Qianshan County and Yuexi County. To the end of the twentieth century, the practice declined sharply, on the verge of extinction. From 2003 to 2004, because the Palace Museum was in dire need of the Koryo Paper to restore the large ensemble murals in Emperor Qianlong's Lodge of Retirement, they investigated and compared mulberry paper of different origins in China and in South Korea, and finally confirmed that mulberry paper in Qianshan and Yuexi was most similar to the paper for the past royal use. Thus they made an order, which was repeatedly reported by the media and brought the recovery of mulberry paper production in Qianshan and Yuexi, and made the reputation of the mulberry paper in the two counties roar. In June, 2008, "mulberry paper making technology" was chosen into the second-batch national intangible cultural heritage protection list. In June 2009, Wang Bailin from Yuexi County and Liu Tongyan from Qianshan County won the title of second-batch national intangible cultural heritage representative inheritors. During the field work in Tanfan Village, Qianshan County in October, 2015 and Banshe Village, Yuexi County in March, 2016, only one papermaking mill in each county was in operation, run by the two national inheritors respectively. It was a proof that although the brand of mulberry paper had enjoyed a reputation, the inheritance status actually was quite frail.

The Souvenir for Restoration of "Juanqinzhai", Emperor Qianlong's Lodge of Retirement

In addition to Xuan paper in Jingxian County and mulberry bark paper in Western Anhui, bast paper also enjoyed a long history in Anhui Province. As early as the Three Kingdoms Period, the paper mulberry bark paper in the southern area of Anhui was quite famous. About 1 800 years ago, Lu Ji from Wu State in the Three Kingdoms Period recorded in his *The Annotations on Animal and Plant Life* mentioned in *The Book of Songs*: "Broussonetia papyrifera, also called paper mulberry tree, was named differently by people from different places. People in south of Yangtze River twisted the bark to make cloth, pestled it to make paper, of several-*zhang* long, white and bright, fine-textured. Fresh leaf of which could be eaten." During the Tang and Song Dynasties, both the well-known Xuanzhou paper and ancient Hui paper and Chi paper, used the paper mulberry bark as main raw material.

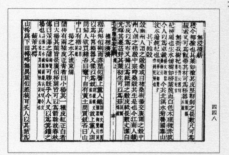
A photo of *The Annotations on Animal and Plant Life* mentioned in *The Book of Songs*

During 2015 to 2017, paper mulberry bark papermaking industry was distributed mainly in Jingxian County, practiced by several bast paper manufacturers, such as Shoujin Bast Paper Factory, Chixing Bast Paper Factory, and Xiongfeng Bast Paper Factory, etc. Some manufacturers made bast paper periodically while producing Xuan paper and painting and calligraphy paper, such as Mingxing Xuan Paper Factory, Jinxing Paper Factory and Yuquan Paper Factory, etc. Two mills in Qianshan and Yuexi counties that usually used paper mulberry bark as raw materials also made bast paper or paper using mixed materials. It's worthy to note a mulberry bark paper mill in Jiang Zushu's family of Mianxi Village, Shendu Town, Shexian County, which had at least hundreds years of history for producing pure paper mulberry bark paper. During our survey, it is revealed that from 1970s to 1980s, this production was once an important source of livelihood for almost every family in the village. Bast paper made in Mianxi Village used the local paper mulberry bark as the raw material, which was generally used for packaging paper. During our survey, the paper was mainly sold to Jinhua in Zhejiang Province to make mosquito-repellent incense paper.

Growing environment for local paper mulberry tree in Mianxi Village of Shexian County

Apart from paper mulberry bark, bark of *Edgeworthia chrysantha* and *Wikstroemia pilosa* Cheng was also used in places such as Jingxian County, Qianshan County and Yuexi County. Oriental Paperbush bark paper and small amount of *Wikstroemia Pilosa* Cheng bark paper could also be found. A special example was Sanxin Paper Co., Ltd. in Xiuning County. Originally it was moved from Taicang of Suzhou to Xiuning County of Huangshan, the headstream area of Xin'an River in 2002. The reason recorded in our investigation was that the Japanese customers wanted to seek better-quality water source to produce better paper. Sanxin used bark of *Wikstroemia pilosa* Cheng as raw material because they made Washi paper exclusively. This is a special case of contemporary development.

Upper Xin'an River near Sanxin Paper Co., Ltd. in Huangshan City

In addition, our investigation from 2014 to 2016 revealed that a special periodical production system was established from 1988 to 1999 when Shexian Four Treasures of the Study Company invested and registered a handmade paper mill with "Chengxintang" as the brand, which produced Xuan paper under the guidance of technicians from Xiaoling Xuan Paper Factory. The mill was located near the new site of National Tax Bureau of Shexian County and possessed six papermaking containers. Most products were processed paper. The mill had the watermark processing workshop and was able to make watermark, gold foil, dyeing, etc. Before the mill was established in 1988, papermaking workers had ever learned papermaking skills in Yangcun Village of Mianxi Town for a short period of time. In 1999, the mill was closed after enterprise restructuring, and the stock paper was sold to South Korean businessmen in bulk.

3.1.3 Severe Challenge for Handmade Paper Production in Anhui Region

As other areas in the country, the inheritance of handmade paper in Anhui region faced severe challenge, which was reflected in the following aspects:

(1) The impact of small-machine-made paper on handmade paper was profound, esp. in Anhui region. The first wave of shock came from the rapid rise of western mechanical wood-pulp machine-made paper during the late 19th century and early 20th century, which caused the mass extinction of widely-distributed handmade paper production sites in Anhui region. Although this was a common national phenomenon, the challenge led to the perishing of 2/3 papermaking mills in Northern and Middle Anhui. The second wave of shock came from the machine-made bamboo paper in 1990s. Although it was acknowledged that handmade bamboo paper was of better quality and was more familiar to local people, the lower price still forced handmade bamboo paper out of Anhui historical stage. The third wave of shock was the rise of mechanized painting and calligraphy paper from Jingxian in Anhui, Jiajiang in Sichuan, Fuyang in Zhejiang and Anyang in Henan, during the late 20th century and early 21st century. They quickly dominated the low-end painting and calligraphy paper market with low prices resulting from the use of cheap straw or bamboo board as pulp resource. Common consumers were not able to distinguish the quality of paper, which allowed the middle and high-end painting and calligraphy paper to be driven out of the market and the rapid decline of the production of real painting and calligraphy paper.

(2) There was a shortage of front-line papermaking practitioners. Because the regional economy of Anhui lagged behind that of other areas in Middle and East China, from 1980s to 1990s, at the first stage of "Reform and Opening up", Anhui had a large number of labor export, which brought the first challenge to the labor force of handmade papermaking industry, and caused the sharp decline of output of handmade paper at that time. After the early 21st century, due to the population stagflation and urbanization, as well as the improvement of economy in Anhui region and improved transportation, the employment opportunities were expanded quickly in the local area. Some conventional industry personnels who used to be farmers found new working opportunities which caused the insufficiency of labor force in traditional local industry. The handmade paper industry was the first to be affected for being practiced in remote areas, taking long-period learning, and being labor-consuming. The interview from 2015 to 2016 revealed: Such labor-consuming procedures in papermaking like scooping paper pulp, drying paper and processing raw materials failed to appeal to the youth and faced the extinction of the skills. In Jingxian County, the average age of front-line handmade paper practitioners was above 45 years old which promised no optimistic future. Similar situation also occurred in mulberry bark paper industry, in which no worker aged below 40 was found still practicing papermaking in Qianshan and Yuexi counties.

(3) The basis of traditional handmade paper practice has evolved. For example, in terms of raw material, due to the promotion of fine varieties of rice, the planting of long-stalk straw grown in the sands, as the core material of standard Xuan paper making, had been gradually reduced. While another core material *Pteroceltis tatarinowii* Maxim. bark had been increasingly out of supply although the local government had taken measures to support the planting by planning planting bases. Lack of bark supply due to the high price also resulted

from the lack of labor force.

(4) Considering the economical efficiency and working-hours benefit, various modern mechanical equipment and chemical products were replacing the original processing equipment and materials, which put Xuan paper traditional skills which were included in human intangible cultural heritage preservation list, under pressure and call for urgent preservation.

(5) Various fake "Xuan paper" or low-quality "Xuan paper" from Jingxian County, Jiajiang of Sichuan, etc. flooded into the market through the network and Wechat, which was sold at a lower price than that of the authentic Xuan paper, and brought pressure on traditional Xuan paper industry. In addition to the sharp fluctuating shrink of painting and calligraphy paper industry market from 2014, and the poor management of papermaking industry have led to, among the 16 Xuan paper national geographical benchmark enterprises, two Xuan paper mills producing the famous brands being closed when our team was investigating in Jingxian County at the beginning of 2016. A series of Xuan paper manufacturers had turned to produce other kinds of paper and would only produce Xuan paper when they obtained such order. Six to seven Xuan paper mills who persisted in producing Xuan paper had been trapped in hard situation.

3.2 Preservation and Research of Handmade Paper in Anhui Province

3.2.1 The Coordination of Macro-policy Promotion, Institutional Construction and Community Participation

1. Existing Laws and Regulations for Protection

(1) *Convention for Protecting the Intangible Cultural Heritage* (hereinafter referred to as *Convention*). In 2009, Xuan paper making skill was selected into "human intangible cultural heritage list of masterpiece" announced by UNESCO (United Nations of Educational, Scientific and Cultural Organization). The preservation of Xuan paper making skill had been included in the protection by all the humankind.

(2) On 21st August, 2014, the 13th session of the Standing Committee of the 12th People's Congress of Anhui Province passed *Regulations on the Protection of Intangible Cultural Heritage*, (hereinafter referred to as *Regulations*) which came into force on 1st, October of the same year. *Regulations* was promulgated and implemented in the whole province officially as local *regulations*. Compared with some provinces who promulgated *Regulations* earlier, the characteristic of implementation way of Anhui Province was that it was in effect after the preservation environment of intangible cultural heritage became more mature. And the main goal was to further ascertain the intangible cultural heritage preservation in the regulation level.

(3) On 31st May, 2013, CPC Anhui Provincial Committee and the People's Government of Anhui Province released *The Outline of Constructing A Powerful Cultural Province*, explicitly arranged special funds to support the work of intangible cultural heritage protection in Anhui region.

(4) In 2005, the hometown of Xuan paper—Jingxian County—promulgated and implemented *Action Plan to Promote Xuan Paper and Xuan Writing-brush Industry*, and accordingly set up Xuan Paper Association. But little was mentioned as to how to overall manage Xuan paper, painting and calligraphy paper and Xuan writing-brush enterprises, not to mention standard operation and supervision mechanisms.

2. Government Guiding, Society and Community Following Up Mode

(1) For Jingxian County, Xuan paper making skill represents a profound local cultural memory with a long history. Multi-level local governments adopted "government subsidies, public participation" to save timely the skills and related information that have varied or lost in inheritance, and to construct public welfare facilities to preserve historical production pictures, data and objects, etc. One of the representative examples is the opening of China Xuan Paper Museum, which is led by China Xuan Paper Co., Ltd.

(2) The People's Government of Jingxian County incorporated the protection of Xuan paper culture into plans for national economy and social development. The government allocated leading protection funds, and encouraged the leading companies such as China Xuan Paper Co., Ltd. to actively participate in, assisted with social donation. The traditional Xuan paper production community Xiaoling and painting and calligraphy paper production base Dingqiao, raised funds and efforts from various channels. Hence, the construction of Xuan paper culture conservation was promoted, and set a good example for the protection of handmade paper in Anhui and other areas in the country.

3. Cultural Resources Promoting Tourism, Spreading Xuan Paper Culture through Brand

(1) From 2006 to 2009, Xuan paper making skill was selected into "national intangible cultural heritage list" and "human intangible cultural heritage list of masterpiece" successively. In 2008, Xuan papermaking skill was performed in Beijing Olympics Opening Ceremony, which strengthened the brand spreading for tourism. First, increasing number of visitor from home and aboard visited and appreciated Xuan paper making skills and its cultural ecology, which raised the spreading of Xuan paper making skill and Xuan paper culture. Second, the academia was motivated to experience and research Xuan paper technology and culture, which brought better sociocultural context and propagation path to the inheritance and protection of Xuan paper making skills. Third, the vitality of Xuan paper making skills and tourism had both been improved.

Certificate of "Human Intangible Cultural Heritage List of Masterpiece"

Handbook for the Protection of Intangible Cultural Heritage in Anhui Province

Performance of Xuan Paper Drying Process in 2008 Beijing Olympics Opening Ceremony (screen shot)

3.2.2 Benchmark Achievements of Technology Inheritance and Protection of Xuan Paper

(1) From 1980s, led by People's Government of Jingxian County with the principle of "money supporting the raw material cultivation", China Xuan Paper Group Company (former Jingxian Xuan Paper Factory of Anhui Province) invested funds or agricultural supplies, built 33.3 km^2 *Pteroceltis tatarinowii* Maxim. tree base in Tingxi, Aimin, Caicun and Beigong in Jingxian County. However, because the land use rights had not been transferred, the enterprise had no authentic right to guarantee the effective operation of the base. Since 2013, China Xuan Paper Group Company (renamed as China Xuan Paper Co., Ltd. when investigated) combined national agricultural comprehensive development projects, to promote the construction of Xuan paper raw material bases. Until the early 2016, when our team investigated the place, the construction of 10 km^2 *Pteroceltis tatarinowii* Maxim. tree base and 0.2 km^2 raw material processing base had been completed, which basically guaranteed the supply and processing of Xuan paper raw material for China Xuan Paper Group Company.

(2) In 2000, led by People's Government of Jingxian County, with Bureau of Quality and Technical Supervision of Jingxian County as the main body for application,

Raw material base of China Xuan Paper Co., Ltd.

China Xuan Paper Group Company finished the Xuan paper origin protection application. In August of the same year, Jingxian County was approved as Xuan paper protected origin by National Geographic Protection Office, the protection base was Jingxian County, and protection name was Xuan paper. One problem reflected in the interviews was that although some enterprises who were included in the list of Protection of Geographical Benchmark had changed their paper products, they were never removed from the list due to the improper operation mechanism.

(3) After Jingxian County was acknowledged as Xuan paper protected origin, in 2002, China Standardization Committee upgraded the recommendatory standards of Xuan paper industry to compulsory national standards, from Standard No. QB/T3515—1999 to GB18739—2002. In 2008, it was revised again, adding more requirements such as source of raw materials.

Cover of *National Standards of Xuan Paper*

The compulsory national standards of Xuan paper industry set the first case in Chinese handmade papermaking industry, which had important demonstration effect on the construction of industry scientific standards.

(4) In 1998, China Xuan Paper Group Company planned to apply for the National Famous Trademark. In January of 1999, the brand of "Hongxing (red star)" Xuan paper was approved as the National Famous Trademark by the Trademark Section of the National Administrative Bureau for Industry and Commerce. From then on, eleven Xuan paper brands such as "Wangliuji" and "Wangtonghe" were included in Famous Trademark List of Anhui Province successively.

(5) In order to facilitate collecting, organizing, and documenting of the making skill data in Xuan paper industry, together with its research and exhibition, in 1993, China Xuan Paper Group Company set up China Xuan Paper Museum, which collected and displayed Xuan Paper and products made by Xuan paper from different dynasties, including exhibition of the making process of Xuan paper and related pictures, calligraphies and paintings. It was closed after being handed over to culture sector of Jingxian County government in 2000. In 2006, China Xuan Paper Group Company constructed China Xuan Paper cultural Park (also served as Xuan Paper Making and Inheriting Base). In 2011, the new Xuan paper museum was replanned and opened in December of 2015. The gross floor area of the completed Xuan Paper Museum was more than 10,000 m^2, consisting of three floors: the first floor was composed of five parts (Xuan paper impression, origin, development, expansion and prosperity), exhibiting the history, handicraft and culture of Xuan paper. The second floor mainly displayed ancient Xuan paper, products made of Xuan paper and works on Xuan paper by famous painters and calligraphers. The third floor was temporary exhibition hall, used for temporary exhibition.

China Xuan Paper Cultural Park

(6) In 2005, Xuan Paper Association was established in Jingxian County, which attempted to introduce the self regulation system to manage the industry. In the next year, the journal of the association *China Xuan Paper* was founded, which carried on research on Xuan paper making skills and inheritance, industry development, etc. The journal was quarterly, which released four issues per year. By the time our team ended our survey in Jingxian County in April of 2016, it had released 41 issues.

Magzine of *China Xuan Paper*

(7) In order to solve the problem of the shortage of inheritors for Xuan paper making skills, China Xuan Paper Co., Ltd., Jingxian County China Xuan Paper Association cooperated with Xuancheng Industrial School, to open Xuan paper making skills training courses, and to explore new mode of inheritance. Since 2012, to stabilize the front-line workers, China Xuan Paper Group Company promulgated and implemented *The Evaluation and Appointment Measures for Technician and Senior Technician in China Xuan Paper Group Company* and *Talent Management and Incentive Measures in China Xuan Paper Group Company* successively, which established the incentive mechanism of talent training, employing and development for the staff who possessed state-recognized college degree or above from full-time higher education institutions, technicians and senior technicians.

3.2.3 A Survey of Anhui Handmade Paper Researches

Today, the modern and contemporary research of Anhui handmade paper is mainly centered on Xuan paper and painting and calligraphy paper making industry in Jingxian County, which conforms to the prosperous development of Xuan paper making in Jingxian County. The research review can be divided into the following phases.

1. Early studies on Xuan paper and exploration of Japanese paper in modern times

(1) In 1923, Hu Yunyu wrote *On Paper* in the third volume of *Series of Puxuezhai* which included an anthology published at his own expenses. In *On Paper*, a specific section focused on Xuan paper, and Chinese academia regards it as the initial research of Xuan paper in China.

(2) In 1878, papermaking department of Japanese Printing Bureau sent Narahara Nobumasa to China. After Narahara returned, he published *The Papermaking Industry in China* which focused on Xuan paper making.

(3) In 1883, Japan published *The Research Journal of Chinese Papermaking* and involved Xuan paper in this book.

(4) In the early 20th century, a Japanese Uchiyamami Satoumon wrote *The Making of Xuan Paper* and published it on the No. 98, 9th Volume of *Japanese Industrial Chemistry Magazine*.

Before the early 20th century, due to the large market demands, Japan began to research and explore the making technique of Xuan paper, meanwhile other foreign countries almost implemented no research on Xuan paper. From 1930s to 1940s, Japan brought some skilled workers from Xiaoling of Jingxian County to Japan, planning to practice Xuan papermaking in Japan. However, since those workers had strong patriotism and sense of technique protection, they refused to provide Japanese with Xuan paper making technique. Also, Japan lacked appropriate natural environment, water, and raw materials, so that quality Xuan paper could not be made in Japan. Finally, Japan released those Chinese workers and sent them back to Jingxian County.

2. Before 1949: studies by local scholars and parties

Precisely, researches on Xuan paper that can be traced started from 1923. Before 1949, the studies and surveys on Xuan paper by local experts include:

(1) The study of Xuan paper making theories and history. In 1923, a celebrated scholar named Hu Yunyu from Jingxian County wrote *On paper* which briefly described the development of Xuan paper and put forward an idea that the raw material is the key of papermaking.

(2) The research on Xuan paper practice. There are two surveys of Xuan paper practice during the Republic of China period. The first one was conducted by Pulp and Paper Research Institute of China in 1936. In the same year, Wei Zhaoqi published an article *The Research of Xuan Paper Making Industry* after this survey. A year later, in 1937, Zhang Yonghui published his article *The Research of Anhui Xuan Paper Industry* based on his field investigation.

Cover of *The research of Anhui Xuan Paper Industry* written by Zhang Yonghui

The research report by Wei Zhaoqi recorded the raw material, making technique, and other aspects of Xuan paper making, such as the distribution of papermaking containers, the status of local paper product, the equipment of paper mills, the method of papermaking, the names of different Xuan paper types, the cost of paper making, and the transporting and marketing of Xuan paper. At the end of the report, Wei pointed out some defects of Xuan paper making: "Firstly, the raw material processing takes such a long time that the papermaking process is not economical for industrial production. Secondly, the application of mucilage does not have a scientific standard. Thirdly, the process of pulp-cooking employs open pot so that much heat was wasted. Also, the process of washing material and bleaching causes much loss of material fibers. Fourthly, since old method of papermaking does not use pulp screen, the pulp is not homogeneous in quality."

In 1936, Zhang Yonghui returned from Germany where he studied. Zhang received an order from the Papermaking Research Institute of China to survey the status of Xuan paper making, and published *The Research of Anhui Xuan Paper Industry* in 1937. According to the research, Zhang pointed out that although native workers kept the Xuan paper making technique a strict secret, the machine-made paper and fake Xuan paper flooded into most parts of market and hit the making and selling of Xuan paper. Those makers of Xuan paper were fully informed that they could not compete against the machine-made paper and counterfeit paper and maintain the development of Xuan paper, unless they would promote their making process to reduce the cost and increase the production. Thus, Xuan paper makers were willing to tell the researchers about their dilemma and hoped they could counterbalance their losses with the help of the officers.

The Review of Xuan Paper Industry in Anhui (hereinafter referred to as *The Review*) records more than ten places in Xiaoling where Xuan paper was made, such as Shuanglingkeng, Fangjiamen, and Xujiawan. About 17 or 18 mills had their own papermaking containers, and those mills possessed more than 40 containers in total. The production of Xuan paper in Xiaoling took about 80% of the total production of Xuan paper in Jingxian County. "*The Review* also records cooperation of Xuan paper makers in Fengkeng. This cooperation aimed to eliminate the selling or competition disputes among mills. The mills in Xiaoling gathered their production together, inspected the quality and fixed the price. Then, Xuan paper could be transported and sold."

In this work, Zhang detailedly described the raw material *Pteroceltis tatarinowii* Maxim. bark, straw, adjuvant, plant mucilage, lime, alkali, and decolorizer of Xuan paper as well as the process and technique in papermaking such as processing raw material, straw processing, drying, pressing and inspecting paper, etc. In the concluding part, Zhang summarized the reason for the quality decline of Xuan paper. The high cost of making raw material, uneven pulp, low rate of finished products, papermaking tools in disrepair, outdated knowledge and technique of mill-owners and lack of skilled workers, overlong material processing, non-standard employment of mucilage, popularity of fake and machine-made paper, all those elements caused the inevitable decline of Xuan paper making. In order to reverse the trend, Zhang suggested that a large factory specializing in alkaline pulping should be established for providing mills with raw material. In the case of the papermaking process, Zhang indicated that the existing handmaking process should be improved so that the paper quality gets better and the cost gets lowered.

The research reports of Zhang and Wei are mutually complementary, and both of them are valuable materials of field study. Starting from these two research reports, the study of Xuan paper eventually possesses its own statistical data with economic significance.

(3) There was a special research of Xuan paper industry status coming from the New Fourth Army. In August of 1938, the New Fourth Army marched into Luoli Village of Jingxian County which was only ten *li* (5 000 meters) away from Xiaoling Village. They established a rural economic committee under the department of the Military command in Zhongcun Village to research on the economic status and political tendency of different social classes, as well as feudal superstition activities and exploitation of farmers by the rich. Then the result of the research was published in the magazine *Kang Di (Fighting Against Enemies)*. Some representative results selected from this research are as follows:

The first paragraph mentioned that "Jingxian County was famous for papermaking. But before World War II (hereinafter referred to as *WWII*), the production of paper except Lianshi paper (a special Xuan paper for painting and calligraphy) in Xiaoling gradually declined. Some paper mills that produced Biaoxin paper which was of poor quality, were shut down because of the challenge of machine-made paper. However, after the war, the price of handmade paper rose rapidly, even the paper of poor quality began to be sold with high prices. For instance, the price of Biaoxin paper (about 14 600 pieces in one load) rose from 7 or 8 Yuan per load to 24 Yuan per load; then, the price stabilized at 14 or 15 Yuan per load. But this boom did not contain the Lianshi paper, for the main market for Lianshi paper (mostly in Beijing, Shanghai, and regions around) was lost." This part pointed out that the reason for the decline of Xuan paper produced by Xiaoling is the loss of market during WWII, while the boom of ordinary handmade paper can be attributed to the decreased import of machine-made paper.

The second paragraph indicated basic reasons for the boom and decline of the production of commodities from the late Qing Dynasty to the time of the Anti-Japanese War. It is said that the invasion of imperialism resulted in the boom of commodities as tobacco, Xuan paper, silk, sack and so on. From the end of the Qing Dynasty and the beginning of the Republic of China to WWI, from WWI to the war of resistance against Japan, the development of Xuan paper was in a short boom which was in accordance with Chinese economic development status at that time. The increased amount of papermaking mills, migration of papermaking mills as well as the increasing investment manifested capitalist relations of production had penetrated into Xuan paper industry and stimulated its development as a commodity. Another reason for the boom is that as a specialty paper, Xuan paper could not be easily substituted by other kinds of paper. This feature, to some extent, helped it resist the challenge from domestic and overseas counterparts. After the September 18th Incident in 1931, the production of Xuan paper began to decline. Then, in the January 28th Incident of 1932, the Commercial Press (in Shanghai) was bombed by Japanese army. Thus, the selling of Xuan paper sharply declined. Before the Lugou Bridge Incident (July 7th Incident), Xuan paper industry in Jingxian County has had a tremendous development from 1935 to 1937; but this boom was broken down by the Anti-Japanese War.

The third paragraph indicated that the main feature of Xuan paper industry in Jingxian County was that the paper mill owners were also paper sellers, and landowners. "Most land in Jingxian County and Xiaoling Village is in the hand of paper sellers and landowners." "Most paper sellers in Xiaoling were also the landowners with great amount of fields." This conclusion demonstrates the unique historical perspective of analysis by CCP (Chinese Communist Party) under that specific historical circumstance, which is of great historical value.

3. Studies on Xuan paper from 1949 to 1976

Compared with the research before the foundation of PRC (People's Republic of China), the researchers of Xuan paper as well as their theses and reports increased during the period from October 1949 to the end of "the Great Cultural Revolution" (1976).

(1) In the early 1950s, Yu Kai has investigated and reported on the recovery of Xuan paper production in Jingxian County. On 29th June 1954, in the article *Time-honored and Well-known Xuan Paper Production* published in *Anhui Daily*, Yu Kai recorded the recovery and expansion of Xuan paper making from the end of 1949 to the middle of 1954, which were valuable data for researching Xuan paper during the early years of New China.

(2) In 1956, Jiang Shixiang published *Discussion on Quality Improvement and Production Method of Xuan Paper*, which was a groundbreaking work for technical innovation of Xuan paper after the establishment of PRC. In this work, the author wrote:

"After the foundation of PRC, the production of Xuan paper failed to meet the social demands, though production and productivity had been annually increased. Its complicated and time-consuming hand-made process resulted in low yields, high cost, great consumption of materials, low efficiency and especially the unstable quality of products. The quality of Xuan paper at that time was even inferior to that at the end of the Qing Dynasty and in the early years of the Republic of China." "The main quality problems of Xuan paper lie in three aspects", based on comments from the VIP users of Xuan paper, e.g. Fine Arts Publishing House affiliated to Ministry of Culture, and Rongbaozhai: "Firstly, the paper was rough with lots of impurities, which caused poor print effect. Secondly, the excessive residual chlorine in paper made paper fragile and the printing discolor. Thirdly, its poor wetting property of Chinese ink affected the quality of replica of the classic works and the performance of watercolor and Chinese ink painting."

The author further described the experimental work of improving paper quality and making method by a joint working group, organized by central government and Anhui Provincial Industrial Bureau. "They concluded that although there still existed problems concerning the experiment of steaming and bleaching straws, other experiments on *Pteroceltis tatarinowii* Maxim. bark boiling, steaming and bleaching were quite successful. The paper made from *Pteroceltis tatarinowii* Maxim. bark surpassed Wangliuji Xuan paper made 30 years ago. After trials by Rongbaozhai and some painters who preferred heavy stroke paintings and enjoyed wetting property of paper, it was considered to be able to meet the need of block-print and peculiar needs of paintings. However, it was not appropriate for those requiring fine lines in painting and low wetting property paper. Only paper made from mulberry bark, or bamboo could meet such demands."

The original record of *Discussion on Improving Quality and Production of Xuan Paper* written by Jiang Shixiang

The original record of *Problems About Xuan Paper* written by Chen Pengnian

(3) The research on reforming Xuan paper production by Chen Pengnian. In 1957, Chen wrote *Problems about Xuan Paper*, and at the beginning of this work, he clearly pointed out: "Nowadays, it will be significant if modern technology can be employed to improve the making process, increase yields, and maintain good quality." "In the making process, the preparation of raw materials and mucilage, making paper, and other complicated processes show the details of the papermaking process. Based on these detailed information, we can draw scientific observations and tips to guarantee its high quality."

Chen raised some fundamental suggestions for improving making process of Xuan paper: to remove those parts from raw materials if they are harmful so as to avoid unnecessary processing steps later on. The pulp must be processed for many times so that the evenness and purity of pulp fibers can be improved. Therefore, the focus should be put on pulp making and beating process.

On making bark and straw pulp, Chen raised constructive advice on production skills and tool employment. Firstly, plant and cultivate *Pteroceltis tatarinowii* Maxim. in a planned way, cooperate with other mills, employ the soaking and fermenting step while skipping the steaming process. Secondly, employ lime liquor for steaming procedure to keep the strength of fibers. Thirdly, in order to improve the old cooking process, Chen designed a steaming loop pot to recycle the lime liquor and keep sufficient liquor for boiling. Thirdly, do researches on the procedures of lime liquor cooking, bleach treatment, dilute lime liquor treatment, and bleach process. Find other alternatives on straw pulp making, Chen proposed to replace water soaking process by water cooking and steaming process. Subdivide straw pulp refining process into four standard processes: lime liquor treatment, bleach treatment, dilute lime liquor treatment, and bleaching. Chen predicted that it is possible to achieve the mechanized pulp production by employing pulping machine, and beating devices could also be used, but the scooping and papermaking step is the most difficult part.

(4) The research of raw material treatment—pulp making by Zhou Naikong. In 1958, Zhou Naikong, a technician of Jingxian Xuan Paper Factory, published *The Raw Material of Xuan Paper—the Method of Making Straw Pulp* in which three processes of producing straw slab, fresh straw, and processed straw were researched and introduced.

(5) Hu Yuxi studied the bark fiber of different ages of *Pteroceltis tatarinowii* Maxim., the raw material of Xuan paper. In 1964, Hu wrote a paper *Comparative Analysis of Pteroceltis tatarinowii Maxim. of Different Ages*, in which his research findings were presented to the public, and this is the first thesis focusing on the bark fiber analysis in Xuan paper making. Based on the anatomy of different-aged bark of *Pteroceltis tatarinowii* Maxim. branches (mostly are phloems), Hu illustrated the distributing and morphological characteristics of the phloem fibers, and explained why the younger or older branches are unfit for the Xuan paper manufacture on the basis of the changing relation of the quantity of phloem fibers and other tissues. According to the well-accepted idea that the two-year-old branches are the best raw material for Xuan paper manufacture, and the younger or older branches cannot be used to make high quality paper, Hu made a comparative study of the phloem fibers in different-aged branches. His thesis attached a transverse section graph of *Pteroceltis tatarinowii* Maxim. bark, "tables of the Comparison of the Different-aged Branch and Bark Tissues in *Pteroceltis tatarinowii* Maxim.", and "the Comparison of the Morphological Characteristics of Phloem Fibers and Wood Fibers in Different-aged Branches of *Pteroceltis tatarinowii* Maxim."

(6) The study of the absorbility, deformability, and durability of Xuan paper was carried out by Chen Zhiwei and other fellows. In the past, the evaluation method of Xuan paper quality lacked scientific methods and specific indicators, depending on the traditional experience acquired from visual inspection and touching. Meanwhile, painters are so acute in Xuan paper evaluation that they can use their terminology to judge the quality of paper as soon as they paint on the paper. However, these terms can merely be understood by insight and cannot become the standards of evaluating the quality of Xuan paper. To change this status, in 1964, Chen and his colleagues collected the suggestions from the painters and combined various test standards of machine-made paper to find out more appropriate quality indicators and test methods, which made it convenient to judge the quality of Xuan paper and further research.

Then, *The study of the Absorbability, Deformability, and Durability of Xuan Paper* was published after experiment and research. Its preface mentions: "From the view of papermaking, the paper with good absorbability, deformability, and durability results from its raw material and production process. Xuan paper uses *Pteroceltis tatarinowii* Maxim. bark and straw as raw materials, and employs the traditional pulping process, such as longtime water soaking and fermenting, plant ash pickling, sun beam bleaching and so on. The scientific basis and process merit of Xuan paper manufacture are worth of further research. In addition, the technique of Xuan paper making can be regarded as a reference for present use of grass fibers and expanding the application field of grass fibers. This reference showed that the past technique could serve the present."

Based on the experiments, they indicated some conclusive factors which affect the absorbability of Xuan paper. Firstly, the larger total surface area that fibers have, the more particles that its cell-wall can absorb. Secondly, within appropriate range, the larger space between each of the fibers, the more ink it absorbs. Meanwhile the space should be of reasonable size. Thirdly, fibers that go through the moderate pulping process have less oxidized cellulose and hydrocellulose, and their absorbility is best. Fourthly, the lower degree of pulp freeness means less hydrocellulose and higher absorbility. Fifth, the vegetable adhesive should be carefully employed. Adjuvant calcium carbonate can help ink diffuse like a halo. But if the particle of calcium carbonate is too rough or the dosage is over 3%, the ink will become light in color.

Deformability is affected by the following factors. Firstly, the key to reduce deformation of paper is to enlarge the space between each of fibers. The more space exists between each of fibers, the slighter the bond and the deformation are. Secondly, as the easiest part to expand, hemicellulose's proportion in pulp should be reduced so that the paper would experience less deformation. Since straw pulp has much hemicellulose, it requires special treatment like soaking and fermenting by water and plant-ash to reduce the proportion of hemicellulose. Thirdly, adjuvant and vegetable adhesive have effects on deformation: the particle of calcium carbonatein pulp can penetrate into the space between fibers so that the change of bond of fibers can be decreased when they shrink. Fourthly, good storage can lessen the drying shrinkage of the paper.

As for the durability, it is affected by the following factors. First one is the damage by worm. Even Xuan paper, without good preservation, will be damaged by worm as well. So the traditional saying that Xuan paper won't be damaged by worm is inaccurate. Secondly, the durability and color-protecting feature of traditional Xuan paper is because the proportion of metal ion in it is low, and it contains tiny calcium salt, all of which made it tend to be alkaline. Furthermore, the less damage the raw material fibers get during the treatment, the better durability the paper has. In this paper, Chen worked out detailed empirical data to support his views, hence made the paper academic and professional.

(7) Maernatsu Rokuro made studies on chemical composition of Xuan paper. Xuan paper has long been Japanese intellectuals' favorite paper for painting and calligraphy. Although Japanese have made Washi paper and constantly made efforts to improve quality of paper, they still deem Xuan paper better. Thus, Japanese gathered information of Xuan paper as much as possible and employed high-tech to analyze it. Maernatsu Rokuro obtained bark sample and paper sample from Takayama Tamotsu and Seki Yoshikuni. Then, lots of researches were carried out based on these samples. In 1962 and 1975, three research reports: *The Chemical Composition of Pteroceltis Tatarinowii Maxim. Bark*, *The Fiber Morphology of Pteroceltis Tatarinowii Maxim.*, and *The Length and Width of Pteroceltis Tatarinowii Maxim. Bark Fibers* were successively published. Through his researches, the chemical composition of *Pteroceltis tatarinowii* Maxim. bark was revealed in a foreign country for the first time. Based on many comparative analyses and sufficient data analysis, they finally got a conclusion that the fibers of Xuan paper usually are 1.5～5.7 mm long (2.9 mm in average), 3～15 μm wide (8.3 μm in average), and the ratio of its length to width is 278%.

Undoubtedly, during this period, the in-depth research of Xuan paper by Japanese scholars played an important role in the history of Xuan paper research.

4. From 1977 to the end of 1999, is a phase in which Xuan paper research became in-depth and comprehensive and lots of experts and research results sprang up. Among them, the representative works are:

(1) The research of Xuan paper making technology by Pan Jixing. His *The History of Papermaking in China* (Cultural Relics Press, 1979), *The History of Chinese Science and Technology: Papermaking and Printing* (Science Press, 1998), *The Four Great Inventions of Ancient China: Their Origin, Development, Spread and Influence in the World* (University of Science and Technology of China Press, 2002), and other famous works discussed the historic status, origin, and raw material of Xuan paper. Specially, Pan deemed that the name "Xuan paper" originated from the record "tribute paper from Xuanzhou" in *The New Book of Tang: Geographical Chronicles*.

A photo of *The History of Papermaking in China*

(2) The research of Xuan paper history by Mu Xiaotian. Mu has worked in Anhui Provincial Museum for a long time. In *The History of Anhui Four Treasures of the Study* (Shanghai Peoples's Fine Arts Publishing House, 1962), he made a preliminary discussion about the history of Xuan paper. After that, Mu authored *Anhui Four Treasures of the Study of China* (Anhui Science and Technology Press, 1983) with Li Minghui, and the first part of this book—"Xuan paper" records his own analysis and statement. In this part, the history of Xuan paper, as well as the materials, making process, making tools are researched and discussed. Additionally, the engraving art on calendered paper, paper screen weaving and the relationship between Xuan paper and Chinese painting and calligraphy are recorded and discussed.

(3) The initial study of weaving paper screen (a papermaking tool) by Ge Zhaoxian. Ge wrote a paper *Preliminary Research of Paper Screen Weaving Technique* and published it in 1981 on the first issue of *Anhui Literature Expo*. He has worked in Jingxian County for a long time, and during his work, he visited Xiaoling Village for many times to interview those workers who wove the paper screen, and recorded its weaving process, materials, standard of size, as well as the family traditions of the workers.

(4) The research of wetting property, durability, deformability, insect resistance of Xuan paper by Liu Renqing. In the aspect of wetting property, Liu and Qv Yaoliang from the State Archives Administration of the PRC co-authored *The Research of Xuan Paper's Wetting Property* on the second issue of *China Pulp and Paper*. This article researched and discussed the reason for wetting effect and best wetting property of Xuan paper from the perspective of its absorbability, fiber morphology, and component. Based on a number of comparative experiments and experimental data, some conclusions were drawn. Firstly, the bark fiber of *Pteroceltis tatarinowii* Maxim. is the optimum material for Xuan paper making, because its evenness, thin cell wall, and soft texture, especially its wrinkles parallel to long axis make a great contribution to the wetting property of Xuan paper, and these wrinkles will take shape after air drying. Secondly, crystalline calcium carbonate and amorphoussilica in Xuan paper are confirmed by scientific method. Then, they found that crystalline calcium carbonateon bark cell-wall is the key to its wetting property and silica just plays a role of assistance. Thirdly, the wetting effect of Xuan paper is mainly presented by the range of ink mark spreading on paper, the clear discrepancy of ink chromaticity and concentration. The superb bark Xuan paper has the best wetting property because it employs more *Pteroceltis tatarinowii* Maxim. bark than regular one.

In addition to the studies on the wetting property, Liu and Qv further made a study on the durability of Xuan paper and published their research results on the 6th issue of *China Pulp and Paper* in 1986. In this article, they recorded the experiment of predicting paper durability and researched different paper types' durability under different temperature conditions in detail. Through the experiments, they found that Xuan paper had the best durability, and it may last over 1 050 years in simulated situation. Because the durability of Xuan paper is related to its PH value, they suggested employing and promoting alkaline making technique to improve the durability of the paper.

After that, Liu successively published many works about Xuan paper and other handmade paper, such as *Xuan Paper and Chinese Painting and Calligraphy* (China Light Industry Press, 1989), *Chinese Calligraphy and Painting Paper* (China Water & Power Press, 2007), *The Story of Papermaking* (China Light Industry Press, 2008), *A Brief Dictionary: General Knowledge of Chinese Handmade Paper and Painting* (China Light Industry Press, 2008), *A Collection of Ancient Chinese Paper* (Intellectual Property Publishing House), *The National Treasure—Xuan Paper* (China Railway Publishing House, 2009) and so on.

(5) The research of improving Xuan paper performance by Yan Jiakuan. Yan was a faculty of Arts Department of Hubei University. It was a new point of view to study Xuan paper from the eyes of an artist and reform it with an artist's wisdom. Yan deemed that the good performance of Xuan paper is due to the fact that it employs abundant long fibers and little short fibers as its material. Chinese artists preferred Xuan paper as their painting material because of its smooth and soft texture, good wetting property, absorbability, and appropriate firmness. As a result, artists have created fine brushwork, free brushwork, and other diversified drawing techniques which are difficult to be used on other types of paper. Since the fibers in Xuan paper are soft with special arrangement, it possesses good hydrophilia. For thousands of years, Chinese artists have confirmed the good performance of Xuan paper by their artistic practice, which means it cannot be replaced by others.

Yan also holds that Xuan paper has defects. Although it has good durability, it still can be affected by the moisture in air and become brittle, rotten, even damaged. Based on his observation, Yan proposed to employ surface sizing in papermaking for improvement, so that it can resist damp, rotting, insect, sun beam, and keep its porous structure and hydrophilia.

In August of 1990, in order to test the effect of paper sizing, Yan painted 30 paintings called *Australian Desert* on improved raw Xuan paper which is a kind of Xuan paper. He exhibited his paintings in October, 1990, in Beijing Fine Art Academy, and in October of 1991, in University of South Queensland, Australia. This experimental artistic creation had drawn attention from domestic and overseas experts.

(6) The research of Xuan paper by Cao Tiansheng. Cao was born in a papermaking family, in Xiaoling Village. He has begun to study Xuan paper since the 1980s. And he published three works about Xuan paper: *Chinese Xuan Paper* (China Light Industry Press, 1993), *Chinese Xuan Paper* (the 2nd Edition, China Light Industry Press, 2000), *History of Xuan Paper* (China Science and Technology Press, 2005). Cao proposed his views of Xuan paper in his books. For example, Xuan paper originated from Hui paper. He put forward the complete definition of Xuan paper and the idea of authentic Xuan paper. Cao also made the study of the history of foreigners stealing Xuan paper making techniques, and defined the start of Xuan paper research. As a descendent of traditional papermaking family, Cao thoroughly sorted the history of Xuan paper research and made a complete study of the historical contribution to Xuan paper making of his clan. His research is favored by academia for being expert and systematic in theory.

(7) The research of Xuan paper by Wu Shixin. Since 1983, Wu published more than 10 articles about Xuan paper research. His articles involve many aspects including history, research status, traditional making technique, modern making technique, raw material, variety, related persons, and anecdotes. *Chinese Xuan Paper* (on the first issue of *Wenxian*, 1986), *A Unique Chinese Art—Xuan Paper: Research of Its Making Technique and Wetting Property* (on *Proceedings of International Seminar on Xuan Paper Art*) are his representative works. In *Chinese Xuan Paper*, Wu proposed that compared to the name "Xuan paper", "Jing paper" or "Jingxian County paper" is a more proper name. In *A Unique Chinese Art—Xuan Paper: Research of Its Making Technique and Wetting Property*, Wu described traditional technique of Xuan paper making and discussed the relation between its traditional making technique and wetting property.

(8) The research of Xuan paper by Xu Guowang. Xu has studied paper for a long time, especially on Xuan paper. In his article *Jingxian County: the Origin of Chinese Xuan Paper*, he used historical data to confirm that "Tang Xuan paper" in some historical records refers to the paper from the region of Xuanzhou in the Tang Dynasty. Xu also defined that Xuan paper must be the paper made from *Pteroceltis tatarinowii* Maxim. bark. Furthermore, from the perspective of the history of papermaking technology, he denied that Xuan paper has existed before the Southern Song Dynasty.

(9) The research of Xuan paper history by Dai Jiangzhang and other authors. In the 8th chapter of *A Brief History of Chinese*

Chinese Calligraphy and Painting Paper and other books written by Liu Renqing

Japanese copy of *A Collection of Ancient Chinese Paper* written by Liu Renqing

Chinese Xuan Paper and other books written by Cao Tiansheng

Papermaking Technology (China Light Industry Press, 1994), an exclusive section is written to discuss the origin, producing place, variety, usage, raw material, and making technology of Xuan paper. It also records authors' understanding of some controversial academic issues.

(10) The research of Xuan paper history by Huang He. Huang He used to be the Deputy Director of Paper History Committe of China Technical Association of Paper Industry. He has also served as the editor in charge of the column of paper history in *China Pulp and Paper* and the editor-in-chief of the special issue *The Study on the History of Paper*. *The Historical Story of Papermaking* (the 2nd edition, Zhonghua Book Co., 1979) and *The Comprehensive Discussion of Ancient and Modern paper and Xuan Paper* (on *Proceedings of International Seminar on Xuan Paper Art*) are two main works of Huang. His views can be concluded as follows: firstly, according to *The New Book of Tang* and *The Old Book of Tang*, Xuan paper originated from Hui paper, and its name is related to the ancient Xuanzhou. Secondly, during the long period of more than a thousand years, with the evolution of material, producing technique, performance and use value, modern Xuan paper is different from the ancient one. Thirdly, the key difference between ancient paper and modern paper is their materials. The ancient paper in Tang and Song Dynasties was made from mulberry bark, and the paper made from this material was smooth and tough, but its wetting property was poor. According to these features of paper, the effect of painting and calligraphy on it is limited. The modern one is made from *Pteroceltis tatarinowii* Maxim. bark (long fibers) and straw (short fibers) and goes through special making processes, like sun bleaching, air drying, and pulp sizing, so that the paper has the property of smooth surface, proper tension, and best wetting property, which performs better for the artistic works. Fourthly, the ancient Xuan paper and modern one are different products of different historical stages. So the ancient Xuan paper should not be confused with the modern one.

(11) The leading researchers and achievements in the 21st century

Cao Tiansheng: *Chinese Xuan Paper* (the 2nd Edition, China Light Industry Press, 2000), *History of Xuan Paper* (China Science and Technology Press, 2005), *A Millenium of Xiaoling Village*, and *The General Knowledge of Xuan Paper* (China Science and Technology Press, 2014).

Zhang Binglun, Fan Jialu: *Papermaking and Printing* (Elephant Press, 2005).

Fan Jialu: *Papermaking II: Writing Brush Manufacturing* (Elephant Press, 2015).

A photo of *Papermaking and Printing*

Wang Xiafei: *Four Traditional Chinese Treasures of the Study* (People's Fine Arts Publishing House, 2005).

Pan Zuyao: *Xuan Paper Making* (China Forestry Publishing House, 2006).

Liu Renqing: *Chinese Calligraphy and Painting Paper* (China Water & Power Press, 2007), and *The National Treasure—Xuan Paper* (China Railway Publishing House, 2009).

Huang Feisong: *A Random Talk about Xuan Paper* (China Federation of Literary and Art Circles Publishing House, 2008), *The Intangible Cultural Heritage Series—Xuan Paper* (Zhejiang People's Publishing House, 2014) and "*Stop and Go—Explore Xuan Paper and Its Story*" (Intellectual Property Publishing House, 2014).

Wu Shixin: *The Historical Story of Chinese Xuan Paper* (China International Press, 2009).

Zhou Naikong: *The Making Technique of Chinese Xuan Paper* (Milkyway Publication (HK) Co., Ltd., 2009).

Tang Shukun: *Illutrated Four Great Inventions of Ancient China: Papermaking* (Zhejiang Education Publishing House, 2015). It introduces Xuan paper and Chengxintang paper. The traditional Chinese character version, English, Malay, Bulgarian, German, Russian and Arabic, German, Russian and Arabic copies have been published during 2016 to 2018.

When the research group's manuscript of Anhui Volume was about to be finalized in

Different versions of "*Illutrated Four Great Inventions of Ancient China: Papermaking*" by Tang Shukun

December of 2017, the research members were informed that are view of *Record of Xuan Paper* by the Chorography Committee of Jingxian County has been completed. This book will be the first comprehensive record of Xuan paper in history. It calls up both ancient documents and modern data, and combine them. Examples, records, biographies, pictures, tables and appenda are all employed in this book. This book consists of 10 volumes which has no chapters but entries. Each volume is subdivided into preface, introduction, chronicle of events, and appendix, etc. According to the information provided by the authors, this book has been published by Publishing House of Local Records in 2019.

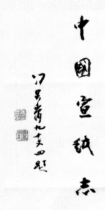

The cover and inscription of "*Record of Xuan Paper*"

图目
Figures

章节	图中文名称	图英文名称
第 一 章	安徽省手工造纸概述	Chapter I Introduction to Handmade Paper in Anhui Province
第 一 节	安徽省手工造纸的历史沿革	Section 1 History of Handmade Paper in Anhui Province
	华佗小像	Portrait of Hua Tuo
	胡正言《十竹斋书画谱》	*Ten Bamboo Studio Manual of Painting and Calligraphy* by Hu Zhengyan
	岳西桑皮纸透光图	A photo of mulberry bark paper made in Yuexi County
	江淮分水岭古代遗存的山间道路	Relic of country road in Jianghuai watershed area
	抄造"红星"三丈三大宣纸	Making 3-zhang-3-chi large-sized Xuan paper of Hongxing (red star) Brand
	潜山县官庄镇的造纸村	A papermaking village in Guanzhuang Town of Qianshan County
	岳西县毛尖山乡的造纸村	A papermaking village in Maojianshan Town of Yuexi County
	《新唐书·地理志》书影	A photo of *The New Book of Tang: Geographical Chronicles*
	《历代名画记》书影	A photo of *Famous Paintings in the Past Dynasties*
	嘉庆《宁国府志》书影	A photo of *The Annals of Ningguo Prefecture during the Reign of Emperor Jiaqing*
	《资治通鉴》书影	A photo of *Comprehensive Mirror to Aid in Government*
	脱脱等撰《宋史》书影	A photo of *The History of Song* by Tuotuo in the Yuan Dynasty
	《建炎以来系年要录》书影	A photo of *Major Events from the First Year of Jianyan Reign*
	古版《江南经略》书影	A photo of *Jiangnan Jinglue* (old version)
	茅元仪在《武备志》中绘制的纸甲	Paper Armour depicted in *A Corpus of Military Equipments* written by Mao Yuanyi
	《六研斋笔记》扫描书影	Scanning record about *Notes from Liuyan Studio*
	《续资治通鉴长编》书影	A photo of *A Sequel to Comprehensive Mirror to Aid in Government*
	旧版《新安志》书影	A photo of *The Annals of Xin'an* (old version)
	弘治年《徽州府志》扫描书影	Scanning records about *The Annals of Huizhou Prefecture* during Hongzhi
	古本《文房四谱》书影	A photo of *Four Treasures* of the Scholar's Study (old version)
	《临川先生文集》扫描书影	Scanning records about *Linchuan Gentleman Anthology*
	清同治十一年(1873年)继善堂刊印《曹氏族谱》书影	Photos of *Genealogy of the Caos* printed by Jishantang in the eleventh year of Tongzhi Reign in the Qing Dynasty (1873)
	明代徽派版画——丰南图	Hui style woodcut in the Ming Dynasty: *Fengnan Illustration*
	清代皖南木活字遗存	The remains of Wooden Movable Words in Southern Anhui in the Qing Dynasty
	旧本查慎行《人海记》书影	A photo of *Renhai Ji* (old version) written by Zha Shenxing in the Qing Dynasty
	旧版南宋祝穆《方舆胜览》书影	Photos of *Fangyu Shenglan* (old version) written by Zhu Mu in Southern Song Dynasty
	"四库"本刘汲《嵩山集》扫描书影	Scanning records about *Songshan Ji* written by Liu Ji in the Song Dynasty (included in *The Complete Library in the Four Branches of Literature*)
	歙县的造纸小村——六合村	A small papermaking village in Shexian County: Liuhe Village
	调查组走访泾县原古艺宣纸厂旧址	Researchers visiting the former Guyi Xuan Paper Factory in Jingxian County
	调查组考察泾县原湖山坑宣纸厂遗址	Researchers visiting the relics of former Hushankeng Xuan Paper Factory in Jingxian County
	调查组在潜山县星杰桑皮纸厂访谈	Researchers interviewing Xingjie Mulberry Bark Paper Factory in Qianshan County
	调查组在岳西县金丝桑皮纸厂访谈	Researchers investigating Jinsi Mulberry Bark Paper Factory in Yuexi County
	调查组在歙县棉溪村构皮纸作坊调查(一)	Researchers investigating Mulberry Bark Paper Mill in Mianxi Village of Shexian County (1)
	调查组在歙县棉溪村构皮纸作坊调查(二)	Researchers investigating Mulberry Bark Paper Mill in Mianxi Village of Shexian County (2)

章节	图中文名称	图英文名称
第 二 节	安徽省手工造纸的当代生产现状	Section 2 Current Production Status of Handmade Paper in Anhui Province
	泾县竹纸旧日的原料林	Bamboo in Jingxian County
	金寨县龙马村废弃的竹纸煮料窑	Abandoned bamboo paper steaming kiln in Longma Village of Jinzhai County
	中国宣纸集团公司"上海世博会纪念宣纸"	Shanghai World Expo Commemorative Xuan Paper, by China Xuan Paper Co., Ltd.
	泾县县城中心区的荷花塘冬景	Winter view of the lotus pond located at the central area of Jingxian County
	旧版《曹氏宗谱》上的曹大三画像	Portrait of Cao Dasan in *Genealogy of the Caos* (old version)
	蔡伦祠遗址(现该地仅剩一座寺庙)	The Relics of Cai Lun Temple (Only One Temple Left)
	棍子歌	Work song: Stick Song
	毛竹接水设施	Bamboo equipment for transporting water

章节	图中文名称	图英文名称
第 三 节	安徽省手工造纸的保护、传承与研究现状	Section 3 Preservation, Inheritance and Current Research of Handmade Paper in Anhui Province
	歙县皮纸作坊的染色纸钱	Dyed joss paper in a bast paper mill of Shexian County
	染过色的加工纸	Dyed processed paper
	泾县"艺宣阁"加工纸的洒金车间	Processing workshop of Gold-sprinkled Xuan Paper in "Yixuange" of Jingxian County
	"掇英轩"的真金手绘粉蜡笺	Powder wax paper decorated with authentic gold in "Duoyingxuan"
	调查组成员在"艺英轩"体验工艺	A researcher experiencing papermaking process in "Yiyingxuan"
	歙县青峰村造纸老人家的访谈	Interviewing old papermakers in Qingfeng Village of Shexian County
	乾隆"倦勤斋"修复工程纪念	The Souvenir for Restoration of "Juanqinzhai", Emperor Qianlong's Lodge of Retirement
	潜山县造纸人刘同烟持国家级非遗传承项目牌匾	National intangible cultural heritage inheritance project plague held by Liu Tongyan, a papermaker in Qianshan County
	岳西县造纸人王柏林的国家级非遗传承人证书	Certificate of National Intangible Cultural Heritage Inheritor by Wang Bolin in Yuexi County
	《毛诗草木鸟兽虫鱼疏》的记载	A photo of *The Annotations on Animal and Plant Life* mentioned in *The Book of Songs*
	歙县棉溪村柳构生长环境	Growing environment for local paper mulberry tree in Mianxi Village of Shexian County
	黄山市三昕纸业有限公司附近的新安江上游	Upper Xin'an River near Sanxin Paper Co., Ltd. in Huangshan City
	"人类非物质文化遗产代表作名录"证书	Certificate of "Human Intangible Cultural Heritage List of Masterpiece"
	《安徽省非物质文化遗产保护手册》	*Handbook for the Protection of Intangible Cultural Heritage in Anhui Province*

	2008年北京奥运会开幕式宣纸晒纸表演（截屏图）	Performance of Xuan Paper Drying Process in 2008 Beijing Olympics Opening Ceremony (screen shot)		宣纸大厦旧照	An old photo of Xuan Paper Building
	中国宣纸股份有限公司原料基地	Raw material base of China Xuan Paper Co., Ltd.		《中国宣纸》会刊	*China Xuan Paper* (a conference proceedings)
	"宣纸国家标准"封面	Cover of *National Standards of Xuan Paper*		宣纸博物馆和宣纸文化园鸟瞰图	An airscape of the Xuan Paper Museum and Xuan Paper Cultural Park
	中国宣纸文化园	China Xuan Paper Cultural Park		中国美术家协会2015年度大会现场	China Artist Association Conference (2015)
	《中国宣纸》杂志	Magzine of *China Xuan Paper*		正在晒纸的邢春荣	Xing Chunrong drying the paper
	张永惠《安徽宣纸工业之调查》原刊书影	Cover of *The Research of Anhui Xuan Paper Industry* written by Zhang Yonghui		正在车间抄纸的周东红	Zhou Donghong making the paper
				正在抄纸的朱建胜	Zhu Jiansheng making the paper
	姜世襄《改进宣纸质量和生产方法的商讨》原刊书影	The original record of *Discussion on Improving Quality and Production of Xuan Paper* written by Jiang Shixiang		正在抄纸的孙双林	Sun Shuanglin making the paper
				正在晒纸的汪息发	Wang Xifa drying the paper
	陈彭年《关于宣纸问题》原刊书影	The original record of *Problems About Xuan Paper* written by Chen Pengnian		正在剪纸的郑志香	Zheng Zhixiang cutting the paper
	《中国造纸史》书影	A photo of *The History of Papermaking in China*		正在检原料的罗鸣	Luo Ming choosing the materials
	刘仁庆著《中国书画纸》等书书影	*Chinese Calligraphy and Painting Paper* and other books written by Liu Renqing		张必跃	Zhang Biyue, a papermaker
				正在揭纸的赵永成	Zhao Yongcheng peeling the paper down
	《中国古纸谱》日文版书影	Japanese copy of *A Collection of Ancient Chinese Paper* written by Liu Renqing		原料基地前的黄迎福	Huang Yingfu at the Raw Material Base
	曹天生著《中国宣纸》等书书影	*Chinese Xuan Paper* and other books written by Cao Tiansheng		曹宁泰旧照	An old photo of Cao Ningtai
	《造纸与印刷》书影	A photo of *Papermaking and Printing*		砍青檀枝条	Cutting *Pteroceltis tatarinowii* Maxim. branch
	《图说中国古代四大发明——造纸术》	Different versions of *Illustrated Four Great Inventions of Ancient China: Papermaking* by Tang Shukun		村里路边剥皮忙	Stripping tree bark in the village
				浸泡剥下的毛皮	Soaking the stripped bark
	《宣纸志（评议稿）》封面和题词	The cover and inscription of *Record of Xuan paper*		快乐踏皮的工人	Workers stamping the bark by feet happily
章节	图中文名称	图英文名称		流水里洗皮坯	Soaking and cleaning the bark materials in a stream
第二章	宣纸	Chapter II Xuan Paper			
第一节	中国宣纸股份有限公司	Section 1 China Xuan Paper Co., Ltd.		晒滩上晾皮坯	Drying the bark on the drying ground
	中国宣纸股份有限公司鸟瞰图	A airscape of China Xuan Paper Co., Ltd.		装皮坯入蒸锅	Putting the bark materials into the steaming pot
	乌溪水景	Waterscape of Wuxi area		撕渡皮	Peeling the bark
	明·嘉靖版《泾县志》书影	A photocopy of *The Annals of Jingxian County* printed in Jiajing Reign of the Ming Dynasty		捏青皮	Tearing the bark apart
				将燎皮背下山	Carrying the dried bark down the hill
	《大清一统志》中"乌溪"的记载	Records of Wuxi area in *Geographical Situation of the Qing Dynasty*		洗皮旧照	An old photo of cleaning the bark
	明·嘉靖版《泾县志》书影	A photocopy of *The Annals of Jingxian County* printed in Jiajing Reign of the Ming Dynasty		女工在选检皮料	Female workers choosing the high quality bark
				碓皮料	Beating the bark with pestle
	"红旗"牌商标图案	Logo of Hongqi (Red Flag) Brand		切皮	Cutting the bark
	1948年极不稳定的宣纸价格（《泾县志》1996年版259页）截图	The unstable prices of Xuan paper in 1948 (printed in *The Annals of Jingxian County* in 1996, page 259)		踩料	Stamping the materials
	打字油印本《泾县宣纸厂志》	A mimeograph of *Records of Xuan Paper Factory in Jingxian County*		袋料扒	Bag used for cleaning the materials
	旧日的泾县宣纸厂工人履历表	Resume of a worker from Xuan Paper Factory in Jingxian County in old days		选稻草	Choosing the straw
	皖南泾县宣纸联营处使用的印戳（上）、徽章（下）	Seal (upper) and badge (lower) of Xuan Paper Joint Business Department in Jingxian County of Southern Anhui		收草坯	Piling the straw
				端料	Soaking the materials in alkali
	泾县宣纸业报呈（左）、冯金池专员回函（右）	Report of Xuan Paper in Jingxian County (left) and Commissioner Feng Jinchi's Reply (right)		草料出锅	Lifting the straw out of pot
				挑草块上摊	Carrying the processed straw on the drying ground
	联营期间工人睡觉吊床旧照	An old photo of hammock for workers during the joint venture period		扯青草	Pulling the straw apart
	联营期间的增产节约总结报告	Final report of increased production and reduced cost during the joint venture period		摊青草	Drying the straw
				翻燎草块	Turning over the processed straw
	公私合营私股领息凭证	Stock interest receipt of the joint venture		背燎草下滩	Carrying the processed straw
	1966年形成国营体制的相关原文件	Original documents of forming state-owned enterprises in 1966		鞭干草	Beating the dried straw
	1966年的安徽省轻工业厅文件	Document issued by Anhui Department of Light Industry in 1966		洗草旧照	An old photo of cleaning the straw
	关于组建中国宣纸集团公司的相关文件	Documents announcing of the formation of China Xuan Paper Co., Ltd.		木碓舂草	Beating the straw with wooden pestle
				袋料	Cleaning the materials with a bag
	"深交所"挂牌上市发布会现场	News Release Conference for listing in Shenzhen Stock Exchange		木槌锤杨桃藤	Beating branches of *Actinidia chinensis* Planch with a wooden hammer
	变更为股份公司的文件	Document announcing the formation of China Xuan Paper Co., Ltd.			
	"红星"品牌商标	The trademark of "Red Star" Brand		划单槽	Stirring the materials
	中国驰名商标（上）与中华老字号（下）标识	Identification of Famous Trademark of China (upper) and China's Time-Honored Brand (lower)			
	国家级非遗生产性保护基地牌匾	Plaque of National Intangible Cultural Heritage Preservation Base			

图中文名称	图英文名称
捞纸	Making the paper
扳榨	Pressing and squeezing the paper
浇帖	Watering the paper pile
牵纸	Pulling the paper on a drying wall
看纸检验	Choosing the high quality paper
剪纸	Cutting the paper
竹篓打包旧照	An old photo of packaging the paper with bamboo basket
单株青檀	*Pteroceltis tatarinowii* Maxim
泾县的沙田稻	Straw grown in the sands in Jingxian County
乌溪河水	River in Wuxi area
生产"红星"宣纸的一处晒滩	Drying ground for making Red Star Xuan paper
生产"红星"宣纸所用的蒸锅	Steaming pot for making Red Star Xuan paper
传统碓旧照	An old photo of the traditonal pestle
纸槽	Papermaking trough
宣纸帘床	Xuan papermaking frame
钢板焙	Iron drying boards
"天下第一剪"	"Shears Second to None"
"红星"古艺宣纤维形态图 (10×)	Fibers of "Red Star" traditional Xuan paper (10× objective)
"红星"古艺宣纤维形态图 (20×)	Fibers of "Red Star" traditional Xuan paper (20× objective)
"红星"古艺宣润墨效果	Writing performance of "Red Star" traditional Xuan paper
"红星"特净纤维形态图 (10×)	Fibers of "Red Star" superb-bark paper (10× objective)
"红星"特净纤维形态图 (20×)	Fibers of "Red Star" superb-bark paper (20× objective)
"红星"特净润墨性效果	Writing performance of "Red Star" superb-bark paper
"红星"净皮纤维形态图 (10×)	Fibers of "Red Star" clean-bark paper (10× objective)
"红星"净皮纤维形态图 (20×)	Fibers of "Red Star" clean-bark paper (20× objective)
"红星"净皮润墨效果	Writing performance of "Red Star" clean-bark paper
"红星"棉料纤维形态图 (10×)	Fibers of "Red Star" Mianliao paper (10× objective)
"红星"棉料纤维形态图 (20×)	Fibers of "Red Star" Mianliao paper (20× objective)
"红星"棉料润墨效果	Writing performance of "Red Star" Mianliao paper
宣纸邮票	Xuan paper stamp
纸箱包装好的成品"红星"宣纸	Packaged "Red Star" Xuan paper product
特制纪念纸的外包装	Package for specially-made memorial paper
北京红星宣纸销售门店	Red Star Xuan Paper Store in Beijing City
张茜写给宣纸厂的信函原件照	A Letter to Xuan Paper Factory written by Zhang Xi
郭沫若的题词原件照	Autograph written by Guo Moruo
李可染赠泾县宣纸厂的牧牛图原件照	Li Keran gives the cowherd as a gift to Xuan Paper Factory in Jingxian County
刘海粟赠泾县宣纸厂题词原件照	Liu Haisu gives autograph as a gift to Xuan Paper Factory in Jingxian County
吴作人赠泾县宣纸厂题词原件照	Wu Zuoren gives autograph as a gift to Xuan Paper Factory in Jingxian County
江泽民赠题词原件照	Jiang Zemin gives autograph as a gift to Xuan Paper Factory in Jingxian County
青檀林基地	*Pteroceltis tatarinowii* Maxim. Forest Base
厂部附近的燎草基地	Processed straw base near the factory area
标准化颁布实施的文件	Documents of Xuan paper standardization
宣纸入选国家非遗牌匾	Intangible Cultural Heritage Plaque of Xuan Paper
1982 年为美国双子星版画社研制的版画纸	Paper made for Gemini Print Group (US) in 1982
红星水库库区	Red Star Reservoir area
塑料丝绳的帘床	Papermaking frame with plastic thread
评聘技师的文件	Professional technicians listed in a document
"红星"古艺宣透光摄影图	A photo of "Red Star" traditional Xuan paper seen through the light
"红星"净皮透光摄影图	A photo of "Red Star" clean-bark paper seen through the light
"红星"棉料透光摄影图	A photo of "Red Star" Mianliao paper seen through the light
"红星"特净透光摄影图	A photo of "Red Star" superb-bark paper seen through the light
"红星"净皮四尺仿古宣透光摄影图	A photo of "Red Star" four-chi vintage Xuan paper (clean-bark) seen through the light

章节	图中文名称	图英文名称
第 二 节	泾县汪六吉宣纸有限公司	Section 2 Wangliuji Xuan Paper Co., Ltd. in Jingxian County
	汪六吉宣纸有限公司大门与内景	View of Wangliuji Xuan Paper Co., Ltd.
	调查组成员访谈李永喜	Researchers interviewing Li Yongxi, a papermaker
	调查组成员访谈李正明	Researchers interviewing Li Zhengming, a papermaker
	"汪六吉"料半采样照片	A sample of "Wangliuji" Liaoban paper
	"汪六吉"黄料采样照片	A sample of "Wangliuji" Huangliao paper
	"汪六吉"料半纤维形态图 (10×)	Fibers of "Wangliuji" Laioban paper (10× objective)
	"汪六吉"料半纤维形态图 (20×)	Fibers of "Wangliuji" Laioban paper (20× objective)
	"汪六吉"料半润墨效果	Writing performance of "Wangliuji" Liaoban paper
	"汪六吉"黄料纤维形态图 (10×)	Fibers of "Wangliuji" Huangliao paper (10× objective)
	"汪六吉"黄料纤维形态图 (20×)	Fibers of "Wangliuji" Huangliao paper (20× objective)
	"汪六吉"黄料润墨效果	Writing performance of "Wangliuji" Huangliao paper
	等待浸泡的青檀毛皮	*Pteroceltis tatarinowii* Maxim. bark for soaking
	入池浸泡杨桃藤	Soaking branches of *Actindia carambola* planch.
	过滤杨桃藤汁	Extracting *Actinidia chinensis* Planch.,
	洗白皮	Cleaning the bark
	洗净的白皮	Clean bark
	选拣台	Table for choosing high-quality bark
	选白皮	Picking out the impurities
	晒滩 (燎草滩)	Drying field for drying the processed straw
	摊晒	Drying the straw
	洗燎草	Cleaning the processed straw
	石碾	Stone roller
	洗漂机	Machine for cleaning and bleaching the materials
	加浆料	Adding the pulp materials
	搞槽 (搅拌)	Stirring the papermaking materials
	捞纸	Making the paper
	放帘	Turning the papermaking screen upside down on the board
	盖纸板	Putting the papermaking board on paper
	压榨	Squeezing and pressing the paper
	压榨完成	Squeezed paper
	抬纸帖	Carrying the squeezed paper to the drying workshop
	整块纸帖烘烤	Drying the paper pile on a drying wall
	浇帖	Watering the paper
	鞭帖	Beating the paper
	纸焙	Drying wall
	晒纸	Drying the paper
	检验	Choosing the high-quality paper

章节	图中文名称	图英文名称
	剪纸	Cutting the paper
	待包装的纸	Piles of paper for packaging
	"汪六吉"产品仓库	Storehouse of "Wangliuji" product
	纸帘	Papermaking screen
	帘床	Papermaking frame for supporting the papermaking screen
	扒头	Tool for stirring the pulp
	浇帖架	Frame for supporting the paper pile
	切帖刀	Paper cutter
	包装好的松毛刷	Packaged brush for pasting the paper
	晒纸墙	Drying wall
	剪刀	Shears
	压纸石	Stone for pressing the paper
	额枪	Tool for separating the paper
	电瓶车	Three-wheeled battery motor vehicle for carrying the squeezed paper pile
	"汪六吉"宣纸产品	"Wangliuji" Xuan paper product
	"汪六吉"展厅的名家题字	Celebrities' autographs in the "Wangliuji" exhibition hall
	环保流程导览图	Flowsheet of environmental protection process
	厂区的环保设备	Environmental protection equipments in the factory area
	"汪六吉"料半透光摄影图	A photo of "Wangliuji" Liaoban paper seen through the light
	"汪六吉"仿古橘红色宣透光摄影图	A photo of "Wangliuji" vintage orange Xuan paper seen through the light
	"汪六吉"黄料透光摄影图	A photo of "Wangliuji" Huangliao paper seen through the light

章节	图中文名称	图英文名称
第三节	安徽恒星宣纸有限公司	Section 3 Anhui Hengxing Xuan Paper Co., Ltd.
	恒星宣纸有限公司门牌与厂区内景	The doorplate of Hengxing Xuan paper Co., Ltd. and a view of the factory area
	后山村旁清澈的小河	River near Houshan Village
	调查组成员在厂区办公室访谈张明喜	Researchers interviewing Zhang Mingxi, a papermaker, in an office of the factory
	"恒星"宣纸成品纸	"Hengxing" Xuan paper
	"恒星"手工纸包装箱成品	Packaged "Hengxing" handmade paper
	"恒星"净皮纤维形态图 (10×)	Fibers of "Hengxing" clean-bark paper (10× objective)
	"恒星"净皮纤维形态图 (20×)	Fibers of "Hengxing" clean-bark paper (20× objective)
	"恒星"净皮润墨效果	Writing performance of "Hengxing" clean-bark paper
	"恒星"御品贡宣纤维形态图 (10×)	Fibers of "Hengxing" tribute Xuan paper (10× objective)
	"恒星"御品贡宣纤维形态图 (20×)	Fibers of "Hengxing" tribute Xuan paper (20× objective)
	"恒星"御品贡宣润墨效果	Writing performance of "Hengxing" tribute Xuan paper
	"红日"精品书画纸纤维形态图 (10×)	Fibers of "Hongri" fine calligraphy and painting paper (10× objective)
	"红日"精品书画纸纤维形态图 (20×)	Fibers of "Hongri" fine calligraphy and painting paper (20× objective)
	"红日"精品书画纸润墨效果	Writing performance of "Hongri" fine calligraphy and painting paper
	厂区内种植的青檀树	Pteroceltis tatarinowii Maxim. planted in the Factory area
	厂区库房里存放的毛皮	Bark stored in the storehouse of the factory
	厂区车间里的燎草	Processed straw in the workshop of the factory
	青弋江边环境	Senery of Qingyi River
	蓄水池	Reservoir
	已处理好的白皮	Processed bark
	挑选白皮	Choosing high-quality bark
	滚筒筛	Roller sifter for sieving the materials
	厂区的捞纸车间大门	Gate of the papermaking workshop in the factory area
	混合浆	Mixing papermaking materials
	搅拌浆料	Stirring the papermaking pulp
	捞纸	Making the paper
	放帘	Turning the papermaking screen upside down on the board
	压榨	Pressing the paper
	揭纸与晒纸	Peeling the paper down and drying the paper
	检验女工	A female worker choosing high-quality paper
	女工在剪纸	A female worker cutting the paper
	"恒星"宣纸印章	Seal of "Hengxing" Xuan paper
	包装机	Packaging machine
	包装好的纸	Packaged paper
	龙须草浆板	*Eulaliopsis binata* pulp board used as papermaking materials
	精品书画纸手工捞纸工序	Procedures of making fine calligraphy and painting paper
	御品贡宣半自动喷浆捞纸工序	Procedures of making tribute Xuan paper by semi-automatic equipment
	压榨	Pressing the paper
	揭纸、晒纸工序	Procedures of peeling the paper down and drying the paper
	检验车间	Workshop for checking the paper
	女工在检验	A female worker Choosing high quality paper
	裁剪后的纸垛	A pile of paper after cutting
	包装车间一角	A view of packaging workshop
	库房里包装好的成品纸	Piles of packaged paper in the storehouse
	石碾	Stone roller
	打浆机	Beating machine
	纸槽	Papermaking trough
	纸帘	Papermaking screen
	帖架	Papermaking shelf for watering the paper pile
	切帖刀	Paper cutter
	刷子	Brush for pasting the paper
	木槌	Mallet for beating the paper
	晒纸墙	Drying wall
	大剪刀	Shears
	扒子	Stirring harrow
	"恒星"成品纸	Final product of "Hengxing" Xuan paper
	后山村里的张氏宗祠	Family Ancestral Hall of the Zhang in Houshan Village
	恒星宣纸有限公司厂区图	A view of Hengxing Xuan Paper Co., Ltd.
	"恒星"净皮透光摄影图	A photo of "Hengxing" clean-bark paper seen through the light
	"恒星"精品书画纸透光摄影图	A photo of "Hengxing" fine calligraphy and painting paper seen through the light
	"恒星"御品贡宣透光摄影图	A photo of "Hengxing" tribute Xuan paper seen through the light
	"恒星"特净扎花透光摄影图	A photo of "Hengxing" superb-bark Zhahua paper seen through the light

章节	图中文名称	图英文名称
第四节	泾县桃记宣纸有限公司	Section 4 Taoji Xuan Paper Co., Ltd. in Jingxian County
	桃记宣纸有限公司厂区全景	A view of Taoji Xuan Paper Co., Ltd.
	苏红村的燎草生产基地[晒摊]	Production base of processed straw in Suhong Village (drying field)
	调查组走访原古艺宣纸厂旧址	Researchers visiting the former site of Traditional Xuan Paper Factory

	图中文名称	图英文名称
	原古艺宣纸厂旧址厂区内景	A view of former Traditional Xuan Paper Factory
	原古艺宣纸厂纸库	Storehouse of former Traditional Xuan Paper Factory
	调查组考察慈坑原漕溪宣纸厂旧址	Researchers visiting former papermaking site of Caoxi Xuan Paper Factory in Cikeng Village of Jingxian County
	原漕溪宣纸厂造纸旧设施	Old tools in the former papermaking site of Caoxi Xuan Paper Factory
	调查组考察泥坑原苏红宣纸厂旧址	Researchers visiting the former papermaking site Suhong Xuan Paper Factory in Nikeng Village
	泥坑原苏红宣纸厂旧址	Former Suhong Xuan Paper Factory in Nikeng Village
	调查组成员访谈胡青山	Researchers interviewing Hu Qingshan, a factory manager
	创始人胡青山	Hu Qingshan, founder of the factory
	现任法人胡凯	Hu Kai, present corporate representative of the factory
	"桃记"宣纸部分产品	Some products of "Taoji" Xuan paper
	"桃记"棉料纤维形态图（10×）	Fibers of "Taoji" Mianliao paper (10× objective)
	"桃记"棉料纤维形态图（20×）	Fibers of "Taoji" Mianliao paper (20× objective)
	"桃记"棉料润墨性效果	Writing performance of "Taoji" Mianliao paper
	放在箩筐里的青檀皮料	Pteroceltis tatarinowii Maxim. bark materials in bamboo basket
	袋中的白皮料	Bark materials in the bags
	处理过的杨桃藤枝	Processed branches of Actinidia chinensis Planch.
	桃记宣纸有限公司厂区旁边的溪流	Stream near Taoji Xuan Paper Co.,Ltd.
	桃记宣纸有限公司生产车间的挑拣台	Table for picking and choosing high-quality materials in Taoji Xuan Paper Co., Ltd.
	平筛设备	Flat filter
	待碾压的草料与石碾	Straw to be rolled and the stone roller
	漂洗设备	Machine for bleaching and cleaning the materials
	工人正将浆料倒入捞纸槽	Pouring the pulp materials into papermaking trough
	捞纸车间	Workshop for papermaking
	已烘干的纸帖	Dried paper pile
	工人在浇帖	A worker watering the paper pile
	晒纸房	Workshop for drying the paper
	女工在逐张检验	Female workers choosing the high-quality paper
	检验完成待包装的宣纸	Xuan paper to be packaged after checking
	捞纸槽	Papermaking trough
	不同规格的帘床	Papermaking frame of various sizes
	纸帘	Papermaking screen
	浇帖架	Frame for supporting the paper pile
	额枪	Tool for separating the paper
	松毛刷	Brush made of pine needles
	大剪刀	Shears
	桃记宣纸有限公司内景	Interior view of Taoji Xuan Paper Co., Ltd.
	原古艺宣纸厂厂区后的檀皮林和晒滩	Pteroceltis tatarinowii forest and drying field behind former Xuan Paper Factory
	《小岭青檀溪水傍 曹氏宣纸天下扬——泾县小岭宣纸历代成名记》一文	A published essay: Pteroceltis tatarinowii near the Stream in Xiaoling Village,Caoshi Xuan Paper Here is World-famous:The Biography of Xuan Paper in Xiaoling Village of Jingxian County
	胡青山正给调查组成员绘原古艺宣纸厂商标	Hu Qingshan illustrating the trademark of former Xuan Paper Factory to the researchers
	胡青山所绘原古艺宣纸厂商标	Trademark of former Xuan Paper Factory (drawn by Hu Qingshan)
	调查组成员考察中与胡青山聊"挑战"	A researcher talking with Hu Qingshan about the "challenges" of papermaking
	"桃记"棉料宣纸透光摄影图	A photo of "Taoji" Mianliao Xuan paper seen through the light
	"桃记"棉料棉连宣透光摄影图	A photo of "Taoji" Mianlian Xuan paper (Mianliao) seen through the light
	"桃记"特净透光摄影图	A photo of "Taoji"superb-bark paper seen through the light

章节	图中文名称	图英文名称
第五节	泾县汪同和宣纸厂	Section 5 Wangtonghe Xuan Paper Factory in Jingxian County
	流经古坝村的漕溪河	Caoxi River flowing through Guba Village
	汪同和宣纸厂区路标	Road sign towards Wangtonghe Xuan Paper Factory area
	汪同和宣纸厂区内景	Interior view of Wangtonghe Xuan Paper Factory area
	汪同和宣纸有限公司所获奖牌	Certificates won by Wangtonghe Xuan Paper Factory
	"墨记汪同和堆栈"旧匾	Former plaque of "Moji Wangtonghe Storehouse"
	"鸡球"牌宣纸	"Jiqiu" Xuan paper
	访谈中侃侃而谈的程彩辉	Interviewing Cheng Caihui
	调查组成员在程涛导引下考察厂区（左一为程涛）	Researchers visiting the factory under the guidance of Cheng Tao (first from the left)
	"国内专用宣纸"类宣纸	"Special Xuan paper for domestic use" Xuan paper
	"墨记"类宣纸	"Moji" Xuan paper
	"精制"类宣纸	"Refined" Xuan paper
	"极品"类宣纸	"Superb" Xuan paper
	"汪同和"特净纤维形态图（10×）	Fibers of "Wangtonghe" superb-bark paper (10× objective)
	"汪同和"特净纤维形态图（20×）	Fibers of "Wangtonghe" superb-bark paper (20× objective)
	"汪同和"特净润墨效果	Writing performance of "Wangtonghe" superb-bark paper
	汪同和宣纸厂厂区门口自栽的青檀树林	Self-planted Pteroceltis tatarinowii Maxim forest at the entrance to Wangtonghe Xuan Paper Factory
	草坯堆（苏红燎草厂）	Piles of processed straw (Suhong Processed Straw Factory)
	流经厂区的溪水	Stream flowing through the factory area
	檀皮蒸煮	Steaming the Pteroceltis tatarinowii bark
	拣黑皮	Picking out the impurities
	漂白好的白皮	Bleached bark
	清洗后放入木榨	Putting the bark materials in wooden pressing device after cleaning
	扳榨	Pressing the paper
	拣白皮	Picking and choosing the high-quality bark
	碾草	Grinding the straw
	打浆	Beating the materials
	捞纸	Making the paper
	压榨工具	Pressing device
	靠帖	Drying the paper pile
	晒纸	Drying the paper
	检纸	Checking the paper quality
	剪纸	Cutting the paper
	石碾	Stone rollers
	打浆机	Beating machine
	车间里的纸槽	Papermaking trough in the workshop
	挂在墙上的纸帘	Papermaking screen hanging on the wall
	帘床	Frame for supporting the papermaking screen
	搅拌泵	Stirring device
	剪档	Jiandang, papermaking tool (for fixing and aligning the wet paper)
	衬档	Chendang, papermaking tool (for fixing and aligning the wet paper)
	浇帖架	Frame for supporting the paper pile

	刷子	Brush made of pine needles
	晒纸焙	Drying wall
	晒纸架	Frame for drying the paper
	额枪	Tool for separating the paper
	剪刀	Shears
	包装工	A packer
	盖印工	A worker stamping on paper
	汪同和"古槽屋"牌匾	"Gu Cao Wu", the plaque of the ancient Wangtonghe Paper Mill
	汪同和古纸庄留下的遗迹	Remains of the ancient Wangtonghe Paper Mill
	汪同和宣纸厂厂区内景	Interior view of the factory area of Wangtonghe Xuan Paper Facory
	穆孝天的题字	Autograph by calligrapher Mu Xiaotian
	厂区外美丽洁净的漕溪河	Beautiful and clean Caoxi River outside the factory
	《汉语大词典》书影	A Photo of *Chinese Dictionary*
	"汪同和"特净透光摄影图	A photo of "Wangtonghe" superb-bark paper seen through the light

章节	图中文名称	图英文名称
第 六 节	泾县双鹿宣纸有限公司	Section 6 Shuanglu Xuan Paper Co., Ltd. in Jingxian County
	双鹿宣纸有限公司正门	Main entrance of Shuanglu Xuan Paper Co., Ltd.
	双鹿宣纸有限公司生产厂区内景	Interior view of Shuanglu Xuan Paper Co., Ltd.
	原双鹿宣纸厂党支部标牌	Party branch sign in the former Shuanglu Xuan Paper Factory
	百岭坑原宣纸作坊遗址	Former Xuan paper mill in Bailingkeng area
	百岭坑宣纸厂遗址	Remains of the former Bailingkeng Xuan Paper Factory
	旧"魁星"宣纸商标	Former trademark of "Kuixing" Xuan paper
	在生产厂区办公室访谈张先荣	Interviewing Zhang Xianrong in the factory office
	在曹光华县城的店中访谈	Interviewing Cao Guanghua at his store in Jingxian County
	"双鹿"净皮宣纸纤维形态图 (10×)	Fibers of "Shuanglu" clean-bark Xuan paper (10× objective)
	"双鹿"净皮宣纸纤维形态图 (20×)	Fibers of "Shuanglu" clean-bark Xuan paper (20× objective)
	"双鹿"净皮宣纸润墨效果	Writing performance of "Shuanglu" clean-bark Xuan paper
	"曹光华"牌"白鹿"宣纸纤维形态图 (10×)	Fibers of "Caoguanghua" Brand "Bailu" Xuan paper (10× objective)
	"曹光华"牌"白鹿"宣纸纤维形态图 (20×)	Fibers of "Caoguanghua" Brand "Bailu" Xuan paper (20× objective)
	"曹光华"牌"白鹿"宣纸润墨效果	Writing performance of "Caoguanghua" Brand "Bailu" Xuan paper
	"曹光华"本色特净宣纸纤维形态图 (10×)	Fibers of "Caoguanghua" superb-bark Xuan paper in original color (10× objective)
	"曹光华"本色特净宣纸纤维形态图 (20×)	Fibers of "Caoguanghua" superb-bark Xuan paper in original color (20× objective)
	"曹光华"本色特净宣纸润墨效果	Writting performance of "Caoguanghua" superb-bark Xuan paper in original color
	已加工的青檀皮	Processed *Pteroceltis tatarinowii* Maxim. bark
	捶打法制作纸药	Making papermaking mucilage by beating
	蒸煮	Steaming the bark
	压榨	Pressing the paper
	选检	Choosing the high-quality bark
	洗涤	Cleaning the materials
	选检	Choosing the high-quality bark
	跳筛设备	Sieving device
	盛放纸药的缸	Papermaking mucilage container
	捞纸	Making the paper
	放纸并揭帘	Putting the paper on the board and taking off the papermaking screen

	扳榨	Pressing the paper
	烘帖	Drying the paper pile by baking
	浇帖	Watering the paper pile
	鞭帖	Patting the paper pile
	晒纸	Drying the paper
	揭纸	Peeling the paper down
	供热装置	Heating device
	检验、剪纸	Checking and cutting the paper
	装箱打包	Packaging the paper
	药榔头	Hammer for beating the papermaking mucilage
	石碾	Stone roller
	打浆机	Beating machine
	捞纸槽	Papermaking trough
	纸帘	Papermaking screen
	挽钩(左), 钉耙(右)	Hook (left), nail rake (right)
	松毛刷	Brush made of pine needles
	晒纸焙	Drying wall
	过滤池	Filter pool
	氧化池	Oxidation pool
	回收仓	Recycle room
	"双鹿"宣纸外包装	Package of "Shuanglu" Xuan paper
	"四宝堂"牌匾和内景	Plaque of "SiBaotang" and the interior view
	获得的荣誉	The certificate
	百岭坑的山水	Landscape of Bailingkeng area
	"曹光华""白鹿"特净透光摄影图	A photo of "Caoguanghua" "Bailu" superb-bark paper seen through the light
	"曹光华"仿古半生熟加工纸透光摄影图	A photo of "Caoguanghua" vintage semi-processed paper seen through the light
	"曹光华"本色特净透光摄影图	A photo of "Caoguanghua" superb-bark paper in original color seen through the light
	"双鹿"净皮透光摄影图	A photo of "Shuanglu" clean-bark paper seen through the light
	"双鹿"四尺棉连透光摄影图	A photo of "Shuanglu" four-chi Mianlian paper seen through the light

章节	图中文名称	图英文名称
第 七 节	泾县金星宣纸有限公司	Section 7 Jinxing Xuan Paper Co., Ltd. in Jingxian County
	金星宣纸有限公司厂区外景	Exterior view of Jinxing Xuan Paper Co., Ltd.
	金星宣纸有限公司厂区内种植的青檀树	*Pteroceltis tatarinowii* Maxim. planted in Jinxing Xuan Paper Co., Ltd.
	第二任厂长黄永堂	Huang Yongtang, the second factory director
	张必福	Zhang Bifu
	写在墙上的金星宣纸有限公司部分荣誉	Part of honor written on the wall of Jinxing Xuan Paper Co., Ltd.
	调查组成员访谈张汉荣	Researchers interviewing Zhang Hanrong
	"金星"宣纸代表纸样	Representative samples of "Jinxing" Xuan paper
	金星宣纸有限公司厂区里种植的青檀树	*Pteroceltis tatarinowii* Maxim. planted in Jinxing Xuan Paper Co., Ltd.
	"金星"净皮宣纸纤维形态图 (10×)	Fibers of "Jinxing" clean-bark Xuan paper (10× objective)
	"金星"净皮宣纸纤维形态图 (20×)	Fibers of "Jinxing" clean-bark Xuan paper (20× objective)
	"金星"净皮宣纸润墨效果	Writing performance of "Jinxing" clean-bark Xuan paper
	"金星"构皮纤维形态图 (10×)	Fibers of "Jinxing" Mulberry bark paper (10× objective)
	"金星"构皮纤维形态图 (20×)	Fibers of "Jinxing" Mulberry bark paper (20× objective)

	图中文名称	图英文名称
	"金星"构皮纸润墨效果	Writing performance of "Jinxing" mulberry bark paper
	檀皮仓库一角	A corner of *Pteroceltis tatarinowii* bark warehouse
	堆放着的燎草	Stacked processed straw
	人工选拣黑皮	Picking out the impurities manually
	皮料甩干机	Dryer for drying the bark materials
	洗草	Cleaning the straw
	张汉荣在介绍碾草工艺	Zhang Hanrong introducing the techniques of grinding the straw
	捞纸车间	Workshop of papermaking
	工人在捞喷浆书画纸	A worker making the calligraphy and painting paper
	喷浆书画纸的放帘动作	Piling process for making calligraphy and painting paper
	浇帖架	Frame for supporting the paper pile
	晒纸	Drying the paper
	工人正在检验纸张	Workers checking the paper quality
	检验合格的纸	Qualified paper
	包装好并入库的纸	Packaged paper for storage
	打浆机	Beating machine
	石碾	Stone roller
	甩干机	Drier for drying the bark materials
	纸槽	Papermaking trough
	纸帘	Papermaking screen
	帖架	Papermaking shelf for watering the paper pile
	刷子	Brush made of pine needles
	额枪	Tool for separating the paper
	纸焙	Drying wall
	压纸石	Pressing stone
	剪刀	Shears
	金星宣纸有限公司网站截图	Website screenshot of Jinxing Xuan Paper Co., Ltd.
	"金星"成品宣纸	The final product of "Jinxing" Xuan paper
	金星宣纸有限公司的荣誉墙	Certificate wall of Jinxing Xuan Paper Co., Ltd.
	金星宣纸有限公司生产车间的警示标语	Warning signs in the production workshop of Jinxing Xuan Paper Co., Ltd.
	金星宣纸有限公司环保设备	Environmental protection equipments of Jinxing Xuan Paper Co., Ltd.
	"金星"构皮纸透光摄影图	A photo of "Jinxing" paper mulberry bark paper seen through the light
	"金星"净皮透光摄影图	A photo of "Jinxing" clean-bark paper seen through the light
	"金星"四尺古云龙纸透光摄影图	A photo of "Jinxing" four-chi traditional Yunlong paper seen through the light
	"金星"楮皮四尺白皮纸透光摄影图	A photo of "Jinxing" four-chi white bark paper (paper mulberry bark) seen through the light

章节	图中文名称	图英文名称
第八节	泾县红叶宣纸有限公司	Section 8 Hongye Xuan Paper Co., Ltd. in Jingxian County
	水西双塔	Shuixi Twin Towers
	流经厂区旁的许家湾溪	Xujiawan Stream near the factory area
	红叶宣纸有限公司厂区正门	Main entrance of Hongye Xuan Paper Co., Ltd.
	调查组成员正在沈学斌家中进行访谈	Researchers interviewing Shen Xuebin at his house
	18 位股东创始人名单	List of 18 initial shareholders
	用"红叶"宣纸创作的书法作品	Calligraphy written on "Hongye" Xuan paper
	红叶宣纸有限公司注册成立证明材料	Establishment proof of Hongye Xuan Paper Co., Ltd.
	沈学斌(左一)给调查组成员展示、讲解纸张特性	Shen Xuebin (first from the left) showing and explaining the characteristics of paper to the researchers
	库存的 70 cm×205 cm 规格罗纹单宣	Ribbing-patterned Xuan paper (specification: 70 cm×205 cm)
	"红叶"净皮纸成品包装	Package of final product of "Hongye" clean-bark paper
	"红叶"净皮宣纤维形态图 (10×)	Fibers of "Hongye" clean-bark Xuan paper (10× objective)
	"红叶"净皮宣纤维形态图 (20×)	Fibers of "Hongye" clean-bark Xuan paper (20× objective)
	"红叶"净皮宣润墨效果	Writing performance of "Hongye" clean-bark Xuan paper
	沙田种植的水稻	Rice grown in sand field
	制浆设备	Pulp-making equipment
	红叶宣纸有限公司厂区旁边的流水	River near Hongye Xuan Paper Co., Ltd.
	洗草旧照	An old photo of cleaning the straw
	捞纸旧照	An old photo of making the paper
	晒纸—八尺宣	Drying the eight *chi* Xuan paper (248 cm× 129 cm)
	晒纸图	A photo of drying the paper
	捞纸槽	Papermaking trough
	帘床	Frame for supporting the papermaking screen
	"红叶"牌"建国八十周年纪念"专用纸纸帘	Special papermaking screen of "Hongye" Brand for "Celebrating the 80th Anniversary of the Founding of the People's Republic of China"
	原红叶宣纸有限公司正门	Former main entrance of Hongye Xuan Paper Co., Ltd.
	"红叶"牌宣纸的荣誉奖牌墙	Honor wall in Hongye Xuan Paper Co., Ltd.
	红叶品牌标识	Logo of Hongye Brand
	停业中的红叶宣纸有限公司厂区内景	Interior view of the closed Hongye Xuan Paper Co., Ltd.
	著名画家冯大中为红叶宣纸厂题写的厂名	Name of Hongye Xuan Paper Factory written by the famous painter, Feng Dazhong
	"红叶"净皮透光摄影图	A photo of "Hongye" clean-bark paper seen through the light

章节	图中文名称	图英文名称
第九节	安徽曹氏宣纸有限公司	Section 9 Anhui Caos Xuan Paper Co., Ltd.
	曹氏宣纸有限公司的指路牌与外部环境	Signpost and external environment of Caos Xuan Paper Co., Ltd.
	曹氏宣纸有限公司厂门	Gate of Caos Xuan Paper Co., Ltd.
	曹氏宣纸有限公司厂区内的展示空间	Showcase in Caos Xuan Paper Co., Ltd.
	曹健勤	Cao Jianqin, founder of Caos Xuan Paper Co., Ltd.
	曹氏家族内传之《曹氏宗谱》	*Genealogy of the Caos* preserved within the Caos family
	调查组成员在厂内访谈曹建勤	Researchers interviewing Cao Jianqin in the factory
	曹人杰旧照	An old photo of Cao Renjie
	《曹氏宗谱》中曹一清的传承脉系记录	Family tree of Cao Yiqing in *Genealogy of the Caos*
	曹氏宣纸有限公司纸库里的宣纸产品	Xuan paper products in the paper storehouse of Caos Xuan Paper Co., Ltd.
	曹氏宣纸有限公司的两种成品包装样式	Two product packaging styles in Caos Xuan Paper Co., Ltd.
	"曹氏宣纸"玉版宣纤维形态图 (10×)	Fibers of "Caoshi Xuan paper" jade-white Xuan paper (10× objective)
	"曹氏宣纸"玉版宣纤维形态图 (20×)	Fibers of "Caoshi Xuan paper" jade-white Xuan paper (20× objective)
	"曹氏宣纸"玉版宣润墨效果	Writing performance of "Caoshi Xuan paper" jade-white Xuan paper
	"曹氏麻纸"纤维形态图 (10×)	Fibers of "Caos Brand hemp paper" (10× objective)
	"曹氏麻纸"纤维形态图 (20×)	Fibers of "Caos hemp paper" (20× objective)
	"曹氏麻纸"润墨性效果	Writing performance of Cao's hemp paper
	车间里的青檀皮料	*Pteroceltis tatarinowii* Maxim. bark materials in the workshop
	野外摊放的燎草料	Processed straw materials spread outdoors

杨桃藤枝	Branches of *Actinidia chinensis* Planch.	"千年古宣"棉料润墨效果	Writing performance of "Millennium Xuan" Mianliao paper
枫坑河的造纸水源	Fengkeng River, the water source of papermaking	"宣和坊"特净纤维形态图（10×）	Fibers of "Xuanhefang" superb-bark paper (10× objective)
蒸皮料	Steaming the bark materials	"宣和坊"特净纤维形态图（20×）	Fibers of "Xuanhefang" superb-bark paper (20× objective)
清洗皮料	Cleaning the bark materials	"宣和坊"特净润墨效果	Writing performance of "Xuanhefang" superb-bark paper
脱水机	Water extractor	已浸泡的杨桃藤枝	Soaked branches of *Actinidia chinensis* Planch.
车间内工人正在捞纸	Workers making the paper in the workshop	"千年古宣"和"宣和坊"古法造纸用到的水源	Water source used in making "Millennium Xuan" and "Xuanhefang" paper with ancient papermaking techniques
"曹氏宣纸"的压榨设备	Pressing device in Caos Xuan Paper Co., Ltd.	捆好的青檀树枝	Bundled *Pteroceltis tatarinowii* Maxim. branches
晒纸	Drying the paper	加工好的皮坯	Processed bark
牵纸的顺序步骤图示	Showing the procedures of peeling the paper	晒滩上摊晒的皮坯	Drying the bark in the drying field
曹建勤正在演示剪纸	Cao Jianqin showing how to cut the paper	选拣皮	Picking the bark
本色黄麻纸	Jute paper in original color	碓皮	Beating the bark
石碾	Stone roller	"千年古宣"和"宣和坊"燎草库标牌	Signpost of "Millennium Xuan" and "Xuanhefang" processed straw storehouse
打浆机	Beating machine	人工选拣燎草	Picking the processed straw by workers
曹氏宣纸有限公司车间里的捞纸槽	Papermaking trough in a workshop of Caoshi Xuan Paper Co., Ltd.	石碾碾草	Grinding the straw by a stone roller
纸帘	Papermaking screen	"千年古宣"和"宣和坊"宣纸生产用的打浆机	Beating machine for making "Millennium Xuan" paper and "Xuanhefang" Xuan paper
扒子	Stirring stick	"千年古宣"宣纸生产用的平筛	Flat filter for making "Millennium Xuan" paper
刷子	Brush made of pine needles	"千年古宣"宣纸生产用的圆筒筛	Cylinder filter for making "Millennium Xuan" paper
晒纸墙（纸焙）	Drying wall	制纸药	Making the papermaking mucilage
特制剪刀	Special shears	过滤纸药	Filtering the papermaking mucilage
宣城市翻印本《宁国府志》	Reprinted version of *The Annals of Ningguo Prefecture* in Xuancheng City	过滤纸药的药袋	Filter bag of papermaking mucilage
文物出版社版重印本《大藏经》	Reprinted version of *Tripitaka* by Cultural Relics Press	捞纸	Making the paper
"曹氏宣纸"搭载"神六"飞船的相关报道	Reports of Shenzhou VI Spacecraft carrying "Caos Xuan paper"	放纸	Turning the papermaking screen upside down on the board
启功题字"宣纸世家"	Autograph of "Xuan Papermaking Family" by the famous calligrapher Qigong	湿纸帖	Wet paper pile
"曹氏宣纸"直营店在北京琉璃厂大街的店铺	Caos Xuan Paper Shop on Liulichang Street in Beijing City	压榨	Pressing the paper
画家陈家泠正在试纸	Chen Jialing, a painter, testing the paper	烘帖	Drying the paper pile
故宫博物院中的"曹氏宣纸"作品陈设	Display works on "Caos Xuan paper" in the Palace Museum	帖架	Frame for supporting the paper pile
泾川小岭的曹大三塑像	Statue of Cao Dasan in Xiaoling Village of Jingchuan County	鞭帖用的鞭	Stick for beating the paper
"曹氏宣纸"玉版宣透光摄影图	A photo of "Caoshi Xuan Paper" jade-white Xuan paper seen through the light	尚未盖章的纸	Unstamped paper
"曹氏"麻纸透光摄影图	A photo of "Caoshi" hemp paper seen through the light	皮碓	Bark pestle
章节	图中文名称	图英文名称	
第十节	泾县千年古宣宣纸有限公司	Section 10 Millennium Xuan Paper Co., Ltd. in Jingxian County	捞纸槽	Papermaking trough
	通往周坑村民组的山区公路	Country road to Zhoukeng Villagers' Group	纸帘	Papermaking trough
	周坑村民组	Zhoukeng Villagers' Group	松毛刷	Brush made of pine needles
	千年古宣宣纸有限公司厂区外景	Exterior view of Millennium Xuan Paper Co., Ltd.	扒头	Stirring stick
	商标查询截图	Screenshot of trademark information	额枪	Tool for separating the paper
	卢一葵	Lu Yikui	装士戎试纸创作及作品	Pei Shirong testing the paper and his works
	在办公室访谈卢一葵	Interviewing Lu Yikui in the office	"千年古宣"的荣誉牌匾	Honor plaque of "Millennium Xuan"
	曹移程在家中接受调查组访谈	Researchers interviewing Cao Yicheng at his home	小岭村的乡间公路	Country road in Xiaoling Village
	长子曹国胜	Cao Guosheng, Cao Yicheng's eldest son	启功为"千年古宣"题词	Autograph of "Millennium Xuan Paper" by Qigong
	次子曹国才	Cao Guocai, Cao Yicheng's second son	王世襄题字牌匾	Autograph written by Wang Shixiang
	"千年古宣"成品纸	"Millennium Xuan" paper product	调查组成员在厂内访谈曹移程	A researcher interviewing Cao Yicheng in the factory
	"宣和坊"成品纸	"Xuanhefang" paper product	在建中的"千年古宣文化创意产业园"	"Cultural and Creative Industrial Park of Millennium Xuan Paper" under construction
	"千年古宣"棉料纤维形态图（10×）	Fibers of "Millennium Xuan" Mianliao paper (10× objective)	"千年古宣"棉料透光摄影图	A photo of "Millennium Xuan" Mianliao paper seen through the light
	"千年古宣"棉料纤维形态图（20×）	Fibers of "Millennium Xuan" Mianliao paper (20× objective)	"宣和坊"特净透光摄影图	A photo of "Xuanhefang" superb-bark paper seen through the light

	"宣和坊""古槽"宣纸透光摄影图	A photo of "Xuanhefang" "Gucao" Xuan paper seen through the light
章节	图中文名称	图英文名称
第十一节	泾县小岭景辉纸业有限公司	Section 11 Xiaoling Jinghui Paper Co., Ltd. in Jingxian County
	景辉纸业有限公司厂区正门	Main entrance of Jinghui Paper Co., Ltd.
	山道上的周坑村指示牌	Signpost to Zhoukeng Village
	村口的指示牌	Signpost at the entrance of the village
	周坑村民集资建路功德碑	Merit stele of a road in Zhoukeng Villagers' Group (money raised by the villagers)
	周坑村的溪水	Stream of Zhoukeng Village
	金永辉	Jin Yonghui
	刘耀谷(右)在车间与调查组成员深入交流	Liu Yaogu (right) talking with a research
	刘耀谷	Liu Yaogu
	曹四明	Cao Siming
	调查组取样的"泾上白"净皮宣纸	"Jingshangbai" clean-bark Xuan paper sample collected by the research group
	景辉纸业有限公司制作的煮硾纸	Impregnated Xuan paper made by Jinghui Paper Co., Ltd.
	金永辉展示自己收藏的清代煮硾纸	Jing Yonghui's collection of impregnated Xuan paper in the Qing Dynasty
	金永辉用煮硾纸创作的绘画作品	Jing Yonghui's paintings on impregnated Xuan paper
	景辉纸业有限公司特净纤维形态图(10×)	Fibers of superb-bark paper in Jinghui Paper Co., Ltd. (10× objective)
	景辉纸业有限公司特净纤维形态图(20×)	Fibers of superb-bark paper in Jinghui Paper Co., Ltd. (20× objective)
	景辉纸业有限公司特净润墨效果	Writing performance of superb-bark paper in Jinghui Paper Co., Ltd.
	景辉纸业有限公司厂门口的青檀树	*Pteroceltis tatarinowii* Maxim. tree at the entrance of Jinghui Paper Co., Ltd.
	剥皮后的青檀树杆子	Peeled branches of *Pteroceltis tatarinowii* Maxim. branches
	景辉纸业有限公司堆放燎草的库房一角	Processed straw storehouse in Jinghui Paper Co., Ltd.
	景辉纸业有限公司厂区附近的山涧水	Mountain water near Jinghui Paper Co., Ltd.
	蒸皮	Steaming the bark
	毛皮	Unprocessed bark
	泡毛皮	Soaking the unprocessed bark
	踏皮后上堆	Stacking the bark after being stamped
	景辉纸业有限公司自建的皮、草摊晒场	Drying field for drying the bark and straw built by Jinghui Paper Co., Ltd.
	翻青皮	Turning the fresh bark over
	摔打皮坯	Beating the bark materials
	纯碱	Sodium carbonate
	浆灰	Pulp ash
	燎皮	Processed bark
	蒸锅	Pot for steaming the materials
	选拣燎皮	Picking the processed bark
	碓皮(打皮)	Beating the bark
	碓皮(打皮)	Beating the bark
	浆草	Soaking and fermenting straw in the lime pulp
	浸泡碱液	Soaking the straw in alkali
	上堆	Piling the straw
	滤碱液	Filtering the pulp
	蒸煮(蒸草坯、青草过程)	Procedure of steaming and boiling (steaming the straw slab and fresh straw)
	选拣台	Picking table

	春后的燎皮浆	Bark pulp after beating
	除砂机	Machine for removing the residues
	滤水机	Machine for filtering the water
	混合浆池	Container for mixing the pulp
	旋翼筛	Rotor sifter
	跳筛	Jiggling sieve
	捞纸	Making the paper
	螺旋杠杆	Screw lever
	抬帖	Carrying the paper pile
	烘帖	Drying the paper pile
	分张	Separating the paper layers
	晒纸	Drying the paper
	纸架	Frame for watering the paper pile
	浇帖水壶	Kettle for watering the paper pile
	纸石	Stone for pressing the paper
	检验	Checking the paper
	药槊头	Hammer for making the papermaking mucilage
	竹刀	Bamboo knife
	打浆机	Beating machine
	捞纸槽	Papermaking trough
	纸帘	Papermaking screen
	扒头	Stirring stick
	帖架	Frame for carrying the paper pile
	鞭帖鞭	Stick for beating the paper pile
	额枪	Tool for separating the paper
	松毛刷	Brush made of pine needles
	晒纸焙	Drying wall
	压纸石	Stone for pressing the paper
	剪刀	Shears
	自建的晒滩	Self-built drying field
	废水处理系统	Wastewater treatment system
	金永辉讲述"泾上白"的品牌文化	Jin Yonghui relating the brand culture of "Jingshangbai"
	有"泾上白"品牌标记的宣纸	Xuan paper with "Jingshangbai" Brand label
	小岭景辉文化园内外景	Exterior view of Jinghui Cultural Park in Xiaoling Village
	金永辉收藏的古纸	Ancient paper collected by Jin Yonghui
	"泾上白"特净透光摄影图	A photo of "Jingshangbai" superb-bark paper seen through the light
章节	图中文名称	图英文名称
第十二节	泾县三星纸业有限公司	Section 12 Sanxing Paper Co., Ltd. in Jingxian County
	三星纸业有限公司生产厂区大门与内景	Gate of Sanxing Paper Co., Ltd. and its interior view
	调查时厂里栽培的青檀树	*Pteroceltis tatarinowii* Maxim. planted in the factory
	张必良正在观看书法家试纸	Zhang Biliang watching a calligrapher testing the paper
	张必良	Zhang Biliang
	"三星"宣纸品名与规格	Types and specifications of "Sanxing" Xuan paper
	"三星"极品宣	"Sanxing" superb Xuan paper
	"三星"极品宣纤维形态图(10×)	Fibers of "Sanxing" superb Xuan paper (10× objective)
	"三星"极品宣纤维形态图(20×)	Fibers of "Sanxing" superb Xuan paper (20× objective)

"三星"极品宣润墨效果	Writing performance of "Sanxing" superb Xuan paper
三星纸业有限公司厂区内种植的青檀树	*Pteroceltis tatarinowii* Maxim. in Sanxing Paper Co., Ltd.
正在清水里浸泡的青檀皮	Soaking *Pteroceltis tatarinowii* Maxim. bark
经过蒸煮漂白后的青檀皮（又称化学皮）	*Pteroceltis tatarinowii* Maxim. bark after being steamed and bleached (also named chemical bark)
沙田稻草	Straw grown in sands
山涧溪水	Mountain water
收购的黑皮	Purchased black bark (the unprocessed bark)
运输黑皮	Transporting the black bark
浸泡中的黑皮	Soaking the black bark
蒸煮	Steaming and boiling the bark
拣黑皮	Picking the black bark
调查组成员在车间观察漂白后的白皮	Researchers observing the white bark (the bleached bark)
拣白皮	Picking white bark
打浆机	Beating machine
洗涤	Cleaning the bark
三星纸业有限公司的燎草仓库	Processed straw storehouse in Sanxing Paper Co., Ltd.
石碾正在碾草料	Grinding the straw materials with a stone roller
捞纸	Making the paper
浇帖架	Frame for supporting the paper pile
晒纸	Drying the paper
检验	Checking the paper
剪纸	Cutting the paper
盖印	Sealing the paper
三星纸业有限公司系列产品	Series products made by Sanxing Xuan Paper Co., Ltd.
捞纸车间	Papermaking workshop
纸帘	Papermaking screen
浇帖架	Frame for supperting the paper pile
松毛刷	Brush made of pine needles
压纸石	Stone for pressing the paper
剪刀	Shears
调研人员与厂方人员核实信息	Researchers verifying details with the factory workers
三星纸业有限公司营销网络	Marketing network of Sanxing Xuan Paper Co., Ltd.
1987年在云岭乡建立的百亩原料加工基地—晒滩	Drying feild, a hundred *mu* of raw materials processing base built in Yunling Town in 1987
三星纸业有限公司自建的环保设施	Environmental protection equipments built in Sanxing Paper Co., Ltd.
书画家试纸	Calligraphers testing the paper
"三星"极品宣透光摄影图	A photo of "Sanxing" superb Xuan paper seen through the light
"三星"桑皮纸透光摄影图	A photo of "Sanxing" mulberry bark paper seen through the light
"三星"黄料阀古宣透光摄影图	A photo of "Sanxing" Huangliao vintage Xuan paper seen through the light

章节	图中文名称	图英文名称
第十三节	安徽常春纸业有限公司	Section 13 Anhui Changchun Paper Co., Ltd.
	在厂区接待室访谈姚忠华	Interviewing Yao Zhonghua at the reception room of the factory
	明星宣纸厂正门	Gate of Mingxing Xuan Paper Factory
	明星宣纸厂创始人姚文明	Yao Wenming, founder of Mingxing Xuan Paper Factory
	明星纸厂前的青弋江	Qingyi River in front of Mingxing Xuan Paper Factory
	现任厂长姚忠华	Yao Zhonghua, present factory director
	"明星"宣纸的包装样式	Packaging of "Mingxing" Xuan paper
	"明星"特净纤维形态图（10×）	Fibers of "Mingxing" superb-bark paper (10× objective)
	"明星"特净纤维形态图（20×）	Fibers of "Mingxing" superb-bark paper (20× objective)
	"明星"特净润墨效果	Writing performance of "Mingxing" superb-bark paper
	"明星"喷浆檀皮纸纤维形态图(10×)	Fibers of "Mingxing" pulp-shooting *Pteroceltis tatarinowii* bark paper (10× objective)
	"明星"喷浆檀皮纸纤维形态图(20×)	Fibers of "Mingxing" pulp-shooting *Pteroceltis tatarinowii* bark paper (20× objective)
	"明星"喷浆檀皮纸润墨效果	Writing performance of "Mingxing" pulp-shooting *Pteroceltis tatarinowii* bark paper
	"明星"喷浆雁皮纸纤维形态图(10×)	Fibers of "Mingxing" pulp-shooting *Wikstroemia pilosa* Cheng bark paper (10× objective)
	"明星"喷浆雁皮纸纤维形态图(20×)	Fibers of "Mingxing" pulp-shooting *Wikstroemia pilosa* Cheng bark paper (20× objective)
	"明星"喷浆雁皮纸润墨效果	Writing performance of "Mingxing" pulp-shooting *Wikstroemia pilosa* Cheng bark paper
	明星宣纸厂仓库里储存的檀皮原料	Raw materials of *Pteroceltis tatarinowii* bark at the storehouse of Mingxing Xuan Paper Factory
	拣白皮	Picking the white bark
	挑拣后的白皮	White bark after being picked
	白皮打浆现场	Beating the white bark
	燎草原料	Processed straw
	石碾在碾草	Grinding the straw with a stone roller
	旋翼筛	Rotor sifter
	除砂器	Machine for removing the grit
	捞纸车间	Papermaking workshop
	放纸	Turning the papermaking screen upside down on the board
	提帘	Lifting the papermaking screen
	放纸架	Frame for piling the paper
	压榨	Pressing the paper
	晒纸：揭纸	Drying the paper: peeling the paper down
	晒纸：上墙	Drying the paper: pasting the paper on the wall
	检验女工	A female worker checking the paper
	剪纸	Cutting the paper
	纸巾（将要回笼打浆的废纸、纸边等）	Waste paper to be recycled
	等待包装装箱的宣纸	Xuan paper to be packaged
	明星宣纸厂仓库里堆放的买来的浆板	Pulp board in the storehouse of Mingxing Xuan Paper Factory
	浸泡浆板	Soaking the pulp board
	打浆	Beating the pulp
	捞纸和放纸	Making the paper and turning the papermaking screen upside down on the board
	喷浆捞纸生产线	Production line of papermaking with pulp-shooting method
	检验与剪纸	Checking and cutting the paper
	打浆机	Beating machine
	宣纸手工捞纸槽	Papermaking trough for making Xuan paper
	喷浆捞纸帘床	Papermaking frame for making the pulp-shooting paper
	挂在墙上的纸帘	Papermaking screen hanging on the wall
	切帖刀	Paper cutter
	松毛刷	Brush made of pine needles
	纸焙	Drying wall

剪刀	Shears
压纸石	Stone for pressing the paper
出口产品	Paper products for export
明星宣纸厂的荣誉墙	Honor wall of Mingxing Xuan Paper Factory
"明星"宣纸商标	Trademark of "Mingxing" Xuan Paper
杨仁恺的题词	Yang Renkai's autography for Mingxing Xuan Paper Factory
明星宣纸厂的机械造纸机	Papermaking machine of Mingxing Xuan Paper Factory
姚忠华	Yao Zhonghua
"明星"特净透光图	A photo of "Mingxing" superb-bark paper seen through the light
"明星"半自动喷浆雁皮纸透光摄影图	A photo of "Mingxing" semi-automatic pulp-shooting *Wikstroemia pilosa* Cheng paper seen through the light
"明星"檀皮书画纸透光摄影图	A photo of "Mingxing" *Pteroceltis tatarinowii* Maxim. bark calligraphy and painting paper seen through the light
"明星"构皮云龙纸透光摄影图	A photo of "Mingxing" mulberry bark Yunlong paper seen through the light

章节	图中文名称	图英文名称
第十四节	泾县玉泉宣纸纸业有限公司	Section 14 Yuquan Xuan Paper Co., Ltd. in Jingxian County
	厂区内景	Interior view of the factory
	S322公路旁的玉泉宣纸厂区和标牌	Yuquan Xuan Paper Factory by the Highway S322 with a signpost
	"玉泉"注册商标	Registered trademark of "Yuquan"
	"玉马"注册商标	Registered trademark of "Yuma"
	"百年中行"纪念宣纸荣誉证书	Xuan paper certificate awarded as "The Centennial Anniversary of Bank of China"
	"十八大纪念宣"荣誉证书	Xuan paper certificate awarded as "Commemoration Paper of the 18th National People's Congress"
	高骏(右二)观看画家试纸	Gao Jun (second from the right) watching a painter testing the paper
	高玉生	Gao Yusheng
	"玉泉"净皮纤维形态图(10×)	Fibers of "Yuquan" clean-bark paper (10× objective)
	"玉泉"净皮纤维形态图(20×)	Fibers of "Yuquan" clean-bark paper (20× objective)
	"玉泉"净皮润墨效果	Writing performance of "Yuquan" clean-bark paper
	厂区仓库里备用的毛皮	Unprocessed bark at the factory warehouse
	厂区仓库里备用的燎草	Processed straw at the factory warehouse
	毛皮原料	Unprocessed bark materials
	挑选毛皮	Picking the bark
	白皮料	White bark materials
	甩干机	Water extractor
	挑选白皮	Picking the white bark
	打浆机	Beating machine
	洗草池	Pool for cleaning the straw
	石碾	Stone roller
	圆筒筛	Cylinder sifter
	平筛	Flat sifter
	混合浆池	Pool for the mixing pulp
	除砂设备	Equipment for removing the grit
	圆筒筛	Cylinder filter
	捞纸	Making the Paper
	"玉泉"宣纸的捞纸车间	Papermaking workshop of "Yuquan" Xuan paper
	放纸	Turning the papermaking screen upside down on the board
	尚未烘烤的纸帖	Paper to be dried

烘好的纸帖	Dride paper pile
晒纸	Drying the paper
揭纸	Peeling the paper down
检验与剪纸	Checking and cutting the paper
叠纸	Folding the paper
宣纸成品	Final products of Xuan paper
打浆机	Beating machine
捞纸槽	Papermaking trough
纸帘	Papermaking screen
帘床	Frame for supporting the papermaking screen
擀棍	Tool for patting the paper
松毛刷	Brush made of pine needles
晒纸焙	Drying wall
晒纸夹	Clip for drying the paper
大剪刀	Shears
"玉泉"宣纸的成品纸仓库	Final paper products of "Yuquan" Xuan paper in the warehouse
安徽省著名商标牌匾	Plaque of Famous Trademark in Anhui Province
面对厂区的山泉水	Spring water near the factory
书画名家的题词与创作作品	Autograph and painting from celebrities
"玉泉"四尺净皮宣纸透光摄影图	A photo of "Yuquan" four-*chi* clean-bark paper seen through the light
"玉泉"皮纸透光摄影图	A photo of "Yuquan" clean-bark paper seen through the light

章节	图中文名称	图英文名称
第十五节	泾县吉星宣纸有限公司	Section 15 Jixing Xuan Paper Co., Ltd. in Jingxian County
	吉星宣纸有限公司生产厂区的外围环境	Surrounding environment of the production area of the Jixing Xuan Paper Co., Ltd.
	制浆车间	Workshop for making the pulp
	捞纸车间	Papermaking workshop
	胡成忠给调查组成员指示原湖山坑的旧址	Hu Chengzhong introducing the former site Hushankeng area to the researchers
	胡成忠与调查组一起勘察晚清、民国年间的宣纸作坊遗址	Hu Chengzhong and the researchers prospecting the former site of Xuan paper mill during the late Qing Dynasty and the Republican Era
	宣纸老艺人胡业斌	Hu Yebin, an old Xuan papermaker
	现任负责人胡成忠	Hu Chengzhong, the present manager
	带包的"日星"宣纸	Packaged "Rixing" Xuan paper
	吉星宣纸有限公司的代表产品——特别定制"本色宣"	Representative "Jixing" Xuan paper: custom edition of "Xuan paper in original color"
	"吉星"本色特净纤维形态图(10×)	Fibers of "Jixing" superb-bark paper in original color (10× objective)
	"吉星"本色特净纤维形态图(20×)	Fibers of "Jixing" superb-bark paper in original color (20× objective)
	"吉星"本色特净润墨效果	Writing performance of "Jixing" superb-bark paper in original color
	吉星宣纸厂区附近的青檀树林	*Pteroceltis tatarinowii* Maxim. woods near the Jixing Xuan Paper Factory
	吉星宣纸厂区的燎草仓库	Straw warehouse of Jixing Xuan Paper Factory
	种养在厂部办公室的猕猴桃藤枝叶	*Actinidia chinensis* Planch. planted in the factory
	流经厂区的山溪水	Stream flowing through the factory area
	从山上引水的管道	Pipe transporting water from the mountain
	毛皮仓库	Unprocessed bark in the warehouse
	胡成忠向调查组展示并讲解白皮制作工艺	Hu Chengzhong displaying and explaining the white bark papermaking techniques
	石碾	Stone roller

章节	图中文名称	图英文名称
	筛选	Filtering the material
	洗料	Cleaning the materials
	草浆	Pulp
	配浆过程中的各环节及设备	Procedures and tools for making the pulp
	浆料	Pulp materials
	捞纸	Making the paper
	捞纸	Making the paper
	捞纸计数器	Counting apparatus
	放纸	Turning the papermaking srceen upside down on the board
	纸帖	Paper pile
	揭纸	Peeling the paper down
	晒纸	Drying the paper
	工人检查纸张	A worker checking the paper
	打浆机	Beating machine
	尺八屏捞纸槽	Papermaking trough (295 cm×175 cm×45 cm)
	绣有"故宫博物院"的纸帘	Papermaking screen embroidered with "The Palace Museum"
	耙子	Rake
	夹晒纸刷板子	Board for clipping and drying the paper
	刷子	Brush made of pine needles
	后山村特制的大剪刀	Shears made in Houshan village
	定制宣纸	Custom edition of Xuan paper
	"日星"宣纸规格与价目表	Specification and price list of "Rixing" Xuan paper
	书画家范扬题写的厂名	Autograph of the name of Jixing Xuan Paper Factory written by the calligrapher and painter Fan Yang
	访谈胡业斌	Interviewing Hu Yebin
	吉星宣纸厂区旁边的原生产遗址	Former papermaking site near the Jixing Xuan Paper Factory
	吉星宣纸厂区后的原晒滩	Former drying field behind Jixing Xuan Paper Factory
	"日星"本色特净透光摄影图	A photo of "Rixing" superb-bark paper in original color seen through the light
	"日星"棉料透光摄影图	A photo of "Rixing" Mianliao paper in original color seen through the light

章节	图中文名称	图英文名称
第十六节	泾县金宣堂宣纸厂	Section 16 Jinxuantang Xuan Paper Factory in Jingxian County
	金宣堂宣纸厂厂区周边环境	Surrounding environment of Jinxuantang Xuan Paper Factory
	皮草加工车间	Workshop for bark processing
	国家地理标志授权证书	Certificate authorized by the National Geographic Symbol
	中国优质产品供应商证书	Certificate of "The Suppliers of High Quality Products in China"
	金宣堂宣纸厂外的浙溪村风光	Landscape of Zhexi Village outside Jinxuantang Xuan Paper Factory
	访谈中的程玉山	Cheng Yushan being interviewed
	调查现场的捞纸工人	Workers making the paper
	调查现场的晒纸工人	Workers drying the paper
	访谈中的程洋	Interviewing Cheng Yang
	剪纸车间堆放的各类待包装宣纸	Different kinds of Xuan paper to be packaged at the cutting workshop
	"金宣堂"特净成品纸	Final product of "Jinxuantang" superb-bark paper
	"星月"净皮四尺单宣成品纸	Final product of "Xingyue" four clean-bark Xuan paper (single-layer)
	"星月"扎花成品宣纸	Final product of "Xingyue" Zhahua Xuan paper
	特种规格（70 cm×280 cm）的特净单宣	Superb-bark Xuan paper (single-layer) with special specification (70 cm×280 cm)
	"金宣堂"特净纤维形态图（10×）	Fibers of "Jinxuantang" superb-bark paper (10× objective)
	"金宣堂"特净纤维形态图（20×）	Fibers of "Jinxuantang" superb-bark paper (20× objective)
	"金宣堂"特净润墨效果	Writing performance of "Jinxuantang" superb-bark paper
	砍伐后来年春天发芽的青檀树	*Pteroceltis tatarinowii* Maxim. with new bud after been cut off
	当地燎草加工户野外堆放的草坯	Straw stack in the outside by local processors
	杨桃藤枝	Branches of Planch.
	纸巾（残次宣纸、纸边）	Defective Xuan paper, paper edges for recycling
	毛皮浸泡	Soaking the bark
	蒸煮	Steaming and boiling the bark
	清洗毛皮	Cleaning the bark
	拣黑皮	Picking out the black bark
	漂白	Bleaching the bark
	清洗白皮	Cleaning the white bark
	燎草堆	Processed straw stack
	清洗燎草	Cleaning the processed straw
	碾压燎草	Grinding the processed paper
	正在碾草的石碾	Stone roller grinding the straw
	水力碎浆机	Hydraulic pulp maker
	除砂机	Machine for removing the grit
	滤水机及装机环境	Machine for filtering water and environment of the machine installment
	打浆机	Beating machine
	浆池	Pulp pool
	撅条后的杨桃藤枝	Branches of *Actinidia chinensis* Planch. after being cut off
	碎条捶药	Hammering the papermaking mucilage
	取液后的杨桃藤枝	Branches of *Actinidia chinensis* Planch. after being pressed
	备用的纸药	Prepared papermaking mucilage
	划单槽	Swirring the papermaking pulp
	捞纸	Making the paper
	放帘	Turning the papermaking screen upside down on the board
	计数器	Counting apparatus
	压榨	Pressing the paper
	鞭帖	Patting the paper pile
	做额	Separating the paper layers
	揭纸	Peeling the paper down
	晒纸	Drying the paper
	收纸	Peeling the paper down
	检纸	Checking the paper
	剪纸	Cutting the paper
	压石	Stone for pressing the paper
	盖印	Sealing the paper
	打包机	Machine for packaging the paper
	包装盒	Box for packaging
	石碾	Stone roller
	打浆设备	Beating equipment
	纸槽	Papermaking trough

	图中文名称	图英文名称
	纸帘	Papermaking screen
	帘床	Frame for the papermaking screen
	浇帖架	Frame for supporting the paper pile
	鞭帖鞭	Papermaking stick for patting the paper pile
	额枪	Tool for separating the paper
	刷子	Brush made of pine needles
	焙笼	Drying wall
	压纸石	Stone for pressing the paper
	后山村所制宣纸剪刀	Shears in Houshan Village
	"金宣堂"直营店内的宣纸产品	Xuan paper products in the direct-sale "Jinxuantang" Xuan Paper Factory Store
	正在工作的剪纸车间	Paper-cutting workshop
	"星月"宣纸指导价	Referential price list of "Xingyue" Xuan paper
	赖少其所题"泾县浙溪金宣堂"	"Jinxuantang in Zhexi Village of Jingxian County" written by Lai Shaoqi
	浙溪村的稻田	Scenery of rice field in Zhexi Village
	金宣堂宣纸厂的污水处理站	Sewage treatment station in Jinxuantang Xuan Paper Factory
	金宣堂宣纸厂附近的晒滩	Drying field near Jinxuantang Xuan Paper Factory
	"金宣堂"牌"中共十八大和十九大纪念宣纸"(内外包装样式)	"Jinxutang" Brand "Commemorating Xuan Paper for the 18th National People's Congress" (exterior and interior look)
	"金宣堂"特净透光摄影图	A photo of "Jinxuantang" superb-bark paper seen through the light
	"金宣堂"棉料透光摄影图	A photo of "Jinxuantang" Mianliao paper seen through the light
	"金宣堂"槽底宣透光摄影图	A photo of "Jinxuantang" caodi paper seen through the light
	"金宣堂"净皮透光摄影图	A photo of "Jinxuantang" clean-bark paper seen through the light

章节	图中文名称	图英文名称
第十七节	泾县小岭金溪宣纸厂	Section 17 Xiaoling Jinxi Xuan Paper Factory in Jingxian County
	金溪宣纸厂所在金坑村民组路边指示牌	Signpost of Jinkeng Villagers' Group where Jinxi Xuan Paper Factory locates
	小岭金溪宣纸厂厂区内的厂牌	Factory plaque of Xiaoling Jinxi Xuan Paper Factory
	金坑风景	Landscape of Jinkeng Villages' Group
	李松林	Li Songlin
	曹永强(右)与朱正海	Cao Yongqiang (right) and Zhu Zhenghai
	"九岭"宣纸礼盒成品纸	Gift box of "Jiuling" Xuan paper
	"九岭"净皮纤维形态图(10×)	Fibers of "Jiuling" clean-bark paper (10× objective)
	"九岭"净皮纤维形态图(20×)	Fibers of "Jiuling" clean-bark paper (20× objective)
	"九岭"净皮润墨效果	Writing performance of "Jiuling" clean-bark paper
	仓库里储备的青檀皮料	Pteroceltis tatarinowii Maxim. bark in the warehouse
	仓库里储备的燎草	Processed straw in the warehouse
	过滤纸药	Filtering the papermaking mucilage
	群山环抱中的小岭金溪宣纸工艺厂	Xiaoling Jinxi Xuan Paper Factory surrounded by mountains
	洗草专用水池	Pool for cleaning the straw
	木榨榨草	Wooden presser pressing the straw
	榨过的草料	Pressed straw materials
	石碾碾草	Stone rollers grinding the straw
	漂白后入袋洗涤的燎草	Bleached Straw in bags for cleaning
	捞纸	Making the paper
	烘干的纸帖	Dried paper pile
	切帖	Cutting the paper pile
	晒纸	Drying the paper
	检验	Checking the paper
	检验完成待包装的宣纸	Xuan paper to be packaged
	捞纸槽	Papermaking trough
	纸帘	Papermaking screen
	额枪	Tool for separating the paper
	松毛刷	Brush made of pine needles
	杀额刀	Knife for cutting the paper
	"九岭"宣纸获奖证书	Certificates of "Jiuling" Xuan paper
	"泾县小岭金溪宣纸厂"题字	Autograph of "Xiaoling Jinxi Xuan Paper Factory in Jingxian County"
	探寻小岭金溪宣纸厂的品牌文化	Investigating the brand culture of Xiaoling Jinxi Xuan Paper Factory
	厂区内的历史建筑	Historical building in the factory area
	"九岭"净皮宣纸透光摄影图	A photo of "Jiuling" clean-bark paper seen through the light

章节	图中文名称	图英文名称
第十八节	黄山白天鹅宣纸文化苑有限公司	Section 18 Huangshan Baitian'e Xuan Paper Cultural Garden Co., Ltd.
	进入白天鹅宣纸文化苑有限公司老厂区的路	Road to the former factory site of Baitian'e Xuan Paper Cultural Garden Co., Ltd.
	白天鹅宣纸文化苑有限公司老厂区环境	View of the former Baitian'e Xuan Paper Cultural Garden Co., Ltd.
	白天鹅宣纸文化苑有限公司新厂区环境	View of the new Baitian'e Xuan Paper Cultural Garden Co., Ltd.
	白天鹅宣纸文化苑有限公司新喷浆纸车间	View of the new pulp-shooting workshop in Baitian'e Xuan Paper Cultural Garden Co., Ltd.
	徐邦达为"黄山纪元宣纸厂"题字	Autograph of Xu Bangda for "Jiyuan Xuan Paper Factory in Huangshan City"
	曹阳明(中)与调查组成员在老厂交流	Cao Yangming (middle) communicating with the researchers in the former factory
	曹阳明	Cao Yangming
	"白天鹅"四尺净皮荣誉证书	Honor certificate of "Baitian'e" four chi clean-bark paper
	"白天鹅"净皮纤维形态图(10×)	Fibers of "Baitian'e" clean-bark paper (10× objective)
	"白天鹅"净皮纤维形态图(20×)	Fibers of "Baitian'e" clean-bark paper (20× objective)
	"白天鹅"净皮润墨效果	Writing performance of "Baitian'e" clean-bark paper
	采购的燎草	Purchased straw
	黄泥坑的山溪水	Stream in Huangnikeng area
	洗燎草	Cleaning the processed straw
	石碾碾草	Grinding the straw with a stone roller
	打浆车间各设备控制开关	Control switches of the equipments in the beating workshop
	振动筛	Sifting device
	配好的浆料	Prepared pulp materials
	捞纸	Making the paper
	湿纸放在槽架上	Wet paper on the frame
	烘帖完成的纸帖	Paper pile after drying
	纸架上的纸帖	Paper pile on the frame
	晒纸上焙	Pasting the paper on the drying wall
	未裁剪的纸	Uncut paper
	裁剪好放入仓库的纸	Cut paper in the storehouse
	有缺陷的纸一	Defective paper I
	有缺陷的纸二	Defective paper II
	有缺陷的纸三	Defective paper III

图中文名称	图英文名称
打浆机	Beating machine
帖架	Frame for supporting the paper pile
松毛刷	Brush made of pine needles
夹晒纸刷板子	Tool for clipping and drying the paper
铁焙	Iron drying device
白天鹅宣纸文化苑有限公司老厂厂房	Former site of Baitian'e Xuan Paper Cultural Garden Co., Ltd.
曹阳明的安徽省级"非遗"传承人证书	Intangible Cultural Heritage Inheritor Certificate of Anhui Province owned by Cao Yangming
《书圣王羲之》拍摄现场	The shooting scene of Wang Xizhi, the Supreme Calligrapher
曹阳明正在做的老化实验	Cao Yangming doing the aging experiment
秸秆烧煮制浆新工艺说明	New papermaking techniques of boiling the straw and making the pulp
秸秆烧煮制浆新工艺实验现场	Testing the new papermaking techniques of boiling the straw and making the pulp
曹阳明与《书圣王羲之》拍摄人员合影	Cao Yangming and the cast of Wang Xizhi, the Supreme Calligrapher
荣誉牌匾	Various honor plaques
"白天鹅"净皮透光摄影图	A photo of "Baitian'e" clean-bark paper seen through the light

章节	图中文名称	图英文名称
第三章	书画纸	Chapter III Calligraphy and Painting Paper
第一节	泾县载元堂工艺厂	Section 1 Zaiyuantang Craft Factory in Jingxian County
	载元堂工艺厂大门	Gate of Zaiyuantang Craft Factory
	厂区外部环境	External view of the factory
	厂区内部环境图	Internal view of the factory
	喷浆技艺	Pulp shooting technique
	手工捞纸技艺	Handmade papermaking technique
	接受访谈的沈维正	Interviewing Shen Weizheng
	调查组成员访谈沈维正	Researchers interviewing Shen Weizheng
	"载元堂"画仙纸	"Zaiyuantang" Huaxian paper
	"载元堂"画仙纸纤维形态图（10×）	Fibers of "Zaiyuantang" Huaxian paper (10× objective)
	"载元堂"画仙纸纤维形态图（20×）	Fibers of "Zaiyuantang" Huaxian paper (20× objective)
	"载元堂"画仙纸润墨效果	Writing performance of "Zaiyuantang" Huaxian paper
	购买来的皮料半成品	Semi-finished bark materials bought from elsewhere
	购买的龙须草、木浆和竹浆浆板	Eulaliopsis binata, wood pulp, and bamboo pulp board bought from elsewhere
	进口分张剂	Imported mucilage for separating the paper layers
	调查组成员在车间探询工艺	A researcher inquiring papermaking procedures
	浸泡浆板	Soaking the pulp board
	打浆	Beating the pulp
	捞纸	Papermaking
	喷浆捞纸车间	Workshop for pulp shooting and papermaking
	工人在压榨纸帖	A worker pressing the paper pile
	浇帖	Watering the paper pile
	杀额	Trimming the deckle edges
	做边	Processing the edge of paper pile
	晒纸	Drying the paper
	检验纸张	Checking the paper
	工人操作切纸机	Workers operating the paper cutting machine
	剪纸工在登记槽单位基本数据	A worker registering basic data of papermaking trough size
	待售的画仙纸产品	Huaxian paper products for selling
	打浆机	Beating machine
	喷浆纸槽	Pulp shooting trough
	手工纸槽	Handmade papermaking trough
	吊帘纸槽	Trough with movable papermaking screen
	纸帘	Papermaking screen
	浆帘床	Screen frame for pulp shooting
	定位桩	Fixed timber pile
	帖架	Frame for supporting the paper pile
	工人在清洗烘焙表面	Workers cleaning the surface of the drying wall
	棒槌	Wooden hammer
	额枪	Stick for separating the paper layers
	刷子	Brush
	大剪刀	Shears
	晒纸过程中的压纸石	Stone for pressing the paper in the procedure of drying the paper
	剪纸过程中的压纸石	Stone for pressing the paper in the procedure of cutting the paper
	纸库局部	A part of paper warehouse
	载元堂工艺厂2015年产品价目表	Product price list of Zaiyuantang Craft Factory in 2015
	制作中的"万年红"春联纸	Making "Wannianhong", red paper for writing spring couplets
	制作中的洒金"万年红"	Making "Wannianhong", red paper with golden dots
	不同工种忙碌的工人	Busy workers in different steps of papermaking
	方增先"安徽省泾县载元堂"题字	Autograph of "Zaiyuantang in Jingxian County of Anhui Province" by Fang Zengxian
	沈维正向调查组成员介绍纸样	Shen Weizheng introducing the paper sample to a researcher
	沈维正向调查组成员展示产品包装	Shen Weizheng showing the product packaging to the researchers
	载元堂工艺厂外景	External view of Zaiyuantang Craft Factory
	技艺娴熟的剪纸女工	A skilled female worker cutting the paper
	"载元堂"画仙纸透光摄影图	A photo of "Zaiyuantang" Huaxian paper seen through the light

章节	图中文名称	图英文名称
第二节	泾县小岭强坑宣纸厂	Section 2 Xiaoling Qiangkeng Xuan Paper Factory in Jingxian County
	通往西山村的乡村公路	Country road to Xishan Village
	强坑宣纸厂内景之一	Internal view of Qiangkeng Xuan Paper Factory
	强坑宣纸厂内景之二	Another internal view of Qiangkeng Xuan Paper Factory
	曹炳集工作过的小岭宣纸厂西山车间（分厂）	Xishan Workshop (branch factory) of Xiaoling Xuan Paper Factory where Cao Bingji once worked
	侄子曹凯（左）与曹友泉（右）	Cao Youquan (right) and his nephew Cao Kai (left)
	"曹友泉宣纸"（书画纸）淘宝网店	Taobao online store of "Cao Youquan Xuan Paper" (calligraphy and painting paper)
	记录曹友泉介绍的纸厂历史	Recording the history of the paper factory that Cao Youquan introduced
	檀皮稻草宣	Xuan Paper made with Pteroceltis tatarinowii Maxim. bark and straw
	檀皮稻草精品宣	High-quality Xuan paper made with Pteroceltis tatarinowii Maxim. bark and straw
	"曹友泉"高级书画纸纤维形态图（10×）	Fibers of "Cao Youquan" advanced calligraphy and painting paper (10× objective)
	"曹友泉"高级书画纸纤维形态图（20×）	Fibers of "Cao Youquan" advanced calligraphy and painting paper (20× objective)
	"曹友泉"高级书画纸润墨效果	Writing performance of "Cao Youquan" advanced calligraphy and painting paper

	混合后的浆料	Mixed pulp materials
	漂白后的檀皮原料	Bleached *Pteroceltis tatarinowii* Maxim. bark
	厂区旁的山泉	Mountain spring alongside the factory
	强坑宣纸厂的水井	Well of Qiangkeng Xuan Paper Factory
	浸泡浆板	Soaking the pulp board
	已配好的浆料	Mixed pulp materials
	放浆	Adding pulp
	捞纸工序的主要环节	Main steps of papermaking procedure
	螺旋杆	Pressing device
	压榨	Pressing the paper
	切额	Trimming the deckle edges
	晒纸	Drying the paper
	检验	Checking the paper
	加盖印章	Stamping the paper
	包装好的书画纸	Packaged calligraphy and painting paper
	捞纸槽	Papermaking trough
	挂在墙上的纸帘	Papermaking screen hanging on the wall
	铁架	Iron frame for carrying the paper
	切帖刀	Knife for cutting the paper pile
	刷夹和松毛刷	Brush holder and pine needle brush
	大剪刀	Shears
	压纸木尺	Wooden ruler for pressing the paper
	鞭帖板	Bamboo whip
	访谈现场	Interviewing a papermaker
	"曹友泉宣纸"天猫旗舰店截屏	Screen shot of "Cao Youquan Xuan Paper" in Tmall flagship store
	工厂周围的檀树	trees around the factory
	小岭强坑宣纸厂的库房	Warehouse of Xiaoling Qiangkeng Xuan Paper Factory
	"曹友泉"书画纸透光摄影图	A photo of "Cao Youquan" calligraphy and painting paper seen through the light

章节	图中文名称	图英文名称
第 三 节	泾县雄鹿宣纸厂	Section 3 Xionglu Xuan Paper Factory in Jingxian County
	雄鹿宣纸厂的路边宣传标牌	Billboard of Xionglu Xuan Paper Factory by the road
	调查组成员访谈董科	A researcher interviewing Dong Ke
	雄鹿宣纸厂产品图	A photo of product of Xionglu Xuan Paper Factory
	雄鹿宣纸厂书画纸纤维形态图（10×）	Fibers of calligraphy and painting paper in Xionglu Xuan Paper Factory (10× objective)
	雄鹿宣纸厂书画纸纤维形态图（20×）	Fibers of calligraphy and painting paper in Xionglu Xuan Paper Factory (20× objective)
	雄鹿宣纸厂书画纸润墨效果	Writing performance of calligraphy and painting paper in Xionglu Xuan Paper Factory
	浸泡前的龙须草浆板	Pulp board of *Eulaliopsis binata* before soaking
	收购的稻草浆原料	Pulp materials of straw bought from elsewhere
	工人在调制化学纸药	A worker modulating chemical papermaking mucilage
	雄鹿宣纸厂从地下井取水	Xionglu Xuan Paper Factory getting water from a underground well
	打浆	Beating the pulp
	清洗后的浆料	Pulp materials after cleaning
	捞纸	Papermaking
	提帘放纸	Turning the papermaking screen upside down on the board

	计数器	Counting apparatus
	压榨	Pressing the paper
	揭纸	Separating the paper layers
	晒纸上墙	Drying the paper on a wall
	收纸	Taking back the paper
	检验纸张	Checking the paper
	剪纸	Cutting the paper
	加盖印章	Stamping the paper
	圆筒筛	Cylinder filter
	纸槽	Papermaking trough
	纸榨	Tools for pressing the paper
	帘床	Frame for supporting the papermaking screen
	千斤顶	Lifting jack
	纸刷	Bush
	调查组成员观察董科试纸	A researcher watching Dong Ke testing the paper
	剪纸车间及成品暂存室	Paper cutting workshop and temporary warehouse for paper products
	产品库	Paper warehouse
	整理剪好的纸张	Sorting the processed paper
	"雄鹿"书画纸透光摄影图	A photo of "Xionglu" calligraphy and painting paper seen through the light

章节	图中文名称	图英文名称
第 四 节	泾县紫光宣纸书画社	Section 4 Ziguang Calligraphy and Painting Xuan Paper Agency in Jingxian County
	丁黄公路	Ding-Huang Highway
	紫光宣纸书画社宣传栏	Billboard of Ziguang Calligraphy and Painting Xuan Paper Agency
	资深宣纸工艺师朱正海访谈吴报景（右）	Zhu Zhenghai, the senior Xuan paper craftsman, interviewing Wu Baojing (right)
	在社区办公室访谈吴报景	Interviewing Wu Baojing in the office of the agency
	"绿杨宝"书画纸纤维形态图（10×）	Fibers of "Lvyangbao" calligraphy and painting paper (10× objective)
	"绿杨宝"书画纸纤维形态图（20×）	Fibers of "Lvyangbao" calligraphy and painting paper (20× objective)
	"绿杨宝"书画纸润墨效果	Writing performance of "Lvyangbao" calligraphy and painting paper
	正在讲解的张世坤	Zhang Shikun introducing papermaking techniques
	购自河南的龙须草浆板	Pulp board of *Fulaliopsis binata* from Henan Province
	半自动喷浆捞纸	Semi-automatic pulp shooting papermaking
	提帘放纸	Turning the papermaking screen upside down on the board
	室内、室外的压榨	Indoor and outdoor pressing device
	晒纸	Drying the paper
	检验纸张	Checking the paper
	待包装的纸	Paper to be packed
	包装好的"绿杨宣"成品	Packaged "Lvyangxuan" paper products
	纸帘	Papermaking screen
	打浆机	Beating machine
	纸槽	Papermaking trough
	帖架	Frame for supporting the paper pile
	晒纸焙	Wall for drying the paper
	额枪	Tool for patting the paper
	刷子	Brush
	压纸石	Stone for pressing the paper

图中文名称	图英文名称
紫光宣纸书画社价目表	Price list of Ziguang Calligraphy and Painting Xuan Paper Agency
紫光宣纸书画社西安店	Store of Ziguang Calligraphy and Painting Xuan Paper Agency in Xi'an City
出口韩国打包好的"画仙纸"	Packaged "Huaxian Paper" exporting to Korea
紫光宣纸书画社厂牌	Plaque of Ziguang Calligraphy and Painting Xuan Paper Agency
吴报景描绘未来发展前景	Wu Baojing forecasting the future development
"绿杨宝"书画纸透光摄影图	A photo of "Lvyangbao" calligraphy and painting paper seen through the light
紫光仿古汉纸透光摄影图	A photo of Ziguang vintage paper seen through the light

章节	图中文名称	图英文名称
第五节	泾县小岭西山宣纸工艺厂	Section 5 Xiaoling Xishan Xuan Paper Craft Factory in Jingxian County
	百顺宣纸工艺品有限公司大门	Gate of Baishun Xuan Paper Artwork Co., Ltd.
	小岭西山宣纸工艺厂边的林荫道	Rood by Xiaoling Xishan Xuan Paper Craft Factory
	曹柏胜	Cao Bosheng
	"曹柏胜"古法檀皮宣	"Cao Bosheng" Xuan Paper made of Pteroceltis tatarinowii Maxim. bark in ancient methods
	"徽家纸号"仿古打印纸	Antique printing paper with "Hui Family Seal"
	"徽家纸号"不干胶书画加工纸标签	Self-adhesive paper label with "Hui Family Seal"
	"曹柏胜"古法檀皮宣纤维形态图 (10×)	Fibers of "Cao Bosheng" Xuan paper made of Pteroceltis tatarinowii Maxim. bark in ancient methods (10× objective)
	"曹柏胜"古法檀皮宣纤维形态图 (20×)	Fibers of "Cao Bosheng" Xuan paper made of Pteroceltis tatarinowii Maxim. bark in ancient methods (20× objective)
	"曹柏胜"古法檀皮宣润墨效果	Writing performance of "Cao Bosheng" Xuan paper made of Pteroceltis tatarinowii Maxim. bark in ancient methods
	古法制作檀皮宣的纸浆	Paper pulp of Pteroceltis tatarinowii Maxim. bark for making Xuan paper in ancient methods
	进口分张剂标签	Label of imported mucilage for separating the paper layers
	流经厂区旁的山泉水	Mountain spring flowing by the factory
	浸泡龙须草浆板	Soaking pulp board of Eulaliopsis binata
	"曹柏胜"古法檀皮宣浆团	Pulp balls of "Cao Bosheng" Xuan paper made of Pteroceltis tatarinowii Maxim. bark in ancient methods
	"徽家纸号"仿古打印纸浆团	Pulp balls of antique printing paper with "Hui Family Seal"
	捞纸	Papermaking
	榨纸	Pressing the paper
	晒纸	Drying the paper
	检纸	Checking the paper
	裁纸机	Machine for cutting the paper
	盖章	Stamping the paper
	打浆机	Beating machine
	捞纸槽	Papermaking trough
	帘床	Frame for supporting the papermaking screen
	切帖刀	Knife for cutting the paper pile
	晒纸刷	Brush for drying the paper
	刷夹	Brush holder
	工人正在往焙墙上刷米汤,使焙面洁净	Workers brushing rice paste to clean the drying wall
	压纸石	Stone for pressing the paper
	剪刀	Shears
	仓库	Warehouse
	小岭西山宣纸工艺厂的常用标签章	Common label seal of Xiaoling Xishan Xuan Paper Craft Factory
	"曹柏胜"古法檀皮宣透光摄影图	A photo of "Cao Bosheng" Xuan paper made of Pteroceltis tatarinowii Maxim. bark in ancient methods seen through the light
	仿古打印纸透光摄影图	A photo of vintage printing paper seen through the light

章节	图中文名称	图英文名称
第六节	安徽澄文堂宣纸艺术品有限公司	Section 6 Chengwentang Xuan Paper Artwork Co., Ltd. in Anhui Province
	百岁坊	Baisui (hundred years) Memorial Gateway
	承流峰	Chengliu Peak
	中国商标网公告信息截图	Screen shot of bulletin information on official website of National Trademark Office
	"澄文堂"厂区大门	Gate of Chengwentang Xuan Paper Artwork Co., Ltd.
	访谈王四海(左)	Interviewing Wang Sihai (left)
	郑礼红协助扳榨	Zheng Lihong assisting with the pressing step
	"红星"书画纸纤维形态图 (10×)	Fibers of "Red Star" calligraphy and painting paper (10× objective)
	"红星"书画纸纤维形态图 (20×)	Fibers of "Red Star" calligraphy and painting paper (20× objective)
	"红星"书画纸润墨效果	Writing performance of "Red Star" calligraphy and painting paper
	浆料	Pulp materials
	眺望九峰水库	Overlooking Jiufeng Reservoir
	浸泡	Soaking the pulp board
	打浆	Beating the pulp
	配浆	Mixing the pulp
	捞纸	Papermaking
	用以压榨的设备	Tools for pressing the paper
	揭纸	Peeling the paper
	晒纸	Drying the paper
	检纸	Checking the paper
	打浆机	Beating machine
	捞纸槽生产线	Production line of papermaking trough
	帘床	Frame for supporting the papermaking screen
	切帖刀	Knife for cutting the paper
	刷把	Brush
	纸焙	Drying wall
	剪刀	Shears
	安徽澄文堂宣纸艺术品有限公司贮纸仓库	Paper storing warehouse of Chengwentang Xuan Paper Artwork Co., Ltd. in Anhui Province
	蔡小汀与王四海	Cai Xiaoting and Wang Sihai
	积压在库的"红星"书画纸	Excess inventory of "Red Star" calligraphy and painting paper
	在库房中若有所思的王四海	Wang Sihai thinking in the warehouse
	"红星"书画纸透光摄影图	A photo of "Red Star" calligraphy and painting paper seen through the light

章节	图中文名称	图英文名称
第四章	皮纸	Chapter IV Bast Paper
第一节	泾县守金皮纸厂	Section 1 Shoujin Bast Paper Factory in Jingxian County
	程家皮纸厂路边指示牌	Road sign of Chengjia Bast Paper Factory
	通往程家皮纸厂的乡道	Country road to Chengjia Bast Paper Factory
	守金皮纸厂周边环境	Surrounding environment of Shoujin Bast Paper Factory
	程守海	Cheng Shouhai
	调查组成员再次访谈程守海	A researcher reinterviewing Cheng Shouhai

	程玮	Cheng Wei
	构皮本色云龙皮纸	Yunlong bast paper made of mulberry bark and tea-leaves
	成品纸仓库	Warehouse of paper products
	构皮本色云龙皮纸纤维形态图（10×）	Fibers of unbleached Yunlong bast paper made of mulberry bark (10× objective)
	构皮本色云龙皮纸纤维形态图（20×）	Fibers of unbleached Yunlong bast paper made of mulberry bark (20× objective)
	构皮本色云龙皮纸润墨效果	Writing performance of unbleached Yunlong bast paper made of mulberry bark
	洗漂构皮料	Cleaning and bleaching mulberry bark materials
	压榨洗漂后的构皮料	Pressing the cleaned and bleached mulberry bark materials
	工人正在拣皮	Workers picking and choosing the bark
	工人在打浆	A worker beating the pulp
	捞纸车间	Papermaking workshop
	捞纸	Papermaking
	压榨	Pressing the paper
	纸帖浇水	Watering the paper pile
	晒纸上墙	Drying the paper on a wall
	检纸	Checking the paper
	拣皮料	Picking and choosing the bark
	调查组成员在拍摄工人打浆	A researcher taking photos of the beating step
	捞纸	Papermaking
	压榨	Pressing the paper
	打浆机	Beating machine
	待打浆的木浆浆板原料	Raw material of wood pulp board to be beaten
	纸槽	Papermaking trough
	纸帘	Papermaking screen
	帘床	Frame for supporting the papermaking screen
	刷子	Brush
	枪棍	Tool for separating the paper layer
	钉耙	Rake for picking up the bark materials
	压纸石	Stone for pressing the paper
	新建办公兼展示区域	New office and display area
	坐落在山脚下的守金皮纸厂外部环境	External view of Shoujin Bast Paper Factory
	程守海和李金梅共同守护的家园	Cheng Shouhai and Li Jinmei's house
	守金皮纸厂外景	External view of Shoujin Bast Paper Factory
	守金皮纸厂周围环境	Surroundings of Shoujin Bast Paper Factory
	构皮本色云龙纸透光摄影图	A photo of Yunlong paper made of paper mulberry bark (original color) seen through the light

章节	图中文名称	图英文名称
第二节	泾县小岭驰星纸厂	Section 2 Xiaoling Chixing Paper Factory in Jingxian County
	驰星纸厂背后的山峰	Mountain peak behind Chixing Paper Factory
	驰星纸厂大门	Entrance of Chixing Paper Factory
	在厂区咨询曹志平的原料加工	Consulting Cao Zhiping about raw materials processing step of paper
	曹志平	Cao Zhiping
	普洱茶包装纸	Packaging paper for Pu'er Tea
	正在蒸煮的中药材	Boiling the Chinese medicinal materials
	驰星纸厂雁皮罗纹纸纤维形态图（10×）	Fibers of *Wikstroemia pilosa* Cheng Luowen paper in Chixing Paper Factory (10× objective)
	驰星纸厂雁皮罗纹纸纤维形态图（20×）	Fibers of *Wikstroemia pilosa* Cheng Luowen paper in Chixing Paper Factory (20× objective)
	驰星纸厂雁皮罗纹纸润墨效果	Writing performance of *Wikstroemia pilosa* Cheng Luowen paper in Chixing Paper Factory
	日本进口的化学纸药	Imported Japanese chemical papermaking mucilage
	周坑村的山涧水	Stream of Zhoukeng Village
	清洗皮料	Cleaning the bark materials
	打浆	Beating the pulp
	拉力剂外包装	Package of paper adhesives
	捞纸车间	Papermaking workshop
	捞纸	Papermaking
	捞纸结束做记号	Making the paper after papermaking
	做好记号等待压榨的纸帖	Paper with marks to be pressed
	压榨工具	Tool for pressing the paper
	压榨	Pressing the paper
	晒纸上墙	Dring the paper on a wall
	揭纸	Peeling the paper down
	仓库中的成品纸	Paper products in the warehouse
	纸帘	Papermaking screen
	铲子	Shovel for separating the paper layers
	压纸石	Stone for pressing the paper
	木头	Wood for pressing the paper
	营业执照	Business license
	"恢复小岭宣纸厂上书"原件照	Original photos of "Asking for Resuming Xiaoling Paper Factory"
	陕西宝鸡的捞纸工	A papermaker from Baoji City of Shaanxi Province
	调查组成员和曹志平等在驰星纸厂大门口合影	A group photo of researchers and Cao Zhiping etc. at the gate of Chixing Paper Factory
	雁皮罗纹纸透光摄影图	A photo of *Wikstroemia pilosa* Cheng Luowen paper seen through the light

章节	图中文名称	图英文名称
第三节	潜山县星杰桑皮纸厂	Section 3 Xingjie Mulberry Bark Paper Factory in Qianshan County
	星杰桑皮纸厂外景	External view of Xingjie Mulberry Bark Paper Factory
	调查组成员在村中与刘同烟交流	Researchers communicating with Liu Tongyan in the village
	坛畈村的农家在收稻谷	Farmers in Tanfan Village harvesting rice
	正在讲解工序的刘同烟（右二）	Liu Tongyan explaining the papermaking procedures
	刘同烟在观察野生桑树皮	Liu Tongyan observing wild mulberry bark
	坛畈村公路上的桑皮纸宣传牌	Billboard of mulberry bark paper by the road of Tanfan Village
	刘同烟与侄子在打槽	Liu Tongyan and his nephew stirring the papermaking materials
	针灸艾条纸纤维形态图（10×）	Fibers of mulberry bark paper for acupuncture and moxibustion (10× objective)
	针灸艾条纸纤维形态图（20×）	Fibers of mulberry bark paper for acupuncture and moxibustion (20× objective)
	针灸艾条纸润墨效果	Writing performance of mulberry bark paper for acupuncture and moxibustion
	水池里浸泡的文化纸边料	Wenhua paper offcut soaking in the sink
	针灸艾条纸（右）与书画纸（白色）	Paper for acupuncture and moxibustion (right) and calligraphy and painting paper (white)
	潜山山间的野桑树	Wild mulberry tree on the hills of Qianshan County
	坛畈村边的野桑树	Wild mulberry trees near Tanfan Village

浸湿的桑皮原料	Wetted mulberry bark
山间洗料的山涧溪水	Stream for cleaning the materials
剥下的桑树皮(前为野桑皮,后为家桑皮)	Mulberry bark (wild ones at the front, cultivated ones at the back)
女工在拣皮	A female worker picking and choosing the bark
入库存放的干皮料	Dried bark materials in the warehouse
蒸锅	Steaming pot
观察炉膛的刘同烟	Liu Tongyan observing the stove heat
刘同烟在洗皮	Liu Tongyan is cleaning the bark
初洗皮料	Cleaning the bark materials for the first time
初捡	Choosing the qualified bark for the first time
中洗	Cleaning the bark materials for the second time
水漂	Bleaching mulberry bark
精选	Choosing the bark materials carefully
踏碓添皮	Beating and adding in mulberry bark
机器搅拌	Machine stirring the bark pulp
灌气	Charge the bag with air
锤袋	Beating the bag
搅拌	Stirring the materials
刘同烟与侄子在划槽	Liu Tongyan and his nephew stirring the materials
下帘	Putting the bamboo papermaking screen into water
捞纸	Papermaking
取帘上塔	Putting the bamboo papermaking screen on the paper pile
榨驼	Tool for squeezing the paper
榨好的纸塔	Paper pile after squeezing
晒纸	Drying the paper
检纸	Checking the paper
切纸	Cutting the paper
包装	Packing the paper
蒸煮锅灶	Pot for steaming and boiling
洗皮篮	Basket for cleaning the bark
正在用皮钩摆散皮料的刘同烟	Liu Tongyan stirring the bark materials with a hook
土筛	Sieve for filtering the bark materials
挑皮篮	Baskets for carrying the bark materials
皮碓	Pestle for beating the bark materials
拌料池	Sink for stirring the materials
过滤池	Sink for filtering the materials
袋料锤	Hammer for beating and cleaning the materials
槽笼	Papermaking trough
竹帘	Bamboo papermaking screen
帘床	Frame for supporting the papermaking screen
植物汁缸与丝布网	Vat holding the plant papermaking mucilage and cloth net
土焙笼	Drying wall
晒纸刷	Brush made of pine needles
四尺针灸艾条纸	4-chi paper for acupuncture and moxibustion
老纸坊	Old paper mill
正在观察桑皮原料蒸煮火候的刘同烟	Liu Tongyan observing stove heat of the steaming and boiling step
刘同烟讲述"神丹皮"传说	Liu Tongyan telling the story of "Shendanpi"
独自洗料的刘同烟	Liu Tongyan cleaning the materials alone
污染的小溪	Polluted stream
2007年调查时的制纸帘户	Interviewing papermaking screen maker in 2007
对潜山桑皮纸参与倦勤斋修复的感谢证明	Proof of thanks to mulberry bark paper in Qianshan County participating in the restoration of Juanqinzhai
桑皮针灸艾条纸透光摄影图	A photo of mulberry bark paper for acupuncture and moxibustion seen through the light
纯桑皮纸(厚)透光摄影图	A photo of pure mulberry bark paper (thick) seen through the light
纯桑皮纸(薄)透光摄影图	A photo of pure mulberry bark paper (thin) seen through the light

章节	图中文名称	图英文名称
第四节	岳西县金丝纸业有限公司	Section 4 Jinsi Paper Co., Ltd. in Yuexi County
	坐落于山上的岳西县金丝纸业生产现场	Production scene of Jinsi Paper Co., Ltd. in Yuexi County on a mountain
	路口的板舍村宣传牌	Billboard of Banshe Village in an intersection
	板舍村中的小河	River in Banshe Village
	王柏林	Wang Bailin
	王柏林(中)、岳西县文化馆官员(右)与调查组负责人	Wang Bailin (middle), officer from the Yuexi County Cultural Centre (right) and the head of researchers
	已晒干的三桠皮原料	Dried mitsumatu bark
	纯桑皮纸	Pure mulberry bark paper
	金丝纸业纯桑皮纸纤维形态图(10×)	Fibers of pure mulberry bark paper in Jinsi Paper Co., Ltd. (10× objective)
	金丝纸业纯桑皮纸纤维形态图(20×)	Fibers of pure mulberry bark paper in Jinsi Paper Co., Ltd. (20× objective)
	金丝纸业纯桑皮纸润墨效果	Writing performance of pure mulberry bark paper in Jinsi Paper Co., Ltd.
	剥下来的野桑皮料	Wild mulberry bark materials
	装在袋中的纸药	Papermaking mucilage in a bag
	板舍村旁的山涧	Stream by the Banshe Village
	岳西县文化馆墙上绘制的生产流程图	Production flow chart on a wall of the Yuexi County Cultural Centre
	已剥好晒干的桑皮料	Dried mulberry bark materials
	漂洗后的皮料	Bleached mulberry bark materials
	蒸煮皮料的纸甑与木柴	Papermaking utensil and firewood for steaming and boiling the bark materials
	沤皮(在陶缸中沤制)	Soaking the bark (in a bag pot)
	揉皮	Rubbing the bark
	洗皮	Cleaning the bark
	榨皮	Pressing the bark
	拣皮	Picking and choosing the bark
	漂洗	Bleaching the bark
	皮碓	Pestle for beating the materials
	袋料	Putting paper pulp in a cloth bag and beating it in water
	王柏林在捞纸	Wang Bailin making the paper
	王柏林在榨纸	Wang Bailin pressing the paper
	揭纸	Peeling the paper down
	焙纸	Drying the paper
	桑皮纸成品	Mulberry bark paper products
	桑皮纸外包装样式	Package of mulberry bark paper
	竹帘	Bamboo papermaking screen

	帘床	Frame for supporting the papermaking screen
	纸刷	Brush
	料筛	Sieve
	拌料池	Sink for stirring the materials
	捞纸池	Papermaking trough
	过滤池	Filtering trough
	调查组成员与王柏林交流市场情况	Researchers exchanging market information with Wang Bailin
	《岳西县志》	*The Annals of Yuexi County*
	调查组成员前往捞纸槽的途中	Researchers on the way to papermaking trough
	文化馆"非遗"牌匾下留影的汪淳	Wang Chun by the Intangible Heritage Certificate plaque in the Cultural Centre
	纯桑皮纸透光摄影图	A photo of pure mulberry bark paper seen through the light
	三桠皮纸透光摄影图	A photo of *Edgeworthia Chrysantha* Lindl. paper seen through the light

章节	图中文名称	图英文名称
第 五 节	歙县深渡镇棉溪村	Section 5 Mianxi Village in Shendu Town of Shexian County
	棉溪村边的新安江	Xin'an river by the Mianxi Village
	冬天"猫冬"休闲的棉溪村民	Relaxing Mianxi Villagers in winter
	深渡镇区景观	View of Shendu Town
	深渡镇政府	Shendu Town Government
	棉溪村外盛开的油菜花田	Rape flowers outside Mianxi Village
	调查组成员在棉溪村村委会查找资料	Researchers searching for information in the Mianxi Village Committee Office
	在江祖术新居院内访谈	Interviewing at the new house of Jiang Zushu
	江祖术家旧宅	Old house of Jiang Zushu
	江祖术在机制信纸上复抄的祖谱	Jiang Zushu's family genealogy transcribed on a machine-made paper
	江祖术与汪寿花旧日合影	An old photo of Jiang Zushu and Wang Shouhua
	调查组成员访谈江祖术	Researchers interviewing Jiang Zushu
	江祖术、三儿子江德成和三儿媳妇汪立菊	Jiang Zushu, his third son Jiang Decheng and daughter-in-law Wang Liju
	汪成棋带队上山寻找"栗树"	Wang Chengqi leading research team to "Li" tree
	棉溪村2007年送检的水质报告	The water quality report of Mianxi Village in 2007
	构树皮	Mulberry bark
	汪成棋示范剥皮流程	Wang Chengqi showing how to strip the bark
	汪立菊示范捆皮	Wang Liju showing how to bind the bark
	煮皮的铁锅	Iron pot for boiling the bark
	蒸煮皮料的灶膛	Stove for steaming and boiling the bark materials
	石臼	Stone mortar
	江祖术示范捞纸动作	Jiang Zushu showing the papermaking procedures
	调查组成员与汪成棋交流捞纸动作	Researchers communicating with Wang Chengqi about papermaking procedures
	汪成棋示范捞纸动作	Wang Chengqi showing the papermaking procedure
	江祖术捞纸（历史照片翻拍）	Jiang Zushu making the paper (retook from an old photo)
	江祖术放纸（历史照片翻拍）	Jiang Zushu piling up the paper (retook from an old photo)
	压榨用的旧石板	An old slate for pressing the paper
	榨床	A stone board for pressing the paper
	江祖术示范刷纸动作	Jiang Zushu showing how to brush the paper
	江祖术揭纸（历史照片翻拍）	Jiang Zushu peeling the paper down (retook from an old photo)
	理纸示范	Showing how to sort the paper
	江祖术家的石板纸槽	Jiang Zushu's stone for papermaking trough
	江祖术家的药兜	Jiang Zushu's basket for cleaning the leaves
	江祖术家的纸帘	Jiang Zushu's papermaking screen
	江祖术家的刷子	Jiang Zushu's brush made of pine needles
	江祖术家的压榨筐	Jiang Zushu's crate for holding the stones
	压榨用的石头	Stone for pressing the paper
	调查组成员与王柏林交流市场情况	Research member exchanging market information with Wang Bailin
	江祖术家构皮纸纤维形态图（10×）	Fibers of mulberry bark paper in Jiang Zushu's house (10× objective)
	江祖术家构皮纸纤维形态图（20×）	Fibers of mulberry bark paper in Jiang Zushu's house (20× objective)
	江祖术家构皮纸润墨效果	Writing performance of mulberry bark paper in Jiang Zushu's house
	江祖术制作的各色构皮纸	Various colors of mulberry bark paper made by Jiang Zushu
	江祖术向调查组成员展示自己当年的记录	Jiang Zushu showing his papermaking records to a researcher
	江祖术保留的销售记录原件	Sale records kept by Jiang Zushu
	从棉溪村看新安江对岸的村落	View of the village on the other side of Xin'an River from Mianxi Village
	调查组成员与汪成棋（前右一）合影	A group photo of researchers and Wang Chengqi (first from the right in the front row)
	向江祖术老人询问造纸术语	Inquiring Jiang Zushu about papermaking terms
	江祖术谈到传承现状	Jiang Zushu talking about current status of papermaking inheritance
	江祖术收藏自用的纸	Paper saved by Jiang Zushu for his own funeral
	江祖术依依不舍卖老纸	Jiang Zushu felt reluctant to sell the paper the saved for his own funeral
	调查组成员与江祖术合影	A group photo of researchers and Jiang Zushu
	棉溪村江祖术家构皮纸透光摄影图	A photo of mulberry bark paper in Jiang Zushu's house of Mianxi Village seen through the light
	棉溪村江祖术家染色构皮纸透光摄影图	A photo of dyed mulberry bark paper in Jiang Zushu's house of Mianxi Village seen through the light

章节	图中文名称	图英文名称
第 六 节	黄山市三昕纸业有限公司	Section 6 Sanxin Paper Co., Ltd. in Huangshan City
	第一次与第二次调查时的新厂区	View of the new Sanxin Paper Factory on our first and second field investigation trips
	布满灰尘的新建平卧式焙纸台群	Newly built tables for drying the paper covered with dust
	在新厂区筹建办公室访谈朱建新	Interviewing Zhu Jianxin in the construction office of the new factory
	新安江畔的万安古镇与古城岩	View of Wan'an Town and ancient city by Xin'an River
	新厂区附近的秀阳古塔	Xiuyang Ancient Tower near the new Sanxin Paper Factory
	朱建新与调查组成员在建设中的抄纸车间	Zhu Jianxin and a researcher in the papermaking workshop
	新厂区已砌好的水池	Newly built water pools at the new Sanxin Paper Factory
	水质纯净的新安江上游	Upstream of Xin'an River
	新厂区正在建设中的纸槽群	Papermaking troughs under construction in the new factory
	在施工现场访谈朱建新	Interviewing Zhu Jianxin at the construction site
	纸品样册	Sample book of decorative papers
	民芸纸样品	Minyun paper samples
	强制纸样品	Qiangzhi paper samples
	三昕纸业有限公司楮皮纸纤维形态图（10×）	Fibers of mulberry bark paper in Sanxin Paper Co., Ltd. (10× objective)
	三昕纸业有限公司楮皮纸纤维形态图（20×）	Fibers of mulberry bark paper in Sanxin Paper Co., Ltd. (20× objective)
	三昕纸业有限公司楮皮纸润墨效果	Writing performance of mulberry bark paper in Sanxin Paper Co., Ltd.
	新厂区内的小水库	Reservoir in the new factory

图中文名称	图英文名称
朱建新在老厂	Zhu Jianxin in the old factory
拣皮	Picking the bark
捞纸	Papermaking
洗料	Cleaning the materials
存放待配料的染料容器	Containers for holding dye
染料配浆	Mixing dye with pulp
捞制民芸纸	Making Minyun paper
刷纸上墙	Drying the paper on a wall
民芸纸样品	Minyun paper samples
烘好的纸张	Dried paper
揉纸	Rubbing the paper
成品纸	Paper products
打浆机	Beating machine
金属制纸槽	Metal papermaking trough
焙笼	Drying wall
三昕纸业有限公司强制纸（白色）	Qiangzhi paper in Sanxin Paper Co., Ltd. (white)
三昕纸业有限公司强制纸（蓝色）	Qiangzhi paper in Sanxin Paper Co., Ltd. (blue)
三昕纸业有限公司民芸纸（黑色）	Minyun paper in Sanxin Paper Co., Ltd. (black)
三昕纸业有限公司民芸纸（红色）	Minyun paper in Sanxin Paper Co., Ltd. (red)
三昕纸业有限公司楮皮纸	Mulberry bark paper in Sanxin Paper Co., Ltd.
摄影画用纸样品	Photographic paper sample
新产区的临时纸库与宿舍	Temporary paper warehouse and dormitory
新安江水	Xin'an River
与朱建新在小水库石坝上聊环保	Talking about environmental protection with Zhu Jianxin on a dam
海阳镇老街	Old street of Haiyang Town
一丝不苟的"三昕"造纸人	Conscientious papermakers of Sanxin Paper Co., Ltd.
手揉纸透光摄影图	A photo of wrinkled paper seen through the light
民芸纸透光摄影图	A photo of Minyun paper seen through the light
染色手揉纸透光摄影图	A photo of dyed wrinkled paper seen through the light

章节	图中文名称	图英文名称
第七节	歙县六合村	Section 7 Liuhe Village in Shexian County
	六合村外的山间公路	Mountain road alongside Liuhe Village
	六合村内的民居	Local residences in Liuhe Village
	流过六合村口的山溪	Mountain stream flowing through Liuhe Village
	《郑氏祭祀簿》	Sacrificial Ceremony Book of the Zhengs
	20世纪70年代后期建造的集中煮皮设施	Facilities for boiling the bark built in the late 1970s
	放置铁锅的灶台	Kitchen range for holding the iron pot
	烧火的石头灶	Stone stove for firing
	村口老祠堂旧址前的古柏	Old cypress in front of the former ancestral hall
	郑义生（右）和郑火土（左）在郑火土家门口合影	A group photo of Zheng Yisheng (right) and Zheng Huotu (left) sitting in Zheng Huotu's doorway
	老屋墙上郑火土的岳父岳母旧照	An old photo of Zheng Huotu's father-in-law and mother-in-law on the wall
	郑火土抄纸工作坊	Zheng Huotu's papermaking mill
	郑为民在自家门口展示用过的纸帘	Zheng Weimin showing papermaking screen in front of his house
	郑火土展示历年所造手工纸纸样	Zheng Huotu showing handmade paper samples he made over the years

图中文名称	图英文名称
郑火土家手工皮纸存放的地方	The container for holding handmade bast paper in Zheng Huotu's house
六合村现存旧日的书写用纸簿	Old writing book of Liuhe Village
郑火土用自制手工纸做的长钱产品	Handmade joss paper products for sacrificial ceremony made by Zheng Huotu
郑火土家手工构皮纸采样	Sampling handmade mulberry bark paper made by Zheng Huotu
郑火土家构皮纸纤维形态图（10×）	Fibers of mulberry bark paper in Zheng Huotu's house (10× objective)
郑火土家构皮纸纤维形态图（20×）	Fibers of mulberry bark paper in Zheng Huotu's house (20× objective)
pH实测比照	pH level test
构树枝	Mulberry tree branch
未加工处理的干构树枝条	Unprocessed dry mulberry tree branches
处理成造纸原料的构树皮	Processed Mulberry bark for papermaking
六合村里种的桑树	Mulberry tree in Liuhe Village
黄子坑河	Huangzikeng River
郑为民展示自家所造加有机制纸辅料的纸张	Zheng Weimin showing his paper made with machine-made paper materials
电动打浆机	Electric beating machine
郑火土演示并讲解捞纸动作	Zheng Huotu showing and explaining papermaking procedures
榨纸床	Bed for pressing the paper
郑火土自己拼合出的"家造"打浆机	Zheng Huotu's home-made beating machine
砌入墙中的脚碓石	Pestle stone in a wall
春臼一	Mortar (1)
存放春臼一的宅院	A courtyard for holding beating mortar (1)
春臼二	Mortar (2)
存放春臼二的造纸坊（已闲置废弃）	Abandoned papermaking mill for storing mortar (2)
郑火土家现用纸槽	Zheng Huotu's papermaking trough currently in use
六合村现存最古老的纸槽	The oldest papermaking trough in Liuhe Village
郑火土家所用纸帘	Zheng Huotu's papermaking screen
郑火土手捏处为帘额	Zheng Huotu holding the top of papermaking screen
六合村其他村民家的纸帘	Papermaking screens from other villagers' houses in Liuhe Village
帘夹正面	The front view of the papermaking screen holder
帘夹反面	The reversed view of the papermaking screen holder
帘床	Frame for supporting the papermaking screen
帘床、纸帘、帘夹放置图	Papermaking screen and its supporting frame and holder
压纸木板	Board for flattening the paper
压纸石	Stone for pressing the paper
闲置的木榨构件	Unused wooden parts
纸刷	Brush
纸凳	Chair for holding the paper
郑火土演示长钱的做法	Zheng Huotu showing how to make joss paper
做纸扎的工具	Tools for making joss paper
郑火土已经做好的纸扎	Completed joss paper made by Zheng Huotu
郑火土用自制手工纸做的长钱	Joss paper made by Zheng Huotu
长钱打开后的形态	Unfolded joss paper
郑火土用机制纸做的珍珠伞	Pearl umbrella made of machine-made paper by Zheng Huotu
珍珠伞打开之后的形态	Unfolded pearl umbrella
村中的古巷	Old alley in the Liuhe Village
双数孔纸钱	Joss paper with even numbers of hole

	纸绳"拔河"	A tug of war using paper rope
	郑火土	Zheng Huotu
	六合村附近的油菜花与山水空间	Landscape of yellow rape flower near the Liuhe Village
	郑火土造染色构皮纸透光摄影图	A photo of dyed mulberry bark paper made by Zheng Huotu seen through the light
	郑火土造染色构皮纸(长)透光摄影图	A photo of dyed mulberry bark paper (long) made by Zheng Huotu seen through the light

章节	图中文名称	图英文名称
第五章	竹纸	Chapter V Bamboo Paper
第一节	歙县青峰村	Section 1 Qingfeng Village in Shexian County
	村里的主干道路	Main road of the village
	通往程忠余家的土路	Dirt road to Cheng Zhongyu's house
	当地称为外山的造纸村	A papermaking village called Waishan by the locals
	废弃的石头老纸槽	Abandoned stone papermaking trough
	程忠余回忆造纸往事	Cheng Zhongyu recalling the past stories of papermaking
	程忠余向调查组展示前些年造的纸	Cheng Zhongyu showing the paper made in previous years to the research team
	刷红纸	Shuahong paper
	表芯纸原纸	Raw paper of Biaoxin paper
	写着乡民贺礼金额的五色纸	Five-color paper recording the gift money amount of the villagers
	歙县青峰村竹纸纤维形态图(10×)	Fibers of bamboo paper in Qingfeng Village of Shexian County (10× objective)
	歙县青峰村竹纸纤维形态图(20×)	Fibers of bamboo paper in Qingfeng Village of Shexian County (20× objective)
	青峰村边的毛竹	Mao bamboo forest next to Qingfeng Village
	青峰村里造纸用的山溪水	Mountain stream used for papermaking in Qingfeng Village
	程忠余家染纸用的化工染料	Cheng Zhongyu's chemical dyes for dyeing paper
	程忠余的儿子演示用"木刀"将竹料劈开	Cheng Zhongyu's son showing how to split the bamboo by a wooden knife
	砍伐下来的竹料	Lopped bamboo materials
	旧日煮山苍子叶的老铁锅	Iron pot used for boiling *Litsea cubeba*'s leaves
	在半停产状态纸坊里捞纸与扣纸演示	Demonstration of papermaking procedures in the papermaking mill nearly stopped production
	半停产纸坊里的木榨	Wooden presser in the papermaking mill nearly stopped production
	数纸	Counting the paper
	乡民用染色纸做成的贺礼簿	Villager's Gift Brochure made of dyed paper
	老纸坊中的纸槽	Papermaking trough in an old papermaking mill
	程忠余家的旧纸帘	Old papermaking screen in Cheng Zhongyu's house
	程忠余的帘床	Frame for supporting the papermaking screen in Cheng Zhongyu's house
	程忠余制作的木刀	Wooden knife made by Cheng Zhongyu
	混料耙	Rake for mixing the materials
	松毛刷	Brush made of pine needles
	压榨用麻绳	Hemp rope for pressing the paper
	压榨用盖板	Wooden board for pressing the paper
	装残破纸的篮子	Basket for carrying the broken paper
	捞筋叉	Fork for picking up the residues
	程忠余招待调查组成员特色粽子餐	Cheng Zhongyu entertaining the researchers with special Zongzi meal
	程忠余老人做点烟的纸媒与燃纸媒	Cheng Zhongyu using paper stick to light a cigarette
	冬天村里烤火炉晒太阳的老人们	Old villagers basking in the sun by the stove
	废弃的泡料池	Abandoned pool for soaking the materials
	程忠余造竹纸透光摄影图	A photo of bamboo paper made by Cheng Zhongyu seen through the light

章节	图中文名称	图英文名称
第二节	泾县孤峰村	Section 2 Gufeng Village in Jingxian County
	泾县的山川林海	Mountains and forests in Jingxian County
	清代嘉庆版《泾县志》书影	*The Annals of Jingxian County* in Jiaqing Reign of the Qing Dynasty
	九华古道泾县小岭段	Jiuhua ancient road in Xiaoling section of Jingxian County
	《泾县土纸样本》	*Samples of Handmade Paper in Jingxian County*
	访谈老мен工	Interviewing an old papermaker
	槌杨桃藤	Beating kiwi fruit cane
	九峰村	Jiufeng Village
	完全废弃的麻塘	Abandoned material pond
	遗弃的石碾	Abandoned stone roller
	捞纸演示	Showing papermaking procedures
	孤峰村的"焙纸笼"遗存	Abandoned paper drying cabin in Gufeng Village
	泾县竹纸纤维形态图(10×)	Fibers of bamboo paper in Jingxian County (10× objective)
	泾县竹纸纤维形态图(20×)	Fibers of bamboo paper in Jingxian County (20× objective)
	《泾县土纸样本》中关于"三六表"纸的信息	Information of "Sanliubiao paper" in *Samples of Handmade Paper in Jingxian County*
	孤峰村竹纸透光摄影图	A photo of bamboo paper in Gufeng Village seen through the light

章节	图中文名称	图英文名称
第三节	金寨县燕子河镇	Section 3 Yanzihe Town in Jinzhai County
	调查组成员与造纸人许修树(左二)、许修林(左三)合影	Researchers with papermaker Xu Xiushu (second from the left) and Xu Xiulin (third from the left)
	龙马村口的路	Road to Longma Village
	薛河制作坊转让协议	Transfer agreement of Xuehe Papermaking Mill
	厂房现场遗留的竹麻	Bamboo materials left in the factory site
	许修安在演示砸破竹子	Xu Xiuan showing how to break the bamboo
	榥甑炉遗存	Abandoned Huangzeng stove
	洗料池旧址	Former site of cleaning pool
	已废弃的舂捣工具	Abandoned tools for pounding the bamboo materials
	废弃的纸槽与槽房	Abandoned papermaking trough and workshop
	许修树家中的旧纸帘	Old papermaking screen in Xu Xiushu's house
	麻凼遗存	Former site of Madang (pond for fermenting the materials)
	旧榥甑	Former Huangzeng
	许修树家的旧纸帘	Old papermaking screen in Xu Xiushu's house
	许修树家的旧捞纸帘架	Old screen frame for supporting the papermaking in Xu Xiushu's house
	打钱模子	Mold for making joss paper
	金寨县竹纸纤维形态图(10×)	Fibers of bamboo paper in Jinzhai County (10× objective)
	金寨县竹纸纤维形态图(20×)	Fibers of bamboo paper in Jinzhai County (20× objective)
	造纸人家中的祭坛供桌	The altar table in a papermaker's house
	用破开的毛竹从山上接泉水的"自来水"	Water tube, using broken bamboo to transport mountain spring water
	山边废弃的造纸遗址	Abandoned papermaking site by the mountain
	旧日用的石头碾盘	Former stone roller
	龙马村竹纸透光摄影图	A photo of bamboo paper in Longma Village seen through the light

章节	图中文名称	图英文名称
第六章	加工纸	Chapter VI Processed Paper

第 一 节	安徽省掇英轩书画用品有限公司	Section 1 Duoyingxuan Calligraphy and Painting Supplies Co., Ltd. in Anhui Province
	"掇英轩"非物质文化遗产荣誉牌匾	Honorary plaques of the intangible cultural heritage of "Duoyingxuan"
	中国科学技术大学人文与社会科学学院与"掇英轩"共建"传统纸笺产学研基地"揭牌仪式	Opening ceremony of the "Traditional Paper Production, Education and Research Base", jointly built by School of Humanities and Social Sciences, USTC and "Duoyingxuan"
	江淮名寺——相隐寺	Famous temple in Jianghuai area—Xiangyin Temple
	张治中故居	Former house of general Zhang Zhizhong
	黄麓镇洪家疃村的乡贤文化展馆	County Sage Cultural Exhibition Hall in Hongjiatuan Village of Huanglu Town
	1985年刘锡宏在日本大玄堂试纸旧照	Old photo of Liu Xihong testing paper in Daxuan Hall in Japan in 1985
	掇英轩文房用品研究所营业执照	Business license of Duoyingxuan Calligraphy and Painting Supplies Research Institute
	研究所阶段的杨宁英旧照(右)	Old photo of Yang Ningying working for the institute (right)
	刘家老屋旧照	Old photo of the aged house of the Liu's family
	老屋里检纸员工旧照	Old photo of staff checking paper
	2000年挂靠脱钩申请	Application for disconnection in 2000
	商标注册证	Registered trademark
	中国艺术研究院聘书	Letter of appointment of the Chinese Academy of Arts
	在老屋小院中接受访谈的刘靖	Interviewing Liu Jing in the yard of old house
	第一批粉蜡笺小样旧照	Old photo of the first batch of Fenla paper produced
	手绘描金粉蜡笺初进"荣宝斋"时旧照	Old photo of hand-painted golden Fenla paper first introduced to "Rongbaozhai"
	"荣宝斋"对"掇英轩"纸笺的反馈材料	Feedback material from "Rongbaozhai" to "Duoyingxuan" paper
	米景扬题"纸中重宝"原件照	Original photo of Mi Jingyang's inscription "Treasure of the Paper"
	日本客户要求经销的信	A letter from a Japanese customer demanding for paper
	杨岗村老屋的"掇英轩"旧址	Former site of "Duoyingxuan" in Yanggang Village
	刘锡宏与师父、师母的合影旧照(左一常秀峰)	Old photo of Liu Xihong with his teacher (the first one on the left) and teacher's wife
	刘干正在学习雕版	Liu Gan is learning graving
	刘晴与杨宁英在店中合影旧照	Old photo of Liu Qing and Yang Ningying in the family store
	方玉红正在洒金(2001年旧照)	Fang Yuhong is sprinkling golden powder on paper (photo taken in 2001)
	在母亲身边观摩的刘子嘉	Liu Zijia is observing papermaking by her mother
	样品室陈列的各色粉蜡笺	Various Fenla paper displayed in the sample room
	琳琅满目的"掇英轩"纸品陈列	Various "Duoyingxuan" paper types
	真金手绘仿明代云龙纹双面粉蜡笺	Golden Double-faced Fenla paper hand-painted with cloud and dragon decorations and imitating the style of Ming Dynasty
	真金雨雪粉蜡笺	Golden Fenla paper with rain and snow decorations
	刘靖向调查组成员讲解洒金粉蜡笺工艺	Liu Jing is explaining the procedure of producing golden Fenla paper to researchers
	手绘描金"龙腾如意"纹粉蜡笺	Golden Fenla paper hand-painted with "Flying Dragon is Good Fortune" decorations
	木版水印绢本宣(石榴图案)	Woodblock printing Silk Xuan paper (with figure of pomegranate)
	绢本宣手卷(八十七神仙图案)	Silk Xuan paper hand scroll (with figure of eighty-seven immortal)
	五色珠光笺(从左往右为银白、金黄、金色、古铜、酒红珠光笺)	Paper with five-color of pearly luster (from left to right: silvery, golden yellow, golden, bronze, claret pearly luster paper)
	木版水印信笺("掇英轩"仿明《十竹斋笺谱》)	Woodblock printing letter paper ("Duoyingxuan" copying of Ming Dynasty)
	"掇英轩"制流沙笺	Liusha paper made by "Duoyingxuan"
	"掇英轩"制银色印花笺	Silver paper with figures of flowers made by "Duoyingxuan"
	各色尺八屏金银印花笺	Various eight-foot gold and silver paper with flower figures
	精品金银印花笺包装	Package of high quality gold and silver paper with flower figures
胡粉、朱砂、胭脂、石绿、石青		Chinese white, cinnabar, rouge, mineral green, azurite
明胶原料		Gelatin material
骨胶原料		Bone glue material
明矾		Alum
购自南京的成品金箔		Gold foil bought from Nanjing
加胶调配好的仿金粉		Gold-like powder mixed with glue
拣纸		Picking the paper
备梏		Sticking the paper
施胶		Soaking the paper in alum
施粉		Painting the paper
挑晾		Drying the paper
托裱		Mounting the paper
上挣		Smoothing the paper
刘靖展示真金手绘粉蜡笺		Liu Jing is showing Fenla paper hand-painted with gold
手绘描金		Hand-painted with golden powder
描金车间的女工们		Female workers in the workshop
卷筒		Reels
真金手绘粉蜡笺包装		Package of Fenla paper hand-painted with gold
仿金手绘粉蜡笺包装		Package of gold-like hand-painted Fenla paper
苏士澍(中国书法家协会主席)在真金手绘粉蜡笺上题轩名		Su Shishu (Chairman of the Chinese Calligrapher's Association) is inscribing on Fenla paper hand-painted with gold
手绘描金常用的毛笔		Brushes used for painting with golden powder
底纹笔		Brush
竹起		Bamboo tool for peeling the paper
棕把		Brush made of palm
多管羊毛排笔		Wool brush with a row of pipes
拖纸盆		Papermaking basin
码放着的挣板		Boards for smoothing the paper
码放着的各色绢		Piles of colored Silk Xuan paper
从泾县定制的宣纸原纸		Base paper of Xuan paper made in Jingxian County
托裱绢用的糨糊		Paste for mounting silken cloth
在挣板上铺绢		Spreading silken cloth on the board for smoothing the paper
托裱		Mounting the paper
绢本宣包装		Package of Silk Xuan paper
仿金材料金黄珠光粉		Gold-like powder for decorating the paper
拖染		Dyeing the paper
涂布		Applying color on the paper
挑晾		Drying the paper
挑晾中的泥金纸		Drying Nijin paper
上挣贴纸条		Sticking tape to fix the paper for smoothing
四尺泥金笺盒		Four feet package of Nijin paper
"马利"牌国画色		Marie's pigment for Chinese painting
雕版		Engraving
雕好的分版		Finished engraving
套印:对版		Overprint: register
套印:上色		Overprint: coloring

套印：覆印	Overprint: overlap print
裁切	Cutting the paper
锦盒包装	Package of brocade box
木刻刀	Woodcut knife
棕老虎	Papermaking tool made by palm
木版水印桌	Woodblock printing table
木版水印架	Woodblock printing shelf
揉面	Kneading dough
变幻多彩的流沙笺系列	Serious of Liusha paper with different figures
调制糨糊	Modulating paste
滴色	Dropping color
覆纸试验	Dyeing paper experiment
提纸试验	Lifting paper experiment
塑料盆	Plastic basin
备楷使用的木棍	Sticks for papermaking
感光胶	Sensitive glue
绘稿	Painting
码放的丝网版	Silk screen plate
丝网印刷	Silk screen printing
收纸	Collecting the paper
检验合格的五色泥金笺	Checking the paper
裁切	Cutting the paper
袋装金银印花笺	Gold and silver paper with flower figures in bags
刮刀	Scraper
气泵	Air pump
合肥城隍庙掇英轩文房用品商店	Duoyingxuan paper stationary store in Town God's Temple of Hefei
北京安徽"四宝堂"对"掇英轩"纸的评价材料	Evaluation of "Duoyingxuan" paper by Beijing Anhui "Sibaotang"
手绘中的粉蜡笺	Hand-painted Fenla paper
规划效果图	Planning effect picture
在建中的"掇英轩"纸笺加工基地新厂区	"Duoyingxuan's" new papermaking factory in building
真金手绘粉蜡笺高端纸品	High quality Fenla paper hand-painted with gold
范曾题"掇英轩"牌匾原件照	Photo of Fan Zeng's inscription "Duoyingxuan" on plaque
与樊嘉禄合影旧照（右樊嘉禄）	Old photo with Fan Jialu (right)
与张秉伦等专家合影旧照（左张秉伦、中华觉明、右潘吉星）	Old photo with Zhang Binglun and other experts (Zhang Binglun is on the left, Hua Jueming is in the middle, Pan Jixing is on the right)
中国历史名纸复原研究中心牌匾	The plaque of Research Centre of Chinese Famous Traditional Handmade Paper
羊脑笺写经折页	Yangnao paper used for copying scriptures
陈列室内的纸笺加工技艺讲座	A lecture on the paper processing skills in the display room
修葺后的老屋后门	The back door of a restored old house
洒金桶	Tool for sprinkling golden powder
《造金银印花笺法实验研究》合作署名论文	Co-authored paper named
"皖省大升纸庄"古纸	Old paper made by Dasheng Paper Factory in Anhui Province
合影：左三陈长智，左四冯大彪，右四杨宁英，右三米景扬，右二唐遒昌	Group photo (the third on the left is Chen Changzhi,the fourth one is Feng Dabiao, the fourth on the right is Yang Ningying, the third one is Mi Jingyang, the second is Tang Naichang)
米景扬为刘靖题的"纸上风云"	Mi Jingyang inscribed "Legends on Paper" for Liu Jing
租用的原长源小学校舍一角	A corner of the rented dormitory of the former Changyuan Primary School
新产品：方格半熟印格纸	New product: checkered Yinge paper
新创制的二十四节气花草纸	New product of The 24 Solar Terms paper decorated with flowers and plants
合肥新办公兼展示空间一角	A corner of the new office and display room
2015年中国台湾"巧手慧心——安徽省传统手工技艺展"演示	The demonstration of "2015 Clever Hand and Wisdom—Anhui Traditional Handicraft Skills Exhibition"
在中央美术学院设计学院的"造纸课"上	"Papermaking lesson" given to Design College students of China Central Academy of Fine Arts
"掇英轩"木版水印笺透光摄影图	A photo of "Duoyingxuan" woodblock printing paper seen through the light
刘靖谈"掇英轩"恢复历史名纸新思路	Liu Jing is talking about "Duoyingxuan's" ideas of recovering famous traditional handmade paper
"掇英轩"植物黄檗染色纸（浓）透光摄影图	A photo of "Duoyingxuan" vegetation Huangbo dyed paper (dense) seen through the light
"掇英轩"植物黄檗染色纸（淡）透光摄影图	A photo of "Duoyingxuan" vegetation Huangbo dyed paper (light) seen through the light
"掇英轩"金粟粉蜡笺透光摄影图	A photo of "Duoyingxuan" Jinsu Fenla paper seen through the light
"掇英轩"绢本宣透光摄影图	A photo of "Duoyingxuan" silk Xuan paper seen through the light
"掇英轩"木版水印笺透光摄影图	A photo of "Duoyingxuan" woodblock printing paper seen through the light
"掇英轩"泥金纸透光摄影图	A photo of "Duoyingxuan" Nijin paper seen through the light

章节	图中文名称	图英文名称
第 二 节	泾县艺英轩宣纸工艺品厂	Section 2 Yiyingxuan Xuan Paper Craft Factory in Jingxian County
	"艺英轩"租用的生产车间——老粮库	An old granary: the production workshop rented by "Yiyingxuan"
	老宣纸二厂所在地外景	Exterior of the Former Second Xuan Paper Factory
	赤滩古镇一角	A corner of Chitan Ancient Town
	朱正海正在调配原料	Zhu Zhenghai is modulating raw materials
	开网店的朱大为	Zhu Dawei is selling the paper online
	水纹笺的视觉效果	The visual effect of Shuiwen paper
	"艺英轩"天然古法草木染色宣	"Yiyingxuan" Vegetation Dyed Xuan paper in ancient natural ways
	染料	Dye
	配色	Color scheming
	褙棍子	Pasting the paper with sticks
	拖染	Dyeing the paper
	挂纸晾干	Hanging and drying the paper
	剪纸	Cutting the paper
	盖章	Sealing the paper
	"艺英轩"天然古法草木染色宣加工使用的植物染料	Plant dye used for "Yiyingxuan" Vegetation Dyed Xuan paper in ancient natural ways
	泡料	Soaking plant dye
	白芨	Bletilla striata
	褙棍子	Sticking the paper to the pole
	试色	Testing the paper
	拖染	Dyeing the paper
	打蜡	Waxing
	研光	Smoothing the paper
	拖盆	Basin for dyeing
	棍子	Stick for hanging the paper
	泡料缸	Tank for soaking plant dye
	剪刀	Shears

章节	图中文名称	图英文名称
	店面经营图	Paper store
	产品价格表	Product price list
	黄淳为"艺英轩"题字的牌匾	A plaque for "Yiyingxuan" inscribed by Huang Chun
	《美术报·艺苑撷英》	China Art Weekly Fine Arts
	朱正海收藏35年的《最忆吴笺照墨光》剪报	The Newspaper clipping on Recall the Calligraphy on Wujian Paper kept by Zhu Zhenghai for 35 years
	"艺英轩"淘宝网店截图	The screen shot of the Taobao online shop of "Yiyingxuan"
	"艺英轩"水纹纸透光摄影图	A photo of "Yiyingxuan" Shuiwen paper seen through the light
	"艺英轩"草木染色纸透光摄影图	A photo of "Yiyingxuan" vegetation dyed paper seen through the light
	"艺英轩"熟煮捶纸透光摄影图	A photo of "Yiyingxuan" Shuzhuchui paper seen through the light

章节	图中文名称	图英文名称
第三节	泾县艺宣阁宣纸工艺品有限公司	Section 3 Yixuange Xuan Paper Craft Co., Ltd. in Jingxian County
	艺宣阁宣纸工艺品有限公司厂区正门	Gate of Yixuange Xuan Paper Craft Co., Ltd.
	百岭坑厂原址内外景观	Internal and external view of the former site of Bailingkeng Factory
	泾县开发区厂区的车间内景	Internal view of the workshop in the factory area of Jingxian Development Zone
	调查组成员在访谈佘贤兵	Researchers are interviewing She Xianbing
	省级工艺美术大师荣誉牌匾	Honor plaque of Provincial Master of Craftsmanship and Art
	书法家试用"艺宣阁"纸品	Calligrapher is testing the products of "Yixuange"
	"艺宣阁"粉彩笺	"Yixuange" Fencai paper
	"艺宣阁"蜡染笺	"Yixuange" Laran paper
	染料配制	Modulating the dye
	上纸	Pasting the paper
	浇染料	Watering the dye
	套色印染	Overprinting the paper
	检验	Checking the quality
	准备裁剪纸	Preparing to cut the paper
	包装	Packing the paper
	购自载元堂工艺厂的楮皮原纸	Mulberry bark paper bought from Zaiyuantang Craft Factory
	水性印染剂浆料	Waterborne printing agent paste
	阴干	Drying the paper in the shadow
	收光与交错研光	Smoothing and polishing the paper
	美工刀裁剪	Cutting the paper by art knife
	盖章	Sealing the paper
	丝网板	Silk screen
	刮刀	Scraper
	研光石	Stone for smoothing the paper
	裁纸机	Paper cutter
	精致粉彩泥金泥银卷筒纸	Delicate colored gold and silver paper
	中国邮政集团邮票册页	The stamp album released by China Post Group
	龙纹纸	Paper with dragon decorations
	纸库	Warehouse of paper
	韩美林题写的"艺宣阁"	The plaque of "Yixuange" inscribed by Han Meilin
	蜡染笺禅意手札	Laran letter paper of Zen style
	仿古蜡染笺	Antique Laran paper
	"艺宣阁"富有特色的门楣	The featured lintel of "Yixuange"

章节	图中文名称	图英文名称
	"艺宣阁"粉彩笺透光摄影图	A photo of "Yixuange" Fencai paper seen through the light
	"艺宣阁"蜡染笺透光摄影图	A photo of "Yixuange" Laran paper seen through the light

章节	图中文名称	图英文名称
第四节	泾县宣艺斋宣纸工艺厂	Section 4 Xuanyizhai Xuan Paper Craft Factory in Jingxian County
	泾县宣艺斋宣纸工艺厂大门	The main entrance of Xuanyizhai Xuan Paper Craft Factory in Jingxian County
	加工厂区车间内景	Internal view of the workshop of the processed paper factory
	"中国质量万里行"颁发的证书	Certificate issued by "China Association for Quality Promotion"
	张金泉	Zhang Jinquan
	厂区一角	A corner of the factory
	生产现场	Production site
	张金泉所获的"加工状元"荣誉证书	"Top in the processed paper industry" certificate of honor received by Zhang Jinquan
	"宣艺斋"粉彩宣	"Xuanyizhai" Fencai Xuan paper
	"宣艺斋"色宣	"Xuanyizhai" Dyed Xuan paper
	"宣艺斋"仿古色宣	"Xuanyizhai" antique Dyed Xuan paper
	洒金纸笺	Sajin paper
	堆放着的青檀皮原料	Stacked raw materials of Pteroceltis tatarinowii Maxim. bark
	龙须草浆板原料	Raw materials of eulaliopsis binata pulp board
	盛在盆中的"胶"	"Glue" in the basin
	铝箔原料	Raw materials of aluminum foil
	地下水	Groundwater
	打浆房	Beating room
	捞纸车间	Workshop for scooping and lifting the papermaking screen out of water and turning it upside down on the board
	捞纸	Scooping and lifting the papermaking screen out of water and turning it upside down on the board
	放纸	Placing the paper
	搭棍子	Pasting the paper with sticks
	上胶	Gluing
	洒银	Sprinkling aluminum foil over the entire sheet of paper
	压纸	Pressing the paper
	悬挂风干	Hanging and drying
	刚理好的纸	Just finished paper
	掸银	Removing the redundant aluminum foil
	待裁剪的纸	Paper to be cut
	包装车间	Packaging workshop
	打浆机	Beating machine
	捞纸槽	Container for scooping and lifting the papermaking screen out of water and turning it upside down on the board
	"一改二"纸帘	Papermaking screen that can make two pieces of paper one time
	帘床	Shelf holding papermaking screen
	电动扒头	An automatic tool used for pulping
	浇帖架	Papermaking screen frame
	刷子	Brush made of pine needles
	纸焙	Paper dryer
	剪刀	Shears
	排笔	brush

章节	图中文名称	图英文名称
	洒银筒	A tube containing aluminum foil
	棕刷	Brown brush
	掸金把子	Brush for removing the redundant aluminum foil
	泾县宣艺斋宣纸工艺厂厂牌	Plate of Xuanyizhai Xuan Paper Craft Factory in Jingxian County
	融现代工艺的"万年红"纸	"Wannianhong" paper made by modern technology
	丝网加工	Processing with screen
	丝网加工万年红纸的刷色	Brushing the color of Wannianhong paper with screen
	曹宝麟为泾县宣艺斋宣纸工艺厂题字	Cao Baolin's inscription to Xuanyizhai Xuan Paper Craft Factory in Jingxian County
	曹宝麟用自己的定制纸写赠张金泉的书法	The calligraphy written by Cao Baolin on his customized paper to Zhang Jinquan
	忙活着的加工纸车间	Processed paper workshop in production
	"宣艺斋"米黄洒银加工纸透光摄影图	A photo of "Xuanyizhai" Cream-colored Sayin processed paper seen through the light

章节	图中文名称	图英文名称
第五节	泾县贡玉堂宣纸工艺厂	Section 5 Gongyutang Xuan Paper Craft Factory in Jingxian County
	贡玉堂宣纸工艺厂指示牌	Road sign towards Gongyutang Xuan Paper Craft Factory
	工人加工墨流宣现场	Workers processing Moliu Xuan paper
	黄村镇风景	Scenery of Huangcun Town
	王学兵	Wang Xuebing
	访谈司绍先老人	Interviewing Si Shaoxian
	工人加工粉蜡笺现场	A worker processing Fenla paper
	紫阳村	Ziyang Village
	不同颜色的虎皮宣	Different colors of Hupi Xuan paper
	司绍先展示墨流宣	Si Shaoxian showing Moliu Xuan paper
	墨流宣及特写效果	Moliu Xuan paper and its close-up effect
	"贡玉"虎皮宣细部效果	Detail effect of "Gongyu" Hupi Xuan paper
	车间里现场悬挂的墨流宣	Moliu Xuan paper hanging in the workshop
	染料(因涉及保密对方不愿告知)	Dye (not willing to inform because of the confidentiality involved)
	制作过程	Working process
	搭完棍子的原纸	Raw paper pasted with sticks
	拖染	Drag the paper through the surface of the pigment for dyeing
	拖染后自然晾干	Drying naturally after dyeing
	烘烤的炭盆	Charcoal brazier for baking
	烘烤	Baking
	撒花	Speckling the paper
	第二次烘烤	Second baking
	纸品仓库的存货架	Stock shelf in paper warehouse
	司绍先展示不同历史时期的宣纸加工纸	Si Shaoxian showing the processed Xuan paper in different historical periods
	棍子	Sticks
	撒把	A pouring tool made of bamboo sticks
	拖纸盆	Wooden cuboid tub to hold dye for paper dyeing
	工作台	Working table
	刷胶刷子	Wool brush for painting gummed paper materials
	丝网	Silk screen
	刮刀	Scraper

章节	图中文名称	图英文名称
	泾县贡玉堂宣纸工艺厂大门	Main entrance of Gongyutang Xuan Paper Craft Factory in Jingxian County
	"贡玉"虎皮宣透光摄影图	A photo of "Gongyu" Hupi Xuan paper seen through the light
	"贡玉"墨流宣透光摄影图	A photo of "Gongyu" Moliu Xuan paper seen through the light

章节	图中文名称	图英文名称
第六节	泾县博古堂宣纸工艺厂	Section 6 Bogutang Xuan Paper Craft Factory in Jingxian County
	"博古堂"生产厂区	"Bogutang" Xuan Paper Factory
	"博古堂"堂名题写原件	The original inscription of the factory name "Bogutang"
	小岭村	Xiaoling Village
	泾县小岭宣纸厂旧址标志牌	The signboard of former Xiaoling Xuan Paper Factory in Jingxian County
	调查组成员采访曹迎春	A researcher interviewing Cao Yingchun
	印染剂	Dye
	液压表	Hydraulic pressure gauge
	套色	Color register
	收纸	Taking back the paper
	收好放置的纸	Collecting the piled paper
	清洗丝网	Washing screen
	检验	Checking the paper
	裁纸	Cutting the paper
	包纸	Wrapping the paper
	打包	Packaging
	丝网板	Screen board
	刮刀	Scraper
	裁纸机	Paper cutter
	生产车间	Production workshop
	仓库	Warehouse
	产品上印制的"博古堂"品牌商标	"Bogutang" brand trademark printed on the product
	"博古堂"粉彩套色笺(冷金)透光摄影图	A photo of "Bogutang" Fencai Taose paper seen through the light

章节	图中文名称	图英文名称
第七节	泾县汇宣堂宣纸工艺厂	Section 7 Huixuantang Xuan Paper Craft Factory in Jingxian County
	在S322省道上的路口指示牌	Road sign on the S322 provincial road
	汇宣堂宣纸工艺厂外围厂名标牌	Name tag outside the Huixuantang Xuan Paper Craft Factory
	调查组成员现场请教汤贵宝	Researchers consulting Tang Guibao
	调查组成员访谈涂振华	A researcher interviewing Tu Zhenhua
	汤贵宝	Tang Guibao
	蜡染笺	Laran paper
	粉彩笺	Fencai paper
	整刀的泥金纸	A set of Nijin paper
	精品宣	High-quality Xuan paper
	复古笺	Xuan paper made by traditional ways
	涂振华带领调查组成员参观车间	Tu Zhenhua leading researchers to visit the workshop
	涂振华在介绍工序	Tu Zhenhua introducing the papermaking procedures
	泥金涂料	Coating material of Nijin paper
	丝网印刷环节	Printing with screen
	印花	Printing

	包装	Packaging		裁边	Trimming the paper
	丝网	Papermaking screen		打蜡	Waxing
	刮刀	Scraper		砑光	Smoothing the paper
	印刷台	Printing table		蓼蓝	Indigo plant
	切纸机	Paper cutter		调制好的颜料	Prepared pigment
	配料桶	Mixing barrel		配色	Color scheming
	工人在配料	Worker mixing ingredients		上杆	Sticking the paper to the pole
	晾晒纸张	Drying the paper		晾干	Drying the paper
	晾晒中的"汇宣堂"加工纸	"Huixuantang" processed paper drying in the air		拖染	Dyeing the paper
	书画家的赞誉	Appreciation of the painter		晾干	Drying the paper
	丝网印刷机器	Screen printer		刷染	Dyeing the paper with a brush
	"汇宣堂"泥金纸透光摄影图	A photo of "Huixuantang" Nijin paper seen through the light		晾干	Drying the paper
	"汇宣堂"泥银纸透光摄影图	A photo of "Huixuantang" Niyin paper seen through the light		托裱（覆背）	Glueing the paper

章节	图中文名称	图英文名称		上墙	Pasting the paper on the wall
第 八 节	泾县凤和堂宣纸加工厂	Section 8 Fenghetang Xuan Paper Factory in Jingxian County		起纸	Peeling the paper down
	翟家村内的"南坛三圣"庙	"Nantan Sansheng" Temple in Zhaijia Village		裁边	Trimming the paper
	凤和堂宣纸加工厂大门	Main entrance of Fenghetang Xuan Paper Factory		打蜡、砑光	Waxing and smoothing the paper
	青弋江翟家村民组段自然环境	Natural environment of Zhaijia Village near the Qingyi River		配色	Color scheming
	凤和堂艺术中心外景	Exterior view of Fenghetang Art Center		上杆	Sticking the paper to the pole
	调查组成员访谈郑智源	Researchers interviewing Zheng Zhiyuan		晾干	Drying the paper
	调查组成员与郑智源夫妻交流	Researchers communicating with Zheng Zhiyuan and his wife		刷染	Dyeing the paper with a brush
	泾县凤和堂宣纸加工厂营业执照	Business license of Fenghetang Xuan Paper Factory in Jingxian County		晾纸	Drying the paper
	泾县凤和堂文化艺术品中心营业执照	Business license of Fenghetang Cultural Arts Center in Jingxian County		碎瓷纹粉蜡笺制作中	Making Fenla paper with broken porcelain figure
	抄经纸	Paper for copying scriptures		打蜡	Waxing
	郑智源向调查组成员介绍产品	Zheng Zhiyuan introducing products to researchers		砑光	Smoothing the paper
	"凤和堂"硬黄纸透光摄影图	A photo of "Fenghetang" Yinghuang paper seen through the light		棕刷	Coir scrub brush
	"凤和堂"磁青竹纸透光摄影图	A photo of "Fenghetang" Ciqing bamboo paper seen through the light		小刷	Small brush
	黄檗与生山栀	Phellodendron amurense and Gardenia Jasmine		大刷	Big brush
	黄连	Coptis		糨糊桶	A bucket for holding glue
	朱砂	Cinnabar		起子	A stick for peeling the paper
	高岭土	Kaolin soil		晾纸杆	A stick for drying the paper
	蜡	Wax		挑纸杆	A stick for picking up the paper
				裁纸尺	Board cutter
	抽水电器设备	Pumping equipment		美工刀（裁纸刀）	Art knife (paper cutter)
	现场测水	On-spot water test		钵与风炉	Bowl and the wind furnace
	配色	Color Scheming		印章石	The seal stone
	上杆	Sticking the paper to the pole		砑光用玉石	Jade for smoothing the paper
	晾干	Drying the paper		洒金筒	Tools for sprinkling gold powder
	刷染	Dyeing the paper with a brush		"凤和堂"粉蜡笺纸品	"Fenghetang" Fenla paper
	晾纸	Drying the paper		"凤和堂"商标注册证	Trademark registration certificate of "Fenghetang"
	刷染	Dyeing the paper with a brush		翟春平与杨晓阳合影	Photo of Zhai Chunping and Yang Xiaoyang
	晾纸	Drying the paper		使用多年的工作围腰	The work wrap used for years
	取杆	Taking sticks down		"凤和堂"脸谱粉蜡笺	The facial makeup drawn on "Fenghetang" Fenla paper
	托裱（覆背）	Glueing the paper			
	上墙	Pasting the paper on the wall		"凤和堂"碎瓷纹粉蜡笺透光摄影图	A photo of "Fenghetang" Fenla paper with broken porcelain figure seen through the light
	起纸	Peeling the paper down		"凤和堂"仿古硬黄纸透光摄影图	A photo of "Fenghetang" antique Yinghuang paper seen through the light

章节	图中文名称	图英文名称
	"风和堂"仿古磁青纸透光摄影图	A photo of "Fenghetang" antique Ciqing paper seen through the light

章节	图中文名称	图英文名称
第 七 章	工具	Chapter VII Tools
第 一 节	泾县明堂纸帘工艺厂	Section 1 Mingtang Papermaking Screen Craft Factory in Jingxian County
	纸帘作坊内景	Internal view of papermaking screen workshop
	工人在作坊编制纸帘	Workers making the papermaking screen in the workshop
	纸帘技师陈明堂	Papermaking screen technician Chen Mingtang
	作坊早年遗留下来的苦竹	Amarus bamboo left in the early years of the workshop
	编帘用的苦竹篾丝	Amarus bamboo splits for making the papermaking screen
	普通线（左）与进口线（右）	Ordinary line (left) and imported line (right)
	制帘用的生漆	Raw lacquer for making the papermaking screen
	剖竹条	Cutting the bamboo strip
	剖竹篾	Cutting the bamboo split
	抽丝	Threading
	正在聚精会神编帘的女工	Women workers making the screen
	芒草秆	Chinese silvergrass straws
	装芒杆	Loading Chinese silvergrass straws
	绷架上的竹帘	Bamboo screen on the stretcher
	用滚球刷漆	Painting with a rolling ball
	用漆把刷漆	Painting with a brush
	上图	Drawing the pattern of bamboo screen
	绣花	Embroidering
	剖竹刀	An iron tool for cutting bamboo strip into bamboo cane
	剖篾台	A worktable for splitting bamboo
	抽丝台	Tool for splitting bamboo
	卡尺	Caliper
	漆把	Painting brush
	调查组成员和陈明堂观看纸帘纹	Researchers and Chen Mingtang watching the pattern of the papermaking screen
	已经废弃的生产纸帘的旧时工具	Abandoned tools for producing the papermaking screen
	访谈陈明堂	Interviewing Chen Mingtang
	新旧生产工具对比（61号为20年前的生产工具）	Comparison of old and new production tools (No. 61 is the production tool 20 years ago)
	历史遗物（苦竹原料）	Historical relics (amarus bamboo as raw material)

章节	图中文名称	图英文名称
第 二 节	泾县全勇纸帘工艺厂	Section 2 Quanyong Papermaking Screen Craft Factory in Jingxian County
	销售店门面	Sales shop facade
	制纸帘人曹康明	Papermaking screen maker Cao Kangming
	半自动化纸帘机器	Semi-automatic papermaking screen machine
	小岭村外生长的苦竹	Amarus bamboo growing outside Xiaoling Village
	晒干后的苦竹竿	Amarus bamboo sticks after drying
	编帘使用的锦线	Thread for making the papermaking screen
	陶盆中待用于漆帘的生漆	Raw lacquer to be used for painting the papermaking screen in pottery pots
	切片后的竹条	Bamboo canes after cutting
	肖全勇展示剖篾	Xiao Quanyong showing how to split the bamboo
	在水中浸泡的竹丝	Bamboo silk soaked in water
	抽丝	Threading
	编帘	Making the papermaking screen
	绷床上刷过漆的纸帘	Painted bamboo screen on the stretcher
	剖竹刀	An iron tool for cutting bamboo strip into bamboo cane
	绷床	Stretcher
	抽丝台	Tool for splitting bamboo
	卡尺	Caliper
	调查成员与肖全勇边走边聊	A researcher and Xiao Quanyong talking while walking
	在曹康明家中进行的访谈	An interview in Cao Kangming's house
	泾县全勇纸帘工艺厂区外围景观	External view of Quanyong Papermaking Screen Factory in Jingxian County
	采访肖全勇	Interviewing Xiao Quanyong
	泾县全勇纸帘工艺厂周围环境	The surroundings of Quanyong Papermaking Screen Factory in Jingxian County

章节	图中文名称	图英文名称
第 三 节	泾县后山大剪刀作坊	Section 3 Houshan Shears Workshop in Jingxian County
	张力伟家庭剪刀作坊外景	External view of Zhang Liwei Family Shears Workshop
	俞宋桃在点火起炉	Yu Songtao making a fire
	大剪刀	Shears
	张力伟	Zhang Liwei
	打制剪刀的铁板	Iron plate for making shears
	已裁切的铁条	Iron bars after cutting
	锻打	Forging
	雕弯	Forging with a small hammer into a certain arc
	镶钢（压钢）	Inlaid steel
	打头片	Forging the blade of shears
	打手柄	Forging the handle of shears
	打眼	Hole drilling in the blade of shears
	砂轮打光	Polishing with grinding wheel
	开口	Making shears edge
	锉片	Rounding off the blade of shears with a file
	淬火	Quenching
	整形	Shaping
	磨口	Sharpening the blade of shears
	制销子	Making shears pin
	铣眼	Milling the notch of shears
	钉铰	Riveting
	试剪	Checking shears
	上油	Applying the oil
	铁钳	Iron clamp
	铁锤	Hammer
	水盆	Basin
	砂轮	Grinding wheel
	铲子	Shovel
	电子秤	Electronic scale
	俞宋桃的作坊	Yu Songtao's mill
	张力伟作坊里的火炉	The stove in Zhang Liwei's mill

表目
Tables

表中英文名称

表1.1 1943年皖西各县造纸情形
Table 1.1 Papermaking status of various counties in Western Anhui in 1943

表1.2 当代安徽手工造纸点分布信息简表（据调查组田野样品采集与实验分析汇总）
Table 1.2 Current Distribution of Handmade Paper in Anhui Province (information gathered from field collected samples and technical analysis)

表2.1 民国三十七年（1948）十月极不稳定的宣纸价格（《泾县志》1996年版第259页）
Table 2.1 The unstable prices of Xuan paper in 1948 (printed in The Annals of Jingxian County in 1996, page 259)

表2.2 隶属关系变化表
Table 2.2 Change of affiliation

表2.3 历任泾县宣纸厂（中国宣纸集团公司）负责人
Table 2.3 Leading executives of Xuan Paper Factory (China Xuan Paper Co., Ltd.) in Jingxian County over the years

表2.4 历任泾县宣纸厂（中国宣纸集团公司）党委主要负责人
Table 2.4 Party Committee Secretary of Xuan Paper Factory (China Xuan Paper Co., Ltd.) in Jingxian County over the years

表2.5 64年红星宣纸企业年产量、产值与利润变化表
Table 2.5 Annual output, output value and profit variation of Red Star Xuan Paper Company in 64 years

表2.6 泾县本地产青檀皮与加工后毛皮的成分构成表
Table 2.6 Ingredients of Local Pteroceltis tatarinowii Maxim. Bark and Processed Bark in Jingxian County

表2.7 不同土壤稻草的化学成分分析
Table 2.7 Chemical composition of different kinds of straw

表2.8 传统纸槽规格数据表
Table 2.8 Specification data of traditional papermaking trough

表2.9 "红星"宣纸常用纸帘规格数据表
Table 2.9 Specification data of Red Star Xuan papermaking screen

表2.10 "红星"宣纸常用帘床规格数据表
Table 2.10 Specification data of Red Star Xuan papermaking frame

表2.11 "红星"古艺宣相关性能参数
Table 2.11 Performance parameters of "Red Star" traditional Xuan paper

表2.12 "红星"特净相关性能参数
Table 2.12 Performance parameters of "Red Star" superb-bark paper

表2.13 "红星"净皮相关性能参数
Table 2.13 Performance parameters of "Red Star" clean-bark paper

表2.14 "红星"棉料相关性能参数
Table 2.14 Performance parameters of "Red Star" Mianliao paper

表2.15 20世纪50～60年代"红星"宣纸的主要品种规格
Table 2.15 Specifications of "Red Star" Xuan paper during 1950~1960

表2.16 历年宣纸价格表
Table 2.16 Price list of Xuan paper over the years

表2.17 宣纸价格表（一）1957年1月
Table 2.17 Price list of Xuan paper Jan. 1957

表2.18 宣纸价格表（二）1983年10月
Table 2.18 Price list of Xuan Paper Oct. 1983

表2.19 宣纸价格表（三）2012年3月
Table 2.19 Price list of Xuan Paper Mar. 2012

表2.20 "汪六吉"宣纸的配料表
Table 2.20 Ingredients of "Wangliuji" Xuan paper

表2.21 "汪六吉"料半相关性能参数
Table 2.21 Performance parameters of "Wangliuji" Liaoban paper

表2.22 "汪六吉"黄料相关性能参数
Table 2.22 Performance parameters of "Wangliuji" Huangliao paper

表2.23 "恒星"宣纸净皮纸样相关性能参数
Table 2.23 Performance parameters of "Hengxing" clean-bark paper

表2.24 "恒星"御品贡宣相关性能参数表
Table 2.24 Performance parameters of "Hengxing" tribute Xuan paper

表2.25 "红日"精品书画纸相关性能参数表
Table 2.25 Performance parameters of "Hongri" fine calligraphy and painting paper

表2.26 "桃记"棉料相关性能参数
Table 2.26 Performance parameters of "Taoji" Mianliao paper

表2.27 "汪同和"特净相关性能参数
Table 2.27 Performance parameters of "Wangtonghe" superb-bark paper

表2.28 汪同和宣纸厂各类纸品价目表
Table 2.28 Price list of different types of paper in Wangtonghe Xuan Paper Factory

表2.29 "双鹿"净皮宣纸相关性能参数
Table 2.29 Performance parameters of "Shuanglu" clean-bark Xuan paper

表2.30 "曹光华"牌"白鹿"宣纸相关性能参数
Table 2.30 Performance parameters of "Caoguanghua" Brand "Bailu" Xuan paper

表2.31 "曹光华"牌本色特净宣纸相关性能参数
Table 2.31 Performance parameters of "Caoguanghua" superb-bark Xuan paper (in original color)

表2.32 "曹光华"半生熟加工纸相关性能参数
Table 2.32 Performance parameters of "Caoguanghua" semi-processed paper

表2.33 "金星"净皮宣纸相关性能参数
Table 2.33 Performance parameters of "Jinxing" clean-bark Xuan paper

表2.34 "金星"构皮纸相关性能参数
Table 2.34 Performance parameters of "Jinxing" mulberry bark paper

表2.35 "金星"成品纸种类、规格和价格
Table 2.35 Paper types, specifications and prices of the final product of "Jinxing" paper

表2.36 "红叶"净皮宣相关性能参数
Table 2.36 Performance parameters of "Hongye" clean-bark Xuan paper

表2.37 "曹氏宣纸"玉版宣相关性能参数
Table 2.37 Performance parameters of "Caoshi Xuan Paper" jade-white Xuan paper

表2.38 "曹氏麻纸"相关性能参数
Table 2.38 Performance parameters of "Caos Brand hemp paper"

表2.39 "千年古宣"牌棉料相关性能参数
Table 2.39 Performance parameters of "Millennium Xuan" Mianliao paper

表2.40 "宣和坊"特净相关性能参数
Table 2.40 Performance parameters of "Xuanhefang" superb-bark paper

表2.41 景辉纸业有限公司特净相关性能参数
Table 2.41 Performance parameters of superb-bark paper in Jinghui Paper Co., Ltd.

表2.42 "景辉纸业"特级净皮相关性能参数
Table 2.42 Performance parameters of superb Xuan paper in Sanxing Paper Co., Ltd.

表2.43 "明星"特净相关性能参数
Table 2.43 Performance parameters of "Mingxing" superb-bark paper

表2.44 "明星"喷浆檀皮纸相关性能参数
Table 2.44 Performance parameters of "Mingxing" pulp-shooting Pteroceltis tatarinowii bark paper

表2.45 "明星"喷浆雁皮纸相关性能参数
Table 2.45 Performance parameters of "Mingxing" pulp-shooting Wikstroemia pilosa Cheng bark paper

表2.46 "玉泉"牌净皮相关性能参数
Table 2.46 Performance parameters of "Yuquan" clean-bark paper

表2.47 "吉星"本色特净相关性能参数
Table 2.47 Performance parameters of "Jixing" superb-bark paper in original color

表2.48 金宣堂宣纸厂特种规格纸品尺寸（2016年4月提供）

Table 2.48 Paper of special specifications in Jinxuantang Xuan Paper Factory (data provided in Apr. 2016)

表2.49 "金宣堂"特净相关性能参数

Table 2.49 Performance parameters of "Jinxuantang" superb-bark Xuan paper

表2.50 "九岭"净宣相关性能参数

Table 2.50 Performance parameters of "Jiuling" clean-bark Xuan paper

表2.51 "白天鹅"净皮相关性能参数

Table 2.51 Performance parameters of "Baitian'e" clean-bark paper

表3.1 "栽元堂"画仙纸的相关性能参数

Table 3.1 Performance parameters of "Zaiyuantang" Huaxian paper

表3.2 "曹友泉"高级书画纸相关性能参数

Table 3.2 Performance parameters of "Cao Youquan" advanced calligraphy and painting paper

表3.3 雄鹿宣纸厂书画纸相关性能参数

Table 3.3 Performance parameters of calligraphy and painting paper in Xionglu Xuan Paper Factory

表3.4 "绿杨宝"书画纸相关性能参数

Table 3.4 Performance parameters of "Lvyangbao" calligraphy and painting paper

表3.5 "曹柏胜"古法檀皮宣相关性能参数

Table 3.5 Performance parameters of "Cao Bosheng" Xuan paper made of *Pteroceltis tatarinowii* bark in ancient methods

表3.6 "红星"书画纸相关性能参数

Table 3.6 Performance parameters of "Red Star" calligraphy and painting paper

表4.1 构皮本色云龙皮纸相关性能参数

Table 4.1 Performance parameters of unbleached Yunlong bast paper made of mulberry bark

表4.2 泾县小岭驰星纸厂雁皮罗纹纸相关性能参数

Table 4.2 Performance parameters of *Wikstroemia pilosa* Cheng Luowen paper of Xiaoling Chixing Paper Factory in Jingxian County

表4.3 潜山星杰针灸艾条纸相关性能参数

Table 4.3 Performance parameters of Qianshan Xingjie Mulberry Bark Paper for acupuncture and moxibustion

表4.4 金丝纸业纯桑皮纸相关性能参数

Table 4.4 Performance parameters of pure mulberry bark paper in Jinsi Paper Co., Ltd.

表4.5 棉溪村江祖术家构皮纸相关性能参数

Table 4.5 Performance parameters of mulberry bark paper in Jiang Zushu's house of Mianxi Village

表4.6 三昕纸业有限公司楮皮纸相关性能参数

Table 4.6 Performance parameters of mulberry bark paper in Sanxin Paper Co., Ltd.

表4.7 三昕纸业有限公司民芸纸相关性能参数

Table 4.7 Performance parameters of Minyun paper in Sanxin Paper Co., Ltd.

表4.8 三昕纸业有限公司强制纸相关性能参数

Table 4.8 Performance parameters of Qianzhi paper in Sanxin Paper Co., Ltd.

表4.9 三昕纸业有限公司纸品市场价格

Table 4.9 Market price list of paper products in Sanxin Paper Co., Ltd.

表4.10 郑火土家构皮纸相关性能参数

Table 4.10 Performance parameters of mulberry bark paper in Zheng Huotu's house

表5.1 歙县青峰村竹纸相关性能参数

Table 5.1 Performance parameters of bamboo paper in Qingfeng Village of Shexian County

表5.2 泾县竹纸相关性能参数

Table 5.2 Performance parameters of bamboo paper in Jingxian County

表5.3 金寨县竹纸相关性能参数

Table 5.3 Performance parameters of bamboo paper in Jinzhai County

表6.1 "掇英轩"泥金笺相关性能参数

Table 6.1 Performance parameters of "Duoyingxuan" Nijin paper

表6.2 "艺英轩"水纹纸相关性能参数

Table 6.2 Performance parameters of "Yiyingxuan" Shuiwen paper

表6.3 "艺英轩"天然古法草木染色宣相关性能参数

Table 6.3 Performance parameters of "Yiyingxuan" Vegetation Dyed Xuan paper in ancient natural ways

表6.4 "艺英轩"熟煮捶纸相关性能参数

Table 6.4 Performance parameters of "Yiyingxuan" Shuzhuchui paper

表6.5 "艺宣阁"粉彩笺相关性能参数

Table 6.5 Performance parameters of "Yixuange" Fencai paper

表6.6 "艺宣阁"蜡染笺相关性能参数

Table 6.6 Performance parameters of "Yixuange" Laran paper

表6.7 泾县宣艺斋宣纸工艺厂主要纸品规格（截至2015年8月在生产的纸品）

Table 6.7 Main paper specifications of Xuanyizhai Xuan Paper Craft Factory in Jingxian County (Paper products produced up to August 2015)

表6.8 "宣艺斋"米黄洒银加工纸相关性能参数

Table 6.8 Performance parameters of "Xuanyizhai" Cream-colored Sayin processed paper

表6.9 "贡玉"虎皮宣相关性能参数

Table 6.9 Performance parameters of "Gongyu" Hupi Xuan paper

表6.10 "贡玉"墨流宣相关性能参数

Table 6.10 Performance parameters of "Gongyu" Moliu Xuan paper

表6.11 "博古堂"粉彩套色笺相关性能参数

Table 6.11 Performance parameters of "Bogutang" Fencai Taose paper

表6.12 "汇宣堂"泥金纸相关性能参数

Table 6.12 Performance parameters of "Huixuantang" Nijin paper

表6.13 "汇宣堂"泥银纸相关性能参数

Table 6.13 Performance parameters of "Huixuantang" Niyin paper

表6.14 "风和堂"硬黄纸相关性能参数

Table 6.8.1 Performance parameters of "Fenghetang" Yinghuang paper

表6.15 "风和堂"宣纸加工厂磁青竹纸相关性能参数

Table 6.15 Performance parameters of "Fenghetang" Ciqing bamboo paper

术语
Terminology

地理名 Places

汉语术语 Term in Chinese	英语术语 Term in English
安丰塘	Anfeng Pool
坝头村	Batou Village
板舍村	Banshe Village
北京凤凰岭书画院	Fenghuangling Painting and Calligraphy Institute, Beijing
曹家村	Caojia Village
茶冲村	Chachong Village
昌桥乡	Changqiao Village
巢湖市	Chaohu City
程家村	Chengjia Village
赤滩街	Chitan Street
大庄村	Dazhuang Village
丁家桥镇	Dingjiaqiao Town
东坑	Dongkeng Village
东乡	Dongxiang Town
董家	Dongjia Village
枫坑村	Fengkeng Village
孤峰村	Gufeng Village
古坝村	Guba Village
关猫山	Guanmao Mountain
官庄镇	Guanzhuang Town
海阳镇	Haiyang Town
荷花塘	Hehua Pool
后山村	Houshan Village
湖山坑	Hushankeng Village
华侨村	Huaqiao Village
黄村镇	Huangcun Town
黄麓镇	Huanglu Town
黄山区	Huangshan District
篁墩	Huangdun Village
金坑	Jinkeng Village
金溪（坑）	Jinxi (keng) Village
金寨县	Jinzhai County
泾川镇	Jingchuan Town
泾县	Jingxian County
九峰村	Jiufeng Village
榔桥镇	Langqiao Town
李家塌	Lijiata Village
李园村	Liyuan Village
六合村	Liuhe Village
毛尖山乡	Maojianshan Town
梅家冲	Meijiachong Village
棉溪村	Mianxi Village
盘坑	Pankeng Village
杞梓里镇	Qizili Town
潜山县	Qianshan County
琴溪镇	Qinxi Town
青峰村	Qingfeng Village
青弋江	Qingyi River
清水塘	Qingshuitang Village
沙丰村	Shafeng Village
上漕村	Shangcao Village
上坊村	Shangfang Village
深渡镇	Shendu Town
深交所	Shenzhen Stock Exchange
寿县	Shouxian County
水山坊	Shuishanfang Village
坛畈村	Tanfan Village
田坊	Tianfang Mill
汀溪乡	Tingxi Town
外浮村	Waifu Village
汪宜坑	Wangyikeng Village
乌溪	Wuxi Village
西山村	Xishan Village
歙县	Shexian County
小岭村	Xiaoling Village
晓角村	Xiaojiao Village
新建村	Xinjian Village
新明乡	Xinming Village
休宁县	Xiuning County
许湾村	Xuwan Village
燕子河镇	Yanzihe Town
殷家坑	Yinjiakeng Village
元龙（牛笼）坑	Yuanlong (Niulong) keng Village
园林村	Yuanlin Village
占云村	Zhanyun Village
张畈村	Zhangfan Village
中郎坑	Zhonglangkeng Village
周家冲	Zhoujiachong Village
周坑村	Zhoukeng Village
朱家村	Zhujia Village
竹岭村	Zhuling Village
庄坑村	Zhuangkeng Village

紫阳村	Ziyang Village

纸 品 名 Paper names

汉语术语	英语术语
Term in Chinese	Term in English
半色皮纸	Bark paper made from half-processed materials
半生宣	half-processed Xuan paper
本色特净宣纸	Superb-bark Xuan paper in original color
本色宣	Xuan paper in original color
表芯纸	Biaoxin paper
"曹柏胜"古法檀皮宣	"Cao Baisheng" mulberry bark Xuan paper made by the ancient method
曹氏麻纸	Caos hemp paper
"曹友泉"高级书画纸	"Cao Youquan" high-grade painting and calligraphy paper
草木染色宣	Vegetation Dyed Xuan paper
册页	Album
"澄文堂"书画纸	"Chengwentang" calligraphy and painting paper
澄心堂纸	Chengxintang paper
楮皮纸	Mulberry bark paper
传统本色类宣纸	Traditional Xuan paper in original color
纯桑皮纸	Pure mulberry bark paper
磁青竹纸	Ciqing bamboo paper
大表纸	Dabiao bamboo paper (for sacrificial purposes)
矾宣	Xuan paper processed with alum
仿古本色皮纸	Bast paper in original color made by traditional method
粉彩笺	Fencai paper
粉蜡笺	Fenla paper
干古纸	Gangu paper with long fiber
高廉纸	Gaolian paper
构皮本色云龙纸	Yunlong mulberry bark paper in original color
构皮茶叶云龙纸	Yunlong mulberry bark and tea leaves paper
构皮纸	Mulberry bark paper
古法檀皮宣	*Pteroceltis tatarinowii* Maxim. bark Xuan paper made with ancient methods
古艺宣	Traditional Xuan paper
骨皮纸	Gupi paper with long fiber
黑壳皮纸	Bast paper (coarse bark kept as raw material)
"红星"古艺宣	Traditional "Hongxing" Xuan paper
"红星"棉料	"Hongxing" papermaking material
"红星"净皮	"Hongxing" clean-bark Xuan paper
"红星"特种净皮	"Hongxing" superb-bark Xuan paper
"红星"书画纸	"Hongxing" calligraphy and painting paper
虎皮宣	Hupi Xuan paper
画仙纸	Huaxian paper
皇家御用纸	Royal paper
黄表纸	Huangbiao paper
黄料	Huangliao paper
黄纸	Yellow paper
"徽家纸号"仿古打印纸	"Hui's Paper", printing paper made with ancient methods
混合书画纸	Painting and calligraphy paper made from multiple raw materials
火纸	Huo paper (bamboo paper)
极品宣	Superb Xuan paper
金粟山写经纸	Jinsu Mountain scripture paper
金银印花笺	Gold and silver paper with flower figures
金寨竹纸	Jinzhai bamboo paper
泾县竹纸	Bamboo paper in Jingxian County
精品锦盒手卷	Refined silk box paper
精品生宣册页	Refined half-processed Xuan paper album
精品套色加工纸	Fine Taose processed paper
精品宣	Refined Xuan paper
精制宣纸	Refined Xuan paper
净皮	Clean-bark paper
蠲纸	Juan paper
绢本宣	Silk xuan paper
库蜡笺	Wax paper
蜡染笺	Wax paper
兰亭蚕纸	Lanting silkworm-egg paper
连史纸	Lianshi paper
流沙笺	Liusha paper
六合村构皮纸	Liuhe mulberry bark paper
六吉黄料	Liuji Huangliao paper
六吉料半	Liuji Liaoban paper
落水纸	Luoshui paper
绿杨宣	Lvyang Xuan paper
绿杨色宣	Lvyang colored Xuan paper
绿杨生宣卡纸	Lvyang half-processed Xuan paper cardboard
棉料	Mianliao paper
棉溪村皮纸	Mianxi bark paper
棉纸	Mian paper
民芸纸	Minyun paper
墨记纸	Moji paper
墨流笺	Moliu Xuan paper
墨流宣	Moliu Xuan paper
木版水印笺	Woodblock printing paper
泥金笺	Nijin paper
泥银笺	Niyin paper
喷浆书画纸	Calligraphy and painting paper made by spraying the pulp on a fixed screen
喷浆檀皮纸	*Pteroceltis tatarinowii* Maxim. bark paper made by spraying the pulp on a fixed screen
喷浆雁皮纸	*Wikstroemia pilosa* Cheng bark paper made by spraying the pulp on a fixed screen
乾隆御贡高丽纸	Tribute Kori paper during Emperor Qianlong reign
强制纸	Qiangzhi paper
染色笺	Colored paper
洒金洒银纸笺	Sajin and Sayin paper
洒金纸	Sajin paper
三六表纸	Sanliubiao bamboo paper
三桠皮纸	Mitsumatu bark paper
桑皮褙纸	Mulberry bark interleaving paper
色宣	Dyed Xuan paper
山水专用纸	Shanshui paper
手工洒金	Handmade Sajin paper
署纸	Shu paper
刷红纸	Shuahong paper
水纹纸(笺)	Shuiwen paper

汉语术语	英语术语
碎瓷纹粉蜡笺	Fenla paper with broken porcelain figure
"桃记"宣纸	"Taoji" Xuan paper
特级净皮	Superb-bark paper
特制绿杨宣	Superb Lvyang Xuan paper
万年红	Wannian (ten thousand years) red paper
五色纸	Five-color paper
歙县青峰村竹纸	Bamboo paper made in Qingfeng Village of Shexian County
"雄鹿"书画纸	"Xionglu" painting and calligraphy paper
宣德贡笺	Xuande tribute paper
薛涛笺	Xuetao paper
雁皮罗纹纸	Wikstroemia pilosa Cheng bark rib pattern paper
雁皮纸	Wikstroemia pilosa Cheng bark paper
印谱	Collection of seal stamps
硬黄纸	Yinghuang paper
玉版宣	Yuban Xuan paper
御品贡宣	Royal tribute Xuan paper
云龙纸	Yunlong paper
针灸艾条桑皮纸	Acupuncture and moxa stick mulberry bark paper
郑火土家构皮纸	Zhenghuo Tujia mulberry bark paper
皱纹纸	Zhouwen paper
煮硾纸	Zhuchui paper
"紫烟"纯桑皮纸	Ziyan pure mulberry bark paper

原料与相关植物名 Raw materials and plants

汉语术语	英语术语
Term in Chinese	Term in English
板栗壳	Chinese chestnut shell
茶叶片	Tea leaves
柴	Wood
楮皮	Mulberry bark
楮皮浆	Mulberry pulp
稻草	Straw
稻草浆	Straw pulp
地下水	Underground water
矾	Alum
仿金粉	Gold-like powder
粉	Powder
感光胶	Sensitive glue
高岭土	Kaolin soil
构皮	Mulberry bark
构皮浆	Mulberry bark pulp
构树皮	Mulberry bark
骨胶	Bone glue
国画色	Traditional Chinese painting pigment
河水	River water
槐米	Sophora japonica
黄檗	Phellodendron amurense
黄连	Coptis
黄麻	Jute
江水	River water
糨糊	Paste
胶	Glue
金箔	Gold foil
金粉	Gold powder
井水	Well water
聚丙烯酰胺	Polyacrylamide
绢	Silk
苦竹	Amarus bamboo
蜡	Wax
狼毒	Stellera chamaejasme (The root of Chinese stellera)
榔叶	Ulmus parvifolia Jacq. leaf
梨木板	Pear wood board
栗树	Chestnut
燎草	Processed straw
燎皮	Bark of the processed straw
蓼蓝	Indigo plant
龙须草	Eulaliopsis binata
龙须草浆板	Eulaliopsis binata pulp board
铝箔	Aluminum foil
毛皮	Unprocessed bark
毛竹	Mao bamboo
煤	Coal
猕猴桃藤	Chinese gooseberry vine
猕猴桃藤汁	Juice of Chinese gooseberry vine
面粉	Flour
墨碇	Ink stick
木浆	Wood pulp
嫩竹	Tender bamboo
黏合剂	Adhesive
糯米	Sticky rice
青丹皮	Pteroceltis tatarinowii Maxim. bark
青花	Minium blue paper
青檀皮	Pteroceltis tatarinowii Maxim. bark
青檀皮浆	Pulp of pteroceltis tatarinowii Maxim. bark
青檀树皮	Pteroceltis tatarinowii Maxim. bark
青桐（中国梧桐）	Green Tung (Chinese parasol)
染料	Dye
三桠皮	Mitsumatu bark
三桠皮浆	Mitsumatu bark plup
桑树皮	Mulberry bark
沙田稻草	Straw grown in sand
山苍子	Litsea cubeba
山苍子树	Litsea cubeba plant
山苍子叶	Litsea cubeba leaves
山涧水	Mountain stream
山泉水	Mountain spring
生漆	Raw lacquer
生山栀	Gardenia jasmine
石灰	Lime
水源	Source of water
丝网	Silk screen
松香	Collophony
藤构	Broussonetia kaempferi Sieb. var. australis Suzuki
藤黄	Gamboge

铁板	Iron plate
桐藤花根	Tongteng Flower Root (Qianshan local name for a papermaking mucilage)
铜金粉	Bronze powder
香叶子	Leaves of *Lindera communis* Hemsl.
橡栗壳	Oak chestnut shell
颜料	Pigment
雁皮	*Wikstroemia pilosa* Cheng bark
雁皮浆	*Wikstroemia pilosa* Cheng bark pulp
杨桃藤	Branches of *Actinidia Chinensis* Planch
杨桃藤汁	Juice of *Actinidia chinensis* Planch Branches
野生猕猴桃枝	Wild branches of *Actinidia chinensis* Planch
野生猕猴桃藤汁	Juice of wild branches of *Actinidia chinensis* Planch
野生桑树皮	Wild mulberry bark
银箔	Silver foil
银粉	Silver powder
印染剂	Printing and dyeing agent
元竹	Yuan Bamboo
增稠剂	Thickener
针叶林木浆	Coniferous wood pulp
纸边	Deckle edge
纸药	Papermaking mucilage
中草药	Chinese herbal medicine
中药白芨	Bletilla striata
朱砂	Cinnabar
竹浆	Bamboo pulp
羊桃藤	Branches of *Actinidia chinensis* Planch

工艺技术和工具设备 Techniques and tools

汉语术语	英语术语
Term in Chinese	Term in English
扒头	Stirring stick
扒子	Stirring stick
拔料	Stirring the papermaking materials
白皮	White / bleached bark
白皮制作	Processing bark
扳榨	Squeezing
拌浆	Mixing pulp
拌料池	Sink for stirring the materials
棒槌	Wooden hammer
包装	Packaging
备楛	Sticking the paper
焙笼	Drying wall
焙纸	Drying the paper
褙棍子	Pasting the paper with sticks
绷绢	Tightening silk on the frame
绷床	Stretcher
编帘	Making the papermaking screen
鞭草	Beating straw materials
鞭干草	Beating dry straw
鞭帖	Patting the paper pile
鞭帖板	Board for beating paper
钵	Bowl
剥草块	Stripping straw balls
剥燎草块	Stripping processed straw balls
剥皮	Peeling
裁边	Trimming the paper
裁剪	Cutting
裁料	Cutting
裁切	Cutting
裁纸	Cutting the paper
裁纸尺	Board cutter
裁纸机	Paper cutter
踩料	Stamping the materials
踩皮	Treading on the materials
槽笼	Papermaking trough
草碓打浆机	Straw beater
草浆	Straw pulp
草坯	Straw pile
搀帖	Carrying a pile of paper by two workers
铲子	Shovel
抄纸	Papermaking
扯青草	Pulling grass
衬档	Chendang, papermaking tool (for fixing and aligning the wet paper)
舂草	Beating the straw
舂捣	Beating
舂臼	Lumpang, beating tool
抽丝	Threading
抽丝台	Tool for splitting bamboo
出锅	Getting the materials out of the pot
出胚	Unpolished scissors
初捡	Choosing the qualified bark for the first time
初洗	Washing for the first time
除砂	Removing the grit
储存	Storing
锤汁	Beating to make pulp
淬皮	Heating the alkaline water in the bucket
淬火	Quenching
锉片	Rounding off the blade of shears with a file
搭棍子	Pasting the paper with sticks
搭纸棍	Pasting the paper with sticks
打包	Packaging
打包装箱	Packing boxes
打槽	Container for beating the materials
打浆	Beating to make pulp
打浆机	Beating machine
打浆装置	Equipment for beating
打捆	Bundling
打蜡	Waxing
打皮	Beating paper bark
打钱模子	Mold for making joss paper
打手柄	Making scissors handles
打头片	Forging the blade of shears
打眼	Hole drilling in the blade of shears
打药	Beating papermaking mucilage
打纸	Pressing the paper

大剪刀	Shears		工作台	Working table
大糨糊刷	Pulp brush in big size		刮刀	Scraper
袋料	Putting paper pulp in a cloth bag and beating it in water		挂晾	Airing
袋料锤	Hammer for beating and cleaming the materials		棍子	Stick for hanging the paper
袋皮	Washing papermaking materials in a bag		过滤	Filtering
掸金(银)把子	Clean the golden or silver decorations on the paper with a hand broom		过滤池	Filter pool
			过滤网	Filtering mesh
掸银	Removing the redun dant aluminum foil		和浆	Mixing pulp
刀纸	One *Dao* (unit measurement of paper) of paper		和色	Mixing color
			烘焙	Baking
稻草	Straw		烘烤	Baking
滴色	Dropping color		烘帖	Drying the paper pile by baking
底纹笔	Brush		烘纸	Baking the paper
第二次拣皮	Picking out the impurities for the second time		划槽	Stirring
电瓶车	Three-wheeled battery motor vehicle for carrying the sequeezed paper pile		划单槽	Swirring the papermaking puulp
			化学皮	Chemically processed bark
电子秤	Electronic scale		楻甑	Steaming pot
雕弯	Forging with a small hammer into a certain arc		回收仓	Recycle room
			绘稿	Drawing and painting
钉铰	Checking shears		混浆	Mixing paper pulp
钉耙	Rake for picking up the bark materials		混料耙	Rake for mixing the materials
定位桩	Fixed timber pile		火叉	Iron fork for setting fire or adding charcoal
抖草坯	Shaking straw			
饾版水印	Watermark		挤皮	Squeezing bark
端料	Soaking the materials in alkali		加湿器	Humidifier
堆放	Piling		夹晒纸刷板子	Board for clipping and drying the paper
碓	Pestling			
碓(碾)料	Milling		拣白皮	Picking and choosing the high quality bark
额枪	Tool for separating the paper			
二次出锅	Getting the materials out after second boiling		拣草	Picking out grass
			拣黑皮	Picking the black bark
二次余皮	Second boiling		拣皮	Picking and choosing the bark
二次煮料	Boiling for the second time		拣选	Picking and choosing
二次装锅	Putting materials in papermaking container for the second boiling		检验	Checking the quality
			检纸	Checking the paper quality
翻草块	Turning over piles of grass		剪(音)档	Jiandang, papermaking tool (for fixing and aligning the wet paper)
翻渡皮	Turning over wet grass			
翻堆	Turning over piles of papermaking materials		剪裁	Cutting
			剪刀	Shears
翻燎草块	Turning over processed straw		剪齐	Trimming the deckle edge
翻燎皮	Turning over processed bark		剪纸	Cutting the paper
翻皮坯	Turning over raw bark		剪纸刀	Scissors
翻青草	Turning over unprocessed grass		浆泵	Pulp pump
翻青皮	Turning the fresh bark over		糨糊桶	A bucker for holding glue
放槽	Releasing water out of the container		浆灰	Pulp ash
放帘	Turning the papermaking screen upside down on the board		浆皮	Putting bark in lime water
			浇帖	Watering the paper pile
放料	Putting materials in the papermaking container		浇帖架	Papermaking screen frame
			脚碓	Foot pestle
放纸	Turning the papermaking screen upside down on the board		搅拌	Stirring the materials
			搅拌泵	Stirring device
分色雕版	Engraving in different colors		解皮	Separating bark
粉碎	Smashing		浸料	Soaking materials
风炉	Wind furnace		浸泡	Soaking the pulp board
封印	Sealing		精选	Choosing the bark materials carefully
复检	Double check			
覆纸	Dyeing paper		净化	Cleaning materials
盖板	Cover plate		撅条	Cutting bark
擀棍	Tool for patting the paper			
搞槽	Stirring the papermaking materials			

卡尺	Calipers	耙子	Rake
开口	Making a cut	拍皮	Beating bark
砍条	Chopping bark	排笔	Brush
砍竹	Cutting bamboo	泡料	Soaking plant dye
看纸	Checking paper	泡料缸	Tank for soaking plant dye
捆麻	Bundling bark	泡皮	Soaking bark
来烧	Burning	配浆	Making pulp
篮子	Basket	配浆料	Blending pulp materials
捞筋叉	Fork for picking up the residues	配胶	Making mucilage
捞纸	Scooping and lifting the papermaking screen out of water and turning it upside down on the board	配料	Ingredients
		配料桶	Mixing barrel
捞纸槽	Container for scooping and lifting the papermaking screen out of water and turning it upside down on the board	配色	Color scheming
		配水	Mixing water
		配汁	Making mucilage
捞纸池	Papermaking trough	喷浆	Pulp shooting
老法	Traditional methods	喷浆帘床	Board for spraying pulp on screen
理齐	Trimming the deckle edge	皮碓	Bark pestle
理纸	Sorting the paper	皮钩	Bark hook
帘床	Shelf holding papermaking screen	皮浆	Bark pulp
帘夹	Papermaking screen holder	皮料拣选	Picking out impurities
帘架	Screen frame	皮坯	Unprocessed bark
晾渡皮	Drying wet bark	皮甑煮锅	Utensil for boiling materials
晾干	Drying	漂白	Bleaching
晾皮坯	Drying unprocessed bark	漂皮	Bleaching bark
晾晒	Drying	漂洗	Rinsing and washing
晾纸杆	Drying pole	破节	Cutting bamboo
燎草	Processed straw	破竹	Cutting bamboo
燎草浆	Pulp of processed straw	剖篾	Splitting bamboo
燎皮	Processed bark	剖篾台	A worktable for splitting bamboo
燎皮浆	Processed bark pulp	剖竹刀	An iron tool for cutting bamboo strip into bamboo cane
料袋	Material bag		
料筛	Material selection	漆把	Painting brush
淋洗	Rinsing and washing	起纸	Peeling the paper down
露漂	Rinsing in the open air	起子	A stick for peeling the paper
滤浆	Filtering pulp	千斤顶	Lifting jack
滤水	Filtering water	牵纸	Peeling paper down
麻凼	Soaking bark in a pool	枪棍	Tool for separating the paper layers
麻绳	Hemp rope	敲口整形	Polishing knife
毛笔	Writing brush	切皮	Cutting the bark
毛刷	Brush	切片	Cutting into sections
美工刀	Art knife	切帖刀	Paper cutter
磨口	Sharpening the blade of shears	切纸	Cutting paper
木版水印架	Woodblock printing shelf	切纸机	Papercutting machine
木版水印桌	Woodblock printing table	青草	Unprocessed grass
木尺	Wooden ruler	青皮	Unprocessed bark
木槌	Wooden hammer	清洗	Cleaning and washing
木刀	Wooden knife	取浆	Taking out pulp
木棍	Wooden stick	取液	Taking out liquid
木刻刀	Wood carving knife	染色	Dyeing
木框	Wooden frame	揉皮	Rubbing bark
木头	Wood	揉纸	Rubbing the paper
碾草	Grinding grass	入库	Getting into the warehouse
碾压	Grinding	洒金桶	Tool for sprinkling golden powder
捏皮	Rubbing bark	洒银	Sprinkling aluminum foil over the entire sheet of paper
沤皮	Soaking the bark		

撒把	A pouring tool made of bamboo sticks	甩干	Spinning dry
撒花	Speckling the paper	甩干机	Drier for drying the bark materals
杀额刀	Knife for cutting the paper	水盆	Water basin
杀青	Processing bamboo	水漂	Bleaching mulberry bark
砂轮	Grinding wheel	丝网	Silk screen
砂轮打光	Polishing with grinding wheel	丝网板	Silk screen board
筛选	Selecting	丝网印刷	Silk screen printing
晒渡皮	Drying wet bark	撕皮	Peeling bark down
晒干	Drying	松毛刷	Brush made of pine needles
晒料	Drying materials	碎条	Shreds
晒皮	Drying bark	踏皮	Stamping bark
晒滩	Drying ground	抬帖	Carrying the paper pile
晒纸	Drying the paper	摊青草	Drying the straw
晒纸焙	Wall for drying the paper	摊青皮	Drying unprocessed bark
晒纸夹	Clip for drying the paper	摊晒	Drying
晒纸架	Frame for drying the paper	檀皮	Wingceltis bark
晒纸墙	Drying wall	檀皮浆	Wingceltis bark pulp
晒纸刷	Brush for drying the paper	套色	Printing and dyeing with a variety of colors
上绷架	Putting paper on the frame	套色印染	Complex process of printing and dyeing with a variety of colors
上杆	Sticking the paper to the pole	添皮	Adding bark
上浆	Adding in rice pulp	挑晾	Drying
上胶	Gluing	挑燎草块	Selecting processed grass piles
上墙	Pasting the paper on the wall	挑皮	Selecting bark
上图	Drawing the pattern of bamboo screen	挑皮篮	Baskets for carrying the bark materials
上油	Applying the oil	挑皮坯	Selecting unprocessed bark
上挣	Smoothing the paper	挑青草	Selecting the straw
烧焙	Baking	挑选	Selecting
施粉	Painting the paper	挑纸杆	A stick for picking up the paper
施胶	Soaking the paper in alum	调皮	Grinding bark
施蜡	Adding in wax	调色	Making pigment
石灰浆池	Lime pulp container	帖架	Paper shelf
石碾	Stone roller	铁焙	Iron baking
试剪	Checking shears	铁锤	Iron hammer
试色	Testing the paper	铁架	Iron frame
收草块	Collecting grass piles	铁钳	Iron clamp
收草坯	Piling the straw	涂布	Applying color on the paper
收渡皮	Keeping wet bark	涂蜡	Waxing
收燎草	Collecting processed grass	土筛	Sieve for filtering the bark materials
收燎皮	Collecting processed bark	退塔	Removing the tools after the pressing process
收皮坯	Collecting raw bark	托裱	Mounting the paper
收青皮	Collecting unprocessed bark	拖盆	Basin for dyeing
收纸	Peeling the paper down	拖染	Dyeing
手工捞纸	Handmade papermaking	拖纸盆	Wooden cuboid tub to hold dye for paper dyeing
手工捞纸槽	Handmade papermaking container	脱水	Dehydration
手绘描金	Hand-painted golden decoration	挽钩	Hook for picking out materials
数棍子	Counting sticks	洗白皮	Washing bleached bark
刷把	Brush	洗草	Washing the straw
刷夹	Brush holder	洗涤	Washing
刷胶刷子	Glue brush	洗浆	Washing to make pulp
刷漆	Painting with brush	洗料	Washing materials
刷染	Dyeing the paper with a brush	洗皮	Washing bark
刷纸	Brushing the paper	洗皮池	Washing pool
刷子	Brush		

汉语术语	英语术语	汉语术语	英语术语
洗皮篮	Washing basket	蒸锅	Steaming pot
洗皮坯	Washing unprocessed bark	蒸料	Steaming
洗漂	Washing and rinsing	蒸皮	Steaming the bark
洗晒	Washing and drying	蒸煮	Procedure of steaming and boiling
铣眼	Milling the notch of shears	整理	Cleaning and sorting
下槽	A variety of steel	挣板	Table for brushing paste
下榾	Removing stick	支皮坯	Processing bark
下帘	Putting the bamboo papermaking screen into water	植物汁缸	Vat holding the plant papermaking mucilage
下挣	Removing stick	纸焙	Drying wall
镶钢（压钢）	Inlaid steel	纸槽	Papermaking trough
小糨糊刷	Small paste brush	纸凳	Chair for holding the paper
校版	Checking wooden watermark plate	纸帘	Papermaking screen
醒帖	Wetting a pile of paper	纸刷	Brush
修版	Revising	纸药	Papermaking mucilage
绣花	Embroidering	纸榨	Tools for pressing the paper
宣纸剪刀	Xuan paper scissors	制版	Making printing mold
宣纸纸槽	Xuan papermaking container	制花	Making flower
悬挂风干	Hanging and drying	制浆	Making pulp
选白皮	Picking out the impurities	制销子	Making shears pin
选草	Selecting grass	制药	Making papermaking mucilage
选黑皮	Selecting unbleached bark	中捡	Another impurity selection
选拣	Selecting	中洗	Cleaning the bark materials for the second time
选皮	Selecting bark	竹刀	Bamboo knife
选条	Selecting materials	竹帘	Bamboo papermaking screen
选纸	Checking the paper	煮料	Boiling materials
削尖	Sharpening	装锅	Putting materials into the pot
压榨	Squeezing and pressing the paper	装芒秆	Putting smooth straw on the screen
压皮	Pressing bark	装箱打包	Packaging the paper
压纸	Pressing the paper	自动烘干套色机	Automatic drying and dyeing machine
压纸石	Stone for pressing the paper	棕把	Brush made of palm
砑光	Smoothing the paper	棕刷	Coir scrub brush
砑光石	Stone for smoothing the paper	做额	Trimming deckle-edge
腌料	Marinading materials	做料	Making materials
腌沤	Soaking and fermenting	做皮	Making bark
药兜	Cloth bag for holding mucilage	做帖	Making paper
药榔头	Hammer for beating the papermaking mucilage	做纸巾	Processing waste paper and deckle-edge
液压机	Hydraulic press machine		
阴干	Drying the paper in the shadow		
印花	Printing		
印刷	Printing		
印刷台	Printing table		
印章石	The seal stone		
玉石	Jade		
圆筒筛	Cylinder sifter		
再选	Re-choosing		
扎捆	Bundling		
扎皮	Bundling bark		
榨干	Squeezing		
榨皮	Pressing the paper		
榨帖	Squeezing a pile of paper		
榨纸	Pressing the paper		
粘皮	Fermenting materials with lime water		
斩料皮	Cutting and grinding bamboo bark		

历 史 文 化 History and culture

汉语术语 Term in Chinese	英语术语 Term in English
《安徽科学技术史稿》	Manuscript of History of Science and Technology in Anhui Province
《安徽历史述要》	History of Anhui
《安徽日报》	Anhui Daily
《安徽宣纸工业之综述》	Research of Anhui Xuan Paper Industry
《安徽政治》	Politics in Anhui
《安庆地区志》	Anqing Chorography
《安庆府志》	Annals of Anqing
八段锦	Ba Duan Jin (a kind of martial arts)
巴拿马	Panama
巴拿马万国博览会	The 1915 Panama Pacific International Exposition
百忍堂	Bai Ren Tang (Zhang's clan hall)
壁贴	Wall sticker

匾	Plaque
伯益族裔	Boyi's descendants
蔡伦庙	Cai Lun Temple
蔡伦术	Cai Lun's papermaking technique
曹锦隆字号	Caojinlong Xuan Paper
《曹氏族谱》	Genealogy of the Caos
《草堂雅集》	Caotang Yaji (Elegant Meetings in Thatched Cottage)
《草堂之灵·说纸篇》	Spirit of Humble Cottage: Introduction to Paper
《程氏墨苑》	Cheng's Engraving on Ink Mould
吃蟹人	Pioneer
吃纸饭	Make a living with papermaking
池纸	Chi Paper (Paper produced in ancient Chizhou)
驰星	Chixing
传男不传女	Male-line succession
春酒	Spring feast
《次韵酬微之赠池纸并诗》	On Giving Chi Paper as Present
《大藏经》	Tripitaka
《大清一统志》	Geographical Situation of the Qing Dynasty
大英博物馆	The British Museum
德厚斋	Dehouzhai
地理标志	Geographical symbol
叮咚踢踏后山张家	The long scissor-making history of in the Zhangs
《冬心画竹题记》	Dongxin Huazhu Tiji (Inscription of Bamboo Painting)
《尔雅》	Er Ya (Chinese Ancient Dictionary)
"帆船"品牌	Fanchuan (sailboat)
《繁昌县志》	Fanchang County Annals
《方氏墨谱》	Notes of The Fangs
《方舆胜览》	Fangyu Shenglan (Geography book compiled during the Southern Song Dynasty)
《飞凫语略》	Feifu Yulue (Calligraphy book compiled during the Ming Dynasty)
"佛"字纸符	Paper amulet with the Chinese character "Fo (Buddha)"
《负暄野录》	Fuxuan Yelu
《格致镜原》	Gezhi Jingyuan (An encyclopedia written by Chen Yuanlong during the Qing Dynasty)
工匠精神	Craftsman's spirit
《工商要闻》	Business News
公私合营	Joint state-private ownership
古法造纸	Ancient papermaking methods
《古今历法通考》	A General View of Ancient and Modern Calendar
关门酒	Feast as closing ceremony
《贵池县志》	Guichi County Annals
《国货研究月刊》	Chinese Goods Monthly
国营	State-owned enterprise
《汉书·地理志》	The Annals of Geography in the Book of Han
《红楼梦》	The Dream of Red Mansion
"红旗"牌	"Hongqi" brand
"红星"牌	"Hongxing" brand
《胡氏篆草》	Hu's Calligraphy
胡侍	Hu Shi
糊篓	Paste basket
《淮南子》	Huai Nan Zi (A philosophy book written during the Western Han Dynasty)
《环翠堂集》	Huancuitang Ji (A book written during the Ming Dynasty)
"徽家纸号"	"Hui Family Seal"
徽派建筑	Huizhou style architecture
徽商	Huizhou merchants
徽纸	Huizhou paper
《徽州府志》	The Annals of Huizhou Prefecture
徽州四雕	Four Carvings in Huizhou
"鸡球"牌	"Jiqiu" brand
祭品	Sacrificial offering
祭祀	Sacrificial ceremony
见面常礼	Greeting etiquette
《建炎以来系年要录》	Major Events from the First Year of Jianyan Reign
《江南经略》	Jiangnan Jinglue (Military book written during the Ming Dynasty)
《江西官报》	Jiangxi Guanbao (Official printing of Jiangxi Government)
《蕉窗九录》	Jiaochuang Jiulu (Notes written during the Ming Dynasty)
"雄鹿"牌	"Xionglu" brand
《金史》	Jin Shi (History of Jin)
《金寨县志》	Jinzhai County Annals
民俗禁忌	Folk taboos
《泾川水东翟氏宗谱》	Zhai's Genealogy of Shuidong Village, Jingchuan Town
泾上白	Jing Shang Bai (paper brand, named after a poem)
泾县连四	Jingxian Liansi Xuan paper
《泾县土纸样本》	Samples of Handmade Paper in Jingxian County
《泾县宣纸厂志》	Records of Xuan Paper Factory in Jingxian County
《泾县志》	The Annals of Jingxian County
旧蔡伦祠所遗碑	Monument relics in ancient Cai Lun Temple
《旧唐书》	Jiu Tang Shu (Old Book of the Tang Dynasty)
倦勤斋	Juan Qin Zhai (a royal palace)
《抗敌》	Kang Di (Magazine on Anti-Japanese War
抗战	War of resistance against Japan
《考槃余事》	Kaopan Yushi (A book on the scholar-officials' life written during the Qing Dynasty)
《科学杂志》	Science
《枯树赋》	Ode to the Dead Tree
《兰亭考》	Lan Ting Kao (Collection of the Orchid Pavilion)
劳动号子	Work songs
老汪六吉	Old Wang Liu Ji Xuan Paper
《历代名画记》	Famous Paintings in the Past Dynasties
《历代诗话》	Notes on Past Famous Poems
联营	Joint venture
《临川先生文集》	Linchuan Gentleman Anthology
《罗纹纸赋》	An ode to Rib Pattern Paper
麻沸散	Ma Fei San (ancient Chinese anesthetics)
漫澍	Manshu
民国	Republic of China (1912~1949)
墨淋	Molin online paper shop
《南陵县志》	Nanning County Annals
南唐	Southern Tang (937~975)
《宁国府志》	The Annals of Ningguo County
《农书》	Book of Agriculture
皮娘	Picking out the impurities
屏风	Screen
《朴学斋丛书》	Collection of Poems from Puxue Studio

72道工序	72 procedures
千年古宣	Millennium Xuan paper
抢饭碗	Grab one's job
《秦汉瓦当文字》	Characters on Eaves Tile of Qin and Han Dynasties
《清国制纸取调巡回日记》	Investigation of Chinese Paper Production During the Qing Dynasty
劝业会	Nanking- Unyre an l'exposition National 1909
《人海记》	Ren Hai Ji (Notes on the Capital City written during the Qing Dynasty)
《日本工业化学杂志》	Japanese Industrial Chemistry Magazine
芍陂	Quebei
"神丹皮"传说	Story of "Shendanpi"
《十国春秋》	Spring and Autumn of Ten Kingdoms
《十竹斋书画谱》	Ten Bamboo Studio Manual of Painting and Calligraphy
世博会	World Exhibition
《世界手工纸大全》	The World's Handmade Paper Collection
《书林清话》	Shulin Qinghua (A Review of Ancient Books)
《书圣王羲之》	Wang Xizhi, the Supreme Calligrapher
《书史》	History of Calligraphy
松萝茶	Songluo Tea
《嵩山集》	Songshan Ji (A book written during the Song Dynasty)
《宋史》	Song Shi (History of the Song Dynasty)
《宋文鉴》	Song Wen Jian (Appreciation of Literature in the Song Dynasty)
《算法统宗》	New Assembly of Mathematics
《唐六典》	Administrative Codes of the Tang Dynasty
"天下第一剪"	"Shears Second to None"
《天中记》	Encyclopedia of Tianzhongshan Mountain
贴墙纸	Paste wallpaper
《通典》	Tong Dian (A history book on politics written during the Tang Dynasty)
屯绿	Tunxi green tea
湾桶	Wooden bucket
皖南区	Southern Anhui
《皖南制纸情形略》	Annals of Papermaking in Southern Anhui
皖纸大升纸庄	Dasheng Wan Paper Shop
汪同和纸庄	Wangtonghe Paper Shop
《王氏族谱》	Genealogy of the Wangs
文房四宝	Four Treasures of the Study
《文房四谱》	Four Treasures of the Scholar's Study
《文脊山记》	Story of Wenji Mountain
芜湖公署	Wuhu Prefectural Administrative Office
芜湖铁画	Wuhu Iron Picture
芜湖专区	Wuhu Region
吴头楚尾	Border between the two kingdoms of Wu and Chu
五禽戏	Wu Qin Xi (Five animal mimic boxing)
《武备志》	A Corpus of Military Equipments
《物理小识》	Common Sense of Physics
惜纸庐	House-shaped burner for burning paper (a custom to show respect to knowledge and paper)
锡铸版	Tin Casting
歙州	She zhou (Hui Zhou)
《新安大族志》	Xin'an Famous Families
新安刻工	Xin'an Carving Craftsman
《新安名族志》	Chronicles of Famous Families in Xin'an
《新安志》	Xin'an Chronicle
《修业堂集》	Xiuye Studio's Collection
《髹饰录》	Xiu Shi Lu (Record of Lacquer Ware)
《续资治通鉴长编》	A Sequel to Comprehensive Mirror to Aid in Government
宣阳都	Xuanyang (Ancient township)
《宣纸的制造》	Xuan Paper Making
宣纸剪	Xuan Papercut
宣纸文化园	Xuan Paper Cultural Park
《宣纸制造工业之调查》	A Survey of Xuan Paper Industry
宣州僧	Xuan Zhou Monks
《学海类编》	Xuehai Leibian (Book series compiled during the Qing Dynasty)
《砚边点滴》	Yanbian Diandi (Notes on Chinese Painting)
药兜	Filter bag
一个师傅带不出好徒弟	One master cannot cultivate a good disciple alone
艺术宣纸研制中心	Art Xuan Paper Research Center
忆宣	Yixuan
《印存玄览初集》	Printed Collection of Metaphysics
永徽年	Gaozong reign during the Tang Dynasty (649~655)
油纸专业公会	You Paper Guild
鱼米之乡	A land flowing with milk and honey
《玉山名胜集》	Ode to the Yushan Scenic Spots
御赐金牌	Gold medal awarded by the emperor
《元和郡县图志》	Illustrated Chronicle of Yuanhe County
元四家	Four representative landscape painters in Yuan Dynasty
原产地域	Original production place
《岳西县乡镇简志》	Brief Chronicle of Yuexi County
《岳西县志》	The Annals of Yuexi County
《造活字印书法》	Movable Type Printing
造金银印花笺法	Gold and silver decorative paper making methods
《长物志》	Zhang Wu Zhi (A book written during the Ming Dynasty on superfluous things)
《长征万里图》	Picture of Long March
浙溪燎草厂	Zhexi Processed Straw Factory
珍珠伞	Pearl umbrella
《郑氏祭祀簿》	Sacrificial Ceremony Book of the Zhengs
《中国纸业》	Chinese Paper
《中国制纸业》	Chinese Paper Industry
纸甲	Paper armour
纸栈	Paper stack
制伞	Making umbrella
中国驰名商标	China well-known trade mark
《中国地方志集成》	Collections of Local Annals in China
《中国宣纸发源地——丁家桥镇故事（第三辑）》	The Birthplace of Chinese Xuan Paper-- Story of Dingjiaqiao Town (vol. III)
"中国宣纸之乡"	"Home of Xuan paper in China"
《中西数学通》	Chinese and Western Mathematics
《装潢志》	Record of Chinese Calligraphy and Painting
《资治通鉴》	Comprehensive Mirror to Aid in Government
《遵生八笺》	Zun Sheng Ba Jian (A book on fitness regimen written during the Ming Dynasty)
做爆引	Making paper fuse
ZAKER新闻	ZAKER News

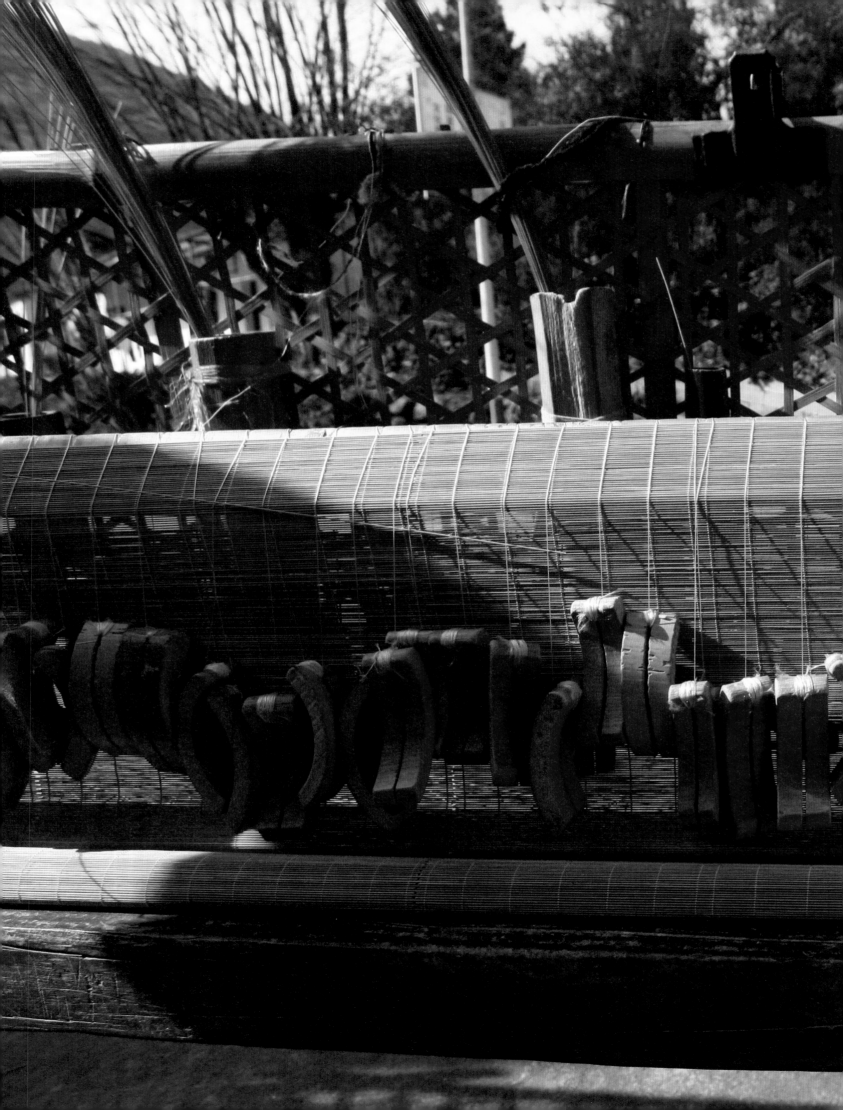

后　记

　　虽然《中国手工纸文库·安徽卷》的撰写工作是在云南、贵州、广西田野调查工作基本完成之后的2014年开始的，但因为《中国手工纸文库》的主持单位中国科学技术大学手工纸研究所在安徽省，中国手工造纸最具代表性的聚集地泾县与云南相比仿佛就在身边，因此不包括团队成员更早的人类学工艺社区调查及文化艺术之旅的积累，对宣纸之都——泾县的田野调查工作实际上从2008年起即已断断续续地在进行着。这部分工作已经带有明确的为《安徽卷》打基础的规划，也采集了若干纸样，形成了若干记录。

　　因为泾县宣纸、书画纸与加工纸业态的密集，《安徽卷》内容的丰富度几乎超过中国其他所有省份，因此分为上、中、下三卷。统稿的工作前后共进行了5轮，这确实源于素材采集与初步完成稿件的多样化和复杂性，以及不可避免地对若干内容需要多轮补充及优化，需要文献与调查现场反复印证。由于田野调查和文献研究基本上是以多位成员不同组合的方式参与的，而且多数章节前后多次的补充修订也不是由一人从头至尾完成的，因而即便工作展开之前制定了田野调查标准、撰稿标准，并提供了示范样稿，全卷的信息采集方式和初稿的表述风格依然存在诸多不统一、不规范之处。

Epilogue

Field investigation of *Library of Chinese Handmade Paper: Anhui* started officially in the year of 2014, after the research team had finished their explorations in Yunnan, Guizhou Provinces and Guangxi Zhuang Autonomous Region. The grantee of the project, Handmade Paper Research Institute of University of Science and Technology of China locates in Anhui Province. So compared to Yunnan Province, Jingxian County which boasts the representative papermaking gathering place, is closer and more convenient for our investigation. Actually, our fieldworks can be traced back as early as 2008, and have lasted ever since intermittently in Jingxian County, the capital of Xuan paper, not to mention our even earlier anthropological survey of techniques, cultural and art investigation of the area. We had purposefully accumulated data and sample paper as the basis of this volume.

Library of Chinese Handmade Paper: Anhui is further divided into three sub-volumes for it includes extensive data on Xuan paper in Jingxian County, calligraphy and painting paper, and processed paper, which is unparalleled for all other provinces in China. The researchers have put into five rounds of sedulous efforts to modify the manuscript, and revisit the papermaking sites for more information and verification due to the diverse and complex materials. Field investigation and literature studies

初稿合成后，统稿与补充调查工作由汤书昆、黄飞松、朱赟、朱正海主持。从2016年3月开始，共进行了5轮统稿，到2019年8月才最终形成了现在的定稿。虽然我们感觉安徽手工造纸调查与研究还有不少需要进一步完善之处，但《安徽卷》的工作从2009年7月汤书昆、王祥、陈彪、黄飞松、周先稠几次组队的预调查开始，已历经10年，从2014年9月正式启动调查至定稿也有5年整。其间，纸样测试、英文翻译、编辑与设计等工作团队成员尽心尽力，使《安徽卷》的品质一天天得到改善，变得更有阅读价值和表达魅力。转眼已到2019年底，各界同仁和团队成员对《安徽卷》均有很高的出版期待，若干未尽事宜只能期待今后有修订缘分时再来完善了。

《安徽卷》书稿的完成和完善有赖于团队成员全心全意的投入与持续不懈的努力，在即将出版付印之际，除了向所有参与成员表达衷心的感谢外，特在后记中对各位同仁的工作做如实的记述。

of each section and chapter are accomplished by the cooperative efforts of multiple researchers, and even the modification was undertaken by many. Therefore, investigation rules, writing norms and format set beforehand may still fail to make amends for the possible deviation in our way of information collection and the writing style of the first manuscript.

Modification and supplementary investigation were headed by Tang Shukun, Huang Feisong, Zhu Yun and Zhu Zhenghai. Ten years have passed since Tang Shukun, Wang Xiang, Chen Biao, Huang Feisong and Zhou Xianchou started the preliminary investigation in July 2009; and another 5 years have passed since the formal investigation started in September 2014. Since March 2016, five rounds of modification contributed to the final version in August 2019. Of course, we admit that the volume should never claim perfection, yet finally, through meticulous works in sample testing, translation, editing and designing, the book actually has been increasingly polished day by day. We can be positive that the book, with fluent writing and intriguing pictures, is worth reading, and ready for publication with best wishes from the academia and our researchers, though we still harbor expectation for further and deeper exploration and modification.

This volume acknowledges the consistent efforts and wholehearted contribution of the following researchers:

一、田野调查、文稿撰写与修订

第一章 安徽省手工造纸概述

撰稿	初稿主执笔：黄飞松、朱赟、汤书昆
	修订与补充完稿：汤书昆、陈敬宇、黄飞松
	参与撰稿：郑久良、孙舰、尹航、何瑗

第二章 宣纸

第一节	中国宣纸股份有限公司（地点：泾县榔桥镇乌溪村）
田野调查	汤书昆、黄飞松、朱赟、陈彪、郑久良、程曦、许骏、王怡青、何瑗
撰稿	初稿主执笔：黄飞松
	修订补稿：汤书昆、黄飞松、朱赟
	参与撰稿：刘伟、王圣融、王怡青
第二节	泾县汪六吉宣纸有限公司（地点：泾县泾川镇茶冲村）
田野调查	朱正海、汤书昆、黄飞松、朱赟、郑久良、罗文伯、程曦、何瑗、王圣融、王怡青、沈佳斐
撰稿	初稿主执笔：朱赟
	修订补稿：黄飞松、汤书昆
	参与撰稿：王圣融
第三节	安徽恒星宣纸有限公司（地点：泾县丁家桥镇后山村）
田野调查	汤书昆、朱正海、黄飞松、朱赟、郑久良、罗文伯、汪梅、程曦、许骏、何瑗、王圣融、王怡青、沈佳斐
撰稿	初稿主执笔：朱赟
	修订补稿：黄飞松、汤书昆
	参与撰稿：郑久良、王圣融
第四节	泾县桃记宣纸有限公司（地点：泾县汀溪乡上漕村）
田野调查	汤书昆、黄飞松、朱正海、朱大为、朱赟、郑久良、程曦、许骏、刘伟、何瑗、沈佳斐
撰稿	初稿主执笔：刘伟
	修订补稿：黄飞松、汤书昆
	参与撰稿：朱赟、沈佳斐
第五节	泾县汪同和宣纸厂（地点：泾县泾川镇古坝村）
田野调查	黄飞松、朱正海、朱大为、何瑗、朱赟、郑久良、程曦、许骏、刘伟、王圣融、王怡青、沈佳斐
撰稿	初稿主执笔：程曦
	修订补稿：黄飞松、汤书昆
	参与撰稿：朱赟、王圣融
第六节	泾县双鹿宣纸有限公司（地点：泾县泾川镇城西工业集中区）
田野调查	汤书昆、黄飞松、朱赟、郑久良、罗文伯、刘伟、何瑗、王圣融、王怡青、沈佳斐
撰稿	初稿主执笔：罗文伯
	修订补稿：黄飞松、汤书昆
	参与撰稿：郑久良、王圣融
第七节	泾县金星宣纸有限公司（地点：泾县丁家桥镇工业园区）
田野调查	黄飞松、朱正海、汤书昆、郑久良、朱赟、程曦、刘伟、何瑗、王圣融、王怡青、沈佳斐
撰稿	初稿主执笔：郑久良
	修订补稿：黄飞松、朱正海、朱赟
	参与撰稿：王圣融、朱赟
第八节	泾县红叶宣纸有限公司（地点：泾县丁家桥镇枫坑村）
田野调查	汤书昆、何瑗、朱赟、刘伟、王圣融、沈佳斐
撰稿	初稿主执笔：刘伟、黄飞松
	修订补稿：汤书昆、黄飞松、朱赟
	参与撰稿：王圣融、王怡青
第九节	安徽曹氏宣纸有限公司（地点：泾县丁家桥镇枫坑村）
田野调查	汤书昆、黄飞松、朱赟、何瑗、郑久良、许骏、刘伟、程曦、王圣融、王怡青、沈佳斐
撰稿	初稿主执笔：许骏、黄飞松、汤书昆
	修订补稿：汤书昆、黄飞松
	参与撰稿：王怡青

第十节	泾县千年古宣宣纸有限公司（地点：泾县丁家桥镇小岭村）	
田野调查	汤书昆、朱赟、刘伟、何瑗、王圣融、王怡青、沈佳斐	
撰稿	初稿主执笔：朱赟、黄飞松	
	修订补稿：黄飞松、汤书昆	
	参与撰稿：王怡青	
第十一节	泾县小岭景辉纸业有限公司（地点：泾县丁家桥镇小岭村）	
田野调查	汤书昆、朱大为、朱赟、郑久良、何瑗、许骏、刘伟、王圣融、王怡青、沈佳斐	
撰稿	初稿主执笔：朱赟、汤书昆	
	修订补稿：汤书昆、黄飞松	
	参与撰稿：刘伟、许骏	
第十二节	泾县三星纸业有限公司（地点：泾县丁家桥镇李园村）	
田野调查	汤书昆、黄飞松、朱正海、朱赟、郑久良、许骏、刘伟、何瑗、王圣融、王怡青、沈佳斐	
撰稿	初稿主执笔：刘伟、黄飞松	
	修订补稿：汤书昆、黄飞松	
	参与撰稿：朱赟	
第十三节	安徽常春纸业有限公司（地点：泾县丁家桥镇工业园区）	
田野调查	汤书昆、黄飞松、朱赟、郑久良、罗文伯、汪梅、何瑗、王圣融、王怡青、沈佳斐	
撰稿	初稿主执笔：朱赟	
	修订补稿：黄飞松、汤书昆	
	参与撰稿：郑久良、王圣融	
第十四节	泾县玉泉宣纸纸业有限公司（地点：泾县丁家桥镇李园村）	
田野调查	汤书昆、黄飞松、朱大为、朱赟、刘伟、郑久良、何瑗、王圣融、王怡青、沈佳斐	
撰稿	初稿主执笔：朱赟	
	修订补稿：黄飞松、汤书昆	
	参与撰稿：王圣融	
第十五节	泾县吉星宣纸有限公司（地点：泾县泾川镇上坊村）	
田野调查	朱正海、黄飞松、汤书昆、朱赟、郑久良、程曦、许骏、刘伟、何瑗、沈佳斐	
撰稿	初稿主执笔：程曦	
	修订补稿：汤书昆、黄飞松	
	参与撰稿：刘伟、朱赟	
第十六节	泾县金宣堂宣纸厂（地点：泾县榔桥镇大庄村）	
田野调查	汤书昆、朱赟、郑久良、罗文伯、汪梅、何瑗、钟一鸣、王圣融、王怡青	
撰稿	初稿主执笔：程曦	
	修订补稿：黄飞松、汤书昆	
	参与撰稿：何瑗、钟一鸣、朱赟	
第十七节	泾县小岭金溪宣纸厂（地点：泾县丁家桥镇小岭村金坑村民组）	
田野调查	黄飞松、汤书昆、郑久良、何瑗、王圣融、朱赟、王怡青、沈佳斐	
撰稿	初稿主执笔：刘伟	
	修订补稿：黄飞松、汤书昆	
	参与撰稿：朱赟、王圣融	
第十八节	黄山白天鹅宣纸文化苑有限公司（地点：黄山市黄山区新明乡、耿城镇）	
田野调查	汤书昆、朱赟、郑久良、许骏	
撰稿	初稿主执笔：朱赟、汤书昆	
	修订补稿：汤书昆	

第三章　书画纸

第一节	泾县载元堂工艺厂（地点：泾县泾川镇城西工业集中区）	
田野调查	朱正海、朱赟、郑久良、罗文伯、程曦、何瑗、沈佳斐	
撰稿	初稿主执笔：郑久良、程曦	
	修订补稿：黄飞松、汤书昆	
	参与撰稿：王圣融、朱赟	

第二节	泾县小岭强坑宣纸厂（地点：泾县丁家桥镇小岭村）
田野调查	刘伟、何瑗、钟一鸣、王圣融、郭延龙、沈佳斐
撰稿	初稿主执笔：刘伟 修订补稿：黄飞松、汤书昆 参与撰稿：何瑗、钟一鸣、郭延龙
第三节	泾县雄鹿纸厂（地点：泾县丁家桥镇李园村）
田野调查	汤书昆、黄飞松、朱赟、郑久良、何瑗、王圣融、沈佳斐
撰稿	初稿主执笔：郑久良 修订补稿：黄飞松、汤书昆 参与撰稿：刘伟、王圣融
第四节	泾县紫光宣纸书画社（地点：泾县丁家桥镇后山村）
田野调查	朱正海、黄飞松、朱赟、郑久良、刘伟、沈佳斐
撰稿	初稿主执笔：郑久良 修订补稿：黄飞松、汤书昆、朱赟
第五节	泾县小岭西山宣纸工艺厂（地点：泾县丁家桥镇小岭村）
田野调查	汤书昆、黄飞松、朱赟、郑久良、程曦、许骏、刘伟、沈佳斐
撰稿	初稿主执笔：朱赟 修订补稿：黄飞松、汤书昆 参与撰稿：何瑗
第六节	安徽澄文堂宣纸艺术品有限公司（地点：泾县黄村镇九峰村）
田野调查	黄飞松、朱赟、王圣融、王怡青、沈佳斐
撰稿	初稿主执笔：黄飞松 修订补稿：汤书昆、黄飞松 参与撰稿：王怡青

第四章　皮纸

第一节	泾县守金皮纸厂（地点：泾县泾川镇园林村）
田野调查	朱正海、黄飞松、朱赟、郑久良、罗文伯、何瑗、王圣融、郭延龙、沈佳斐
撰稿	初稿主执笔：郑久良 修订补稿：黄飞松、汤书昆、朱赟
第二节	泾县小岭驰星纸厂（地点：泾县丁家桥镇小岭村）
田野调查	汤书昆、黄飞松、朱正海、朱赟、郑久良、罗文伯、王圣融、沈佳斐
撰稿	初稿主执笔：黄飞松、朱赟 修订补稿：汤书昆 参与撰稿：罗文伯
第三节	潜山县星杰桑皮纸厂（地点：安庆市潜山县官庄镇坛畈村）
田野调查	汤书昆、朱赟、郑久良、刘伟、程曦
撰稿	初稿主执笔：汤书昆、郑久良 修订补稿：汤书昆
第四节	岳西县金丝纸业有限公司（地点：安庆市岳西县毛尖山乡板舍村）
田野调查	汤书昆、朱赟、汪淳、郑久良、刘伟、王圣融、尹航
撰稿	初稿主执笔：王圣融、汤书昆 修订补稿：汤书昆 参与撰稿：朱赟、刘伟
第五节	歙县深渡镇棉溪村（地点：黄山市歙县深渡镇棉溪村）
田野调查	汤书昆、陈琪、刘靖、朱赟、王秀伟、朱岱、沈佳斐、孙燕、叶婷婷
撰稿	初稿主执笔：汤书昆、朱赟 修订补稿：汤书昆、陈琪
第六节	黄山市三昕纸业有限公司（地点：黄山市休宁县海阳镇晓角村）
田野调查	汤书昆、刘靖、陈政、李宪奇、朱赟、王秀伟、郑久良、陈琪、朱岱、沈佳斐
撰稿	初稿主执笔：汤书昆、郑久良 修订补稿：汤书昆

第七节	歙县六合村（地点：黄山市歙县杞梓里镇六合村）	
田野调查	陈琪、汤书昆、陈政、孙燕、叶婷婷、沈佳斐	
撰稿	初稿主执笔：汤书昆、孙燕	
	修订补稿：汤书昆、陈琪	

第五章　竹纸		
第一节	歙县青峰村（地点：黄山市歙县青峰村）	
田野调查	陈琪、汤书昆、贡斌、李宪奇、沈佳斐	
撰稿	初稿主执笔：汤书昆、王圣融	
	修订补稿：汤书昆、陈琪	
	参与撰稿：陈琪	
第二节	泾县孤峰村（地点：泾县昌桥乡孤峰村、泾川镇古坝村、黄村镇九峰村）	
田野调查	陈彪、黄飞松、朱正海、沈佳斐	
撰稿	初稿主执笔：黄飞松、陈彪	
	修订补稿：黄飞松、汤书昆	
第三节	金寨县燕子河镇（地点：金寨县燕子河镇龙马村／燕溪村）	
田野调查	汤书昆、黄飞松、张静明、蓝强	
撰稿	初稿主执笔：汤书昆、黄飞松	
	修订补稿：汤书昆、黄飞松	

第六章　加工纸		
第一节	安徽省掇英轩书画用品有限公司（地点：巢湖市黄麓镇）	
田野调查	汤书昆、陈彪、李宪奇、朱赟、郑久良、刘伟、王圣融、程曦、叶珍珍、沈佳斐	
撰稿	初稿主执笔：汤书昆、钟一鸣	
	修订补稿：汤书昆、李宪奇	
	参与撰稿：叶珍珍	
第二节	泾县艺英轩宣纸工艺品厂（地点：泾县琴溪镇赤滩街道）	
田野调查	汤书昆、朱赟、刘伟、郑久良、程曦、许骏、王圣融、郭延龙、王怡青、沈佳斐	
撰稿	初稿主执笔：朱赟	
	修订补稿：汤书昆	
第三节	泾县艺宣阁宣纸工艺品有限公司（地点：泾县泾川镇城西工业集中区）	
田野调查	黄飞松、朱大为、朱赟、郑久良、程曦、许骏、刘伟、王圣融、沈佳斐	
撰稿	初稿主执笔：黄飞松、朱赟	
	修订补稿：黄飞松、汤书昆	
	参与撰稿：王圣融	
第四节	泾县宣艺斋宣纸工艺厂（地点：泾县泾川镇城西工业集中区）	
田野调查	黄飞松、郑久良、程曦、朱赟、王圣融、郭延龙、沈佳斐	
撰稿	初稿主执笔：程曦	
	修订补稿：汤书昆、黄飞松	
	参与撰稿：郑久良、刘伟、王圣融	
第五节	泾县贡玉堂宣纸工艺厂（地点：泾县黄村镇紫阳村）	
田野调查	朱正海、朱大为、朱赟、郑久良、程曦、许骏、刘伟、王圣融、郭延龙、沈佳斐	
撰稿	初稿主执笔：刘伟	
	修订补稿：汤书昆、黄飞松	
	参与撰稿：郭延龙	
第六节	泾县博古堂宣纸工艺厂（地点：泾县丁家桥镇小岭村）	
田野调查	汤书昆、朱正海、朱赟、郑久良、王圣融、郭延龙、沈佳斐	
撰稿	初稿主执笔：朱赟	
	修订补稿：汤书昆、黄飞松	
第七节	泾县汇宣堂宣纸工艺厂（地点：泾县泾川镇曹家村）	
田野调查	汤书昆、朱正海、朱赟、郑久良、刘伟、王圣融、郭延龙、沈佳斐	
撰稿	初稿主执笔：郑久良	
	修订补稿：汤书昆、黄飞松	
	参与撰稿：王圣融、朱赟	

第八节	泾县凤和堂宣纸加工厂（地点：泾县泾川镇五星村）
田野调查	汤书昆、黄飞松、沈佳斐
撰稿	初稿主执笔：沈佳斐、汤书昆、黄飞松 修订补稿：汤书昆、黄飞松

第七章 工具

第一节	泾县明堂纸帘工艺厂（地点：泾县丁家桥镇）
田野调查	朱赟、刘伟、王圣融、郭延龙、沈佳斐
撰稿	初稿主执笔：刘伟 修订补稿：黄飞松、汤书昆 参与撰稿：王圣融
第二节	泾县全勇纸帘工艺厂（地点：泾县丁家桥镇工业园区）
田野调查	黄飞松、朱赟、刘伟、王圣融、郭延龙、沈佳斐
撰稿	初稿主执笔：刘伟 修订补稿：汤书昆、黄飞松 参与撰稿：王圣融
第三节	泾县后山大剪刀作坊（地点：泾县丁家桥镇后山村）
田野调查	朱正海、朱赟、黄飞松、刘伟、王圣融、廖莹文、沈佳斐
撰稿	初稿主执笔：黄飞松 修订补稿：黄飞松、汤书昆 参与撰稿：朱赟

二、技术与辅助工作

手工纸分布示意图绘制	主持：郭延龙、陈龑 参与绘制：郭延龙、朱赟、何瑗、姚的卢、叶珍珍
实物纸样测试分析	主持：朱赟、陈龑 测试：朱赟、陈龑、何瑗、郑久良、程曦、汪宣伯、钟一鸣、叶婷婷、郭延龙、王圣融、王怡青、黄立新、赵梦君、王裕玲、宋福星
实物纸样拍摄	黄晓飞
实物纸样整理	汤书昆、朱赟、倪盈盈、郑斌、付成云、蔡婷婷、刘伟、何瑗、王圣融、叶珍珍、陈龑、王怡青
实物纸样透光纤维图制作	朱赟、陈龑、何瑗、刘伟、王怡青、廖莹文

三、总序、编撰说明、附录与后记部分

总序
撰稿	汤书昆

编撰说明
撰稿	汤书昆、朱赟

附录
术语整理编制	朱赟、陈登航、付成云、秦庆
图目整理编制	朱赟、王怡青、王圣融、叶珍珍、廖莹文
表目整理编制	朱赟、王怡青、王圣融、叶珍珍、廖莹文

后记
撰稿	汤书昆

四、统稿与翻译

统稿主持	汤书昆
统稿规划	朱赟、朱正海
翻译主持	方媛媛
其他参与翻译人员	刘丽、汪晓婧、高倩、高丁祎、胡昕、刘惠敏

1. Field investigation, writing and modification

	Chapter I Introduction to Handmade Paper in Anhui Province	
Writers	First manuscript written by: Huang Feisong, Zhu Yun, Tang Shukun Modified by: Tang Shukun, Chen Jingyu, Huang Feisong Zheng Jiuliang, Sun Jian, Yin Hang, He Ai have also contributed to the writing	

	Chapter II Xuan Paper	
Section 1	China Xuan Paper Co., Ltd. (Location: Wuxi Village in Langqiao Town of Jingxian County)	
Field investigators	Tang Shukun, Huang Feisong, Zhu Yun, Chen Biao, Zheng Jiuliang, Cheng Xi, Xu Jun, Wang Yiqing, He Ai	
Writers	First manuscript written by: Huang Feisong Modified by: Tang Shukun, Huang Feisong, Zhu Yun Liu Wei, Wang Shengrong, Wang Yiqing have also contributed to the writing	
Section 2	Wangliuji Xuan Paper Co., Ltd. in Jingxian County (Location: Chachong Village in Jingchuan Town of Jingxian County)	
Field investigators	Zhu Zhenghai, Tang Shukun, Huang Feisong, Zhu Yun, Zheng Jiuliang, Luo Wenbo, Cheng Xi, He Ai, Wang Shengrong, Wang Yiqing, Shen Jiafei	
Writers	First manuscript written by: Zhu Yun Modified by: Huang Feisong, Tang Shukun Wang Shengrong has also contributed to the writing	
Section 3	Anhui Hengxing Xuan Paper Co., Ltd. (Location: Houshan Village in Dingjiaqiao Town of Jingxian County)	
Field investigators	Tang Shukun, Zhu Zhenghai, Huang Feisong, Zhu Yun, Zheng Jiuliang, Luo Wenbo, Wang Mei, Cheng Xi, Xu Jun, He Ai, Wang Shengrong, Wang Yiqing, Shen Jiafei	
Writers	First manuscript written by: Zhu Yun Modified by: Huang Feisong, Tang Shukun Zheng Jiuliang, Wang Shengrong have also contributed to the writing	
Section 4	Taoji Xuan Paper Co., Ltd. in Jingxian County (Location: Shangcao Village in Tingxi Town of Jingxian County)	
Field investigators	Tang Shukun, Huang Feisong, Zhu Zhenghai, Zhu Dawei, Zhu Yun, Zheng Jiuliang, Cheng Xi, Xu Jun, Liu Wei, He Ai, Shen Jiafei	
Writers	First manuscript written by: Liu Wei Modified by: Huang Feisong, Tang Shukun Zhu Yun, Shen Jiafei have also contributed to the writing	
Section 5	Wangtonghe Xuan Paper Factory in Jingxian County (Location: Guba Village in Jingchuan Town of Jingxian County)	
Field investigators	Huang Feisong, Zhu Zhenghai, Zhu Dawei, He Ai, Zhu Yun, Zheng Jiuliang, Cheng Xi, Xu Jun, Liu Wei, Wang Shengrong, Wang Yiqing, Shen Jiafei	
Writers	First manuscript written by: Cheng Xi Modified by: Huang Feisong, Tang Shukun Zhu Yun, Wang Shengrong have also contributed to the writing	
Section 6	Shuanglu Xuan Paper Co., Ltd. in Jingxian County (Location: Chengxi Industrial Park in Jingchuan Town of Jingxian County)	
Field investigators	Tang Shukun, Huang Feisong, Zhu Yun, Zheng Jiuliang, Luo Wenbo, Liu Wei, He Ai, Wang Shengrong, Wang Yiqing, Shen Jiafei	
Writers	First manuscript written by: Luo Wenbo Modified by: Huang Feisong, Tang Shukun Zheng Jiuliang, Wang Shengrong have also contributed to the writing	
Section 7	Jinxing Xuan Paper Co., Ltd. in Jingxian County (Location: Industrial Zone in Dingjiaqiao Town of Jingxian County)	
Field investigators	Huang Feisong, Zhu Zhenghai, Tang Shukun, Zheng Jiuliang, Zhu Yun, Cheng Xi, Liu Wei, He Ai, Wang Shengrong, Wang Yiqing, Shen Jiafei	
Writers	First manuscript written by: Zheng Jiuliang Modified by: Huang Feisong, Zhu Zhenghai, Zhu Yun Wang Shengrong, Zhu Yun have also contributed to the writing	
Section 8	Hongye Xuan Paper Co., Ltd. in Jingxian County (Location: Fengkeng Village in Dingjiaqiao Town of Jingxian County)	
Field investigators	Tang Shukun, He Ai, Zhu Yun, Liu Wei, Wang Shengrong, Shen Jiafei	
Writers	First manuscript written by: Liu Wei, Huang Feisong Modified by: Tang Shukun, Huang Feisong, Zhu Yun Wang Shengrong, Wang Yiqing have also contributed to the writing	
Section 9	Anhui Caoshi Xuan Paper Co., Ltd. (Location: Fengkeng Village in Dingjiaqiao Town of Jingxian County)	
Field investigators	Tang Shukun, Huang Feisong, Zhu Yun, He Ai, Zheng Jiuliang, Xu Juan, Liu Wei, Cheng Xi, Wang Shengrong, Wang Yiqing, Shen Jiafei	
Writers	First manuscript written by: Xu Jun, Huang Feisong, Tang Shukun Modified by: Tang Shukun, Huang Feisong Wang Yiqing has also contributed to the writing	
Section 10	Millennium Xuan Paper Co., Ltd. in Jingxian County (Location: Xiaoling Village in Dingjiaqiao Town of Jingxian County)	
Field investigators	Tang Shukun, Zhu Yun, Liu Wei, He Ai, Wang Shengrong, Wang Yiqing, Shen Jiafei	
Writers	First manuscript written by: Zhu Yun, Huang Feisong Modified by: Huang Feisong, Tang Shukun Wang Yiqing has also contributed to the writing	

Section 11	Xiaoling Jinghui Paper Co., Ltd. in Jingxian County (Location: Xiaoling Village in Dingjiaqiao Town of Jingxian County)
Field investigators	Tang Shukun, Zhu Dawei, Zhu Yun, Zheng Jiuliang, He Ai, Xu Jun, Liu Wei, Wang Shengrong, Wang Yiqing, Shen Jiafei
Writers	First manuscript written by: Zhu Yun, Tang Shukun Modified by: Tang Shukun, Huang Feisong Liu Wei, Xu Jun have also contributed to the writing
Section 12	Sanxing Paper Co., Ltd. in Jingxian County (Location: Liyuan Village in Dingjiaqiao Town of Jingxian County)
Field investigators	Tang Shukun, Huang Feisong, Zhu Zhenghai, Zhu Yun, Zheng Jiuliang, Xu Jun, Liu Wei, He Ai, Wang Shengrong, Wang Yiqing, Shen Jiafei
Writers	First manuscript written by: Liu Wei, Huang Feisong Modified by: Tang Shukun, Huang Feisong Zhu Yun has also contributed to the writing
Section 13	Anhui Changchun Paper Co., Ltd. (Location: Industrial Zone in Dingjiaqiao Town of Jingxian County)
Field investigators	Tang Shukun, Huang Feisong, Zhu Yun, Zheng Jiuliang, Luo Wenbo, Wang Mei, He Ai, Wang Shengrong, Wang Yiqing, Shen Jiafei
Writers	First manuscript written by: Zhu Yun Modified by: Huang Feisong, Tang Shukun Zheng Jiulaing, Wang Shengrong have also contributed to the writing
Section 14	Yuquan Xuan Paper Co., Ltd. in Jingxian County (Location: Liyuan Village in Dingjiaqiao Town of Jingxian County)
Field investigators	Tang Shukun, Huang Feisong, Zhu Dawei, Zhu Yun, Liu Wei, Zheng Jiuliang, He Ai, Wang Shengrong, Wang Yiqing, Shen Jiafei
Writers	First manuscript written by: Zhu Yun Modified by: Huang Feisong, Tang Shukun Wang Shengrong has also contributed to the writing
Section 15	Jixing Xuan Paper Co., Ltd. in Jingxian County (Location: Shangfang Village in Jingchuan Town of Jingxian County)
Field investigators	Zhu Zhenghai, Huang Feisong, Tang Shukun, Zhu Yun, Zheng Jiuliang, Cheng Xi, Xu Jun, Liu Wei, He Ai, Shen Jiafei
Writers	First manuscript written by: Cheng Xi Modified by: Tang Shukun, Huang Feisong Liu Wei, Zhu Yun have also contributed to the writing
Section 16	Jinxuantang Xuan Paper Factory in Jingxian County (Location: Dazhuang Village in Langqiao Town of Jingxian County)
Field investigators	Tang Shukun, Zhu Yun, Zheng Jiuliang, Luo Wenbo, Wang Mei, He Ai, Zhong Yiming, Wang Shengrong, Wang Yiqing
Writers	First manuscript written by: Cheng Xi Modified by: Huang Feisong, Tang Shukun He Ai, Zhong Yiming, Zhu Yun have also contributed to the writing
Section 17	Xiaoling Jinxi Xuan Paper Factory in Jingxian County (Location: Jinkeng Villages' Group of Xiaoling Village in Dingjiaqiao Town of Jingxian County)
Field investigators	Huang Feisong, Tang Shukun, Zheng Jiuliang, He Ai, Wang Shengrong, Zhu Yun, Wang Yiqing, Shen Jiafei
Writers	First manuscript written by: Liu Wei Modified by: Huang Feisong, Tang Shukun Zhu Yun, Wang Shengrong have also contributed to the writing
Section 18	Huangshan Baitian'e Xuan Paper Cultural Garden Co., Ltd. (Location: Xinming Town and Gengcheng Town of Huangshan District in Huangshan City)
Field investigators	Tang Shukun, Zhu Yun, Zheng Jiuliang, Xu Jun
Writers	First manuscript written by: Zhu Yun, Tang Shukun Modified by: Tang Shukun

Chapter III Calligraphy and Painting Paper

Section 1	Zaiyuantang Xuan Paper Factory in Jingxian County (Location: Chengxi Industrial Zone in Jingchuan Town of Jingxian County)
Field investigators	Zhu Zhenghai, Zhu Yun, Zheng Jiuliang, Luo Wenbo, Cheng Xi, He Ai, Shen Jiafei
Writers	First manuscript written by: Zheng Jiuliang, Cheng Xi Modified by: Huang Feisong, Tang Shukun Wang Shengrong, Zhu Yun have also contributed to the writing
Section 2	Xiaoling Qiangkeng Xuan Paper Factory in Jingxian County (Location: Xiaoling Village in Dingjiaqiao Town of Jingxian County)
Field investigators	Liu Wei, He Ai, Zhong Yiming, Wang Shengrong, Guo Yanlong, Shen Jiafei
Writers	First manuscript written by: Liu Wei Modified by: Huang Feisong, Tang Shukun He Ai, Zhong Yiming, Guo Yanlong have also contributed to the writing
Section 3	Xionglu Xuan Paper Factory in Jingxian County (Location: Liyuan Village in Dingjiaqiao Town of Jingxian County)
Field investigators	Tang Shukun, Huang Feisong, Zhu Yun, Zheng Jiuliang, He Ai, Wang Shengrong, Shen Jiafei
Writers	First manuscript written by: Zheng Jiuliang Modified by: Huang Feisong, Tang Shukun Liu Wei, Wang Shengrong have also contributed to the writing

Section 4	Ziguang Xuan Paper Factory in Jingxian County (Location: Houshan Village in Dingjiaqiao Town of Jingxian County)
Field investigators	Zhu Zhenghai, Huang Feisong, Zhu Yun, Zheng Jiuliang, Liu Wei, Shen Jiafei
Writers	First manuscript written by: Zheng Jiuliang Modified by: Huang Feisong, Tang Shukun, Zhu Yun
Section 5	Xiaoling Xishan Xuan Paper Factory in Jingxian County (Location: Xiaoling Village in Dingjiaqiao Town of Jingxian County)
Field investigators	Tang Shukun, Huang Feisong, Zhu Yun, Zheng Jiuliang, Cheng Xi, Xu Jun, Liu Wei, Shen Jiafei
Writers	First manuscript written by: Zhu Yun Modified by: Huang Feisong, Tang Shukun He Ai has also contributed to the writing
Section 6	Chengwentang Xuan Paper Co., Ltd. in Anhui Province (Location: Jiufeng Village in Huangcun Town of Jingxian County)
Field investigators	Huang Feisong, Zhu Yun, Wang Shengrong, Wang Yiqing, Shen Jiafei
Writers	First manuscript written by: Huang Feisong Modified by: Tang Shukun, Huang Feisong Wang Yiqing has also contributed to the writing

Chapter IV Bast Paper

Section 1	Shoujin Bast Paper Factory in Jingxian County (Location: Yuanlin Village in Jingchuan Town of Jingxian County)
Field investigators	Zhu Zhenghai, Huang Feisong, Zhu Yun, Zheng Jiuliang, Luo Wenbo, He Ai, Wang Shengrong, Guo Yanlong, Shen Jiafei
Writers	First manuscript written by: Zheng Jiuliang Modified by: Huang Feisong, Tang Shukun, Zhu Yun
Section 2	Xiaoling Chixing Paper Factory in Jingxian County (Location: Xiaoling Village in Dingjiaqiao Town of Jingxian County)
Field investigators	Tang Shunkun, Huang Feisong, Zhu Zhenghai, Zhu Yun, Zheng Jiuliang, Luo Wenbo, Wang Shengrong, Shen Jiafei
Writers	First manuscript written by: Huang Feisong, Zhu Yun Modified by: Tang Shukun Luo Wenbo has also contributed to the writing
Section 3	Xingjie Mulberry Bark Paper Factory in Qianshan County (Location: Tanfan Village in Guanzhuang Town of Qianshan County in Anqing City)
Field investigators	Tang Shukun, Zhu Yun, Zheng Jiuliang, Liu Wei, Cheng Xi
Writers	First manuscript written by: Tang Shukun, Zheng Jiuliang Modified by: Tang Shukun
Section 4	Jinsi Paper Co., Ltd. in Yuexi County (Location: Banshe Village in Maojianshan Town of Yuexi County in Anqing City)
Field investigators	Tang Shukun, Zhu Yun, Wang Chun, Zheng Jiuliang, Liu Wei, Wang Shengrong, Yin Hang
Writers	First manuscript written by: Wang Shengrong, Tang Shukun Modified by: Tang Shukun Zhu Yun, Liu Wei have also contributed to the writing
Section 5	Mianxi Village in Shendu Town of Shexian County (Location: Mianxi Village in Shendu Town of Shexian County in Huangshan City)
Field investigators	Tang Shukun, Chen Qi, Liu Jing, Zhu Yun, Wang Xiuwei, Zhu Dai, Shen Jiafei, Sun Yan, Ye Tingting
Writers	First manuscript written by: Tang Shukun, Zhu Yun Modified by: Tang Shukun, Chen Qi
Section 6	Sanxin Paper Co., Ltd. in Huangshan City (Location: Xiaojiao Village in Haiyang Town of Xiuning County in Huangshan City)
Field investigators	Tang Shukun, Liu Jing, Chen Zheng, Li Xianqi, Zhu Yun, Wang Xiuwei, Zheng Jiuliang, Chen Qi, Zhu Dai, Shen Jiafei
Writers	First manuscript written by: Tang Shukun, Zheng Jiuliang Modified by: Tang Shukun
Section 7	Liuhe Village in Shexian County (Location: Liuhe Village in Qizili Town of Shexian County in Huangshan City)
Field investigators	Chen Qi, Tang Shukun, Chen Zheng, Sun Yan, Ye Tingting, Shen Jiafei
Writers	First manuscript written by: Tang Shukun, Sun Yan Modified by: Tang Shukun, Chen Qi

Chapter V Bamboo Paper

Section 1	Qingfeng Village in Shexian County (Location: Qingfeng Village in Shexian County of Huangshan City)
Field investigators	Chen Qi, Tang Shukun, Gong Bin, Li Xianqi, Shen Jiafei
Writers	First manuscript written by: Tang Shukun, Wang Shengrong Modified by: Tang Shukun, Chen Qi Chen Qi has also contributed to the writing
Section 2	Gufeng Village in Jingxian County (Location: Gufeng Village in Changqiao Town, Guba Village in Jingchuan Town, Jiufeng Village in Huangcun Town, Jingxian County)
Field investigators	Chen Biao, Huang Feisong, Zhu Zhenghai, Shen Jiafei
Writers	First manuscript written by: Huang Feisong, Chen Biao Modified by: Huang Feisong, Tang Shukun

Section 3	Yanzihe Town in Jinzhai County (Location: Longma / Yanxi Village in Yanzihe Town of Jinzhai County)
Field investigators	Tang Shukun, Huang Feisong, Zhang Jingming, Lan Qiang
Writers	First manuscript written by: Tang Shukun, Huang Feisong Modified by: Tang Shukun, Huang Feisong

Chapter VI Processed Paper

Section 1	Duoyingxuan Calligraphy and Painting Supplies Co., Ltd. in Anhui Province (Location: Huanglu Town of Chaohu City)
Field investigators	Tang Shukun, Chen Biao, Li Xianqi, Zhu Yun, Zheng Jiuliang, Liu Wei, Wang Shengrong, Cheng Xi, Ye Zhenzhen, Shen Jiafei
Writers	First manuscript written by: Tang Shukun, Zhong Yiming Modified by: Tang Shukun, Li Xianqi Ye Zhenzhen has also contributed to the writing
Section 2	Yiyingxuan Xuan Paper Craft Factory in Jingxian County (Location: Chitan Street in Qinxi Town of Jingxian County)
Field investigators	Tang Shukun, Zhu Yun, Liu Wei, Zheng Jiuliang, Cheng Xi, Xu Jun, Wang Shengrong, Guo Yanlong, Wang Yiqing, Shen Jiafei
Writers	First manuscript written by: Zhu Yun Modified by: Tang Shukun
Section 3	Yixuange Xuan Paper Craft Co., Ltd. in Jingxian County (Location: Chengxi Industrial Zone in Jingchuan Town of Jingxian County)
Field investigators	Huang Feisong, Zhu Dawei, Zhu Yun, Zheng Jiuliang, Cheng Xi, Xu Jun, Liu Wei, Wang Shengrong, Shen Jiafei
Writers	First manuscript written by: Huang Feisong, Zhu Yun Modified by: Huang Feisong, Tang Shukun Wang Shengrong has also contributed to the writing
Section 4	Xuanyizhai Xuan Paper Craft Factory in Jingxian County (Location: Chengxi Industrial Zone in Jingchuan Town of Jingxian County)
Field investigators	Huang Feisong, Zheng Jiuliang, Cheng Xi, Zhu Yun, Wang Shengrong, Guo Yanlong, Shen Jiafei
Writers	First manuscript written by: Cheng Xi Modified by: Tang Shukun, Huang Feisong Zheng Jiuliang, Liu Wei, Wang Shengrong have also contributed to the writing
Section 5	Gongyutang Xuan Paper Craft Factory in Jingxian County (Location: Ziyang Village in Huangcun Town of Jingxian County)
Field investigators	Zhu Zhenghai, Zhu Dawei, Zhu Yun, Zheng Liangjiu, Cheng Xi, Xu Jun, Liu Wei, Wang Shengrong, Guo Yanlong, Shen Jiafei
Writers	First manuscript written by: Liu Wei Modified by: Tang Shukun, Huang Feisong Guo Yanlong has also contributed to the writing
Section 6	Bogutang Xuan Paper Craft Factory in Jingxian County (Location: Xiaoling Village in Dingjiaqiao Town of Jingxian County)
Field investigators	Tang Shukun, Zhu Zhenghai, Zhu Yun, Zheng Jiuliang, Wang Shengrong, Guo Yanlong, Shen Jiafei
Writers	First manuscript written by: Zhu Yun Modified by: Tang Shukun, Huang Feisong
Section 7	Huixuantang Xuan Paper Craft Factory in Jingxian County (Location: Caojia Village in Jingchuan Town of Jingxian County)
Field investigators	Tang Shukun, Zhu Zhenghai, Zhu Yun, Zheng Jiuliang, Liu Wei, Wang Shengrong, Guo Yanlong, Shen Jiafei
Writers	First manuscript written by: Zheng Jiuliang Modified by: Tang Shukun, Huang Feisong Wang Shengrong, Zhu Yun have also contributed to the writing
Section 8	Fenghetang Xuan Paper Craft Factory in Jingxian County (Locaion: Wuxing Village in Jingchuan Town of Jingxian County)
Field investigators	Tang Shukun, Huang Feisong, Shen Jiafei
Writers	First manuscript written by: Shen Jiafei, Tang Shukun, Huang Feisong Modified by: Tang Shukun, Huang Feisong

Chapter VII Tools

Section 1	Mingtang Papermaking Screen Craft Factory in Jingxian County (Location: Dingjiaqiao Town in Jingxian County)
Field investigators	Zhu Yun, Liu Wei, Wang Shengrong, Guo Yanlong, Shen Jiafei
Writers	First manuscript written by: Liu Wei Modified by: Huang Feisong, Tang Shukun Wang Shengrong has also contributed to the writing
Section 2	Quanyong Papermaking Screen Craft Factory in Jingxian County (Location: Industrial Zone in Dingjiaqiao Town of Jingxian County)
Field investigators	Huang Feisong, Zhu Yun, Liu Wei, Wang Shengrong, Guo Yanlong, Shen Jiafei
Writers	First manuscript written by: Liu Wei Modified by: Tang Shukun, Huang Feisong Wang Shengrong has also contributed to the writing

Section 3		Houshan Shears Workshop in Jingxian County (Location: Houshan Village in Dingjiaqiao Town of Jingxian County)
Field investigators		Zhu Zhenghai, Zhu Yun, Huang Feisong, Liu Wei, Wang Shengrong, Liao Yingwen, Shen Jiafei
Writers		First manuscript written by: Huang Feisong Modified by: Huang Feisong, Tang Shukun Zhu Yun has also contributed to the writing

2. Technical Analysis and Other Related Works

Handmade paper distribution maps	Headed by: Guo Yanlong, Chen Yan Drawn by: Guo Yanlong, Zhu Yun, He Ai, Yao Dilu, Ye Zhenzhen
Sample paper test	Headed by: Zhu Yun, Chen Yan Members: Zhu Yun, Chen Yan, He Ai, Zheng Jiuliang, Cheng Xi, Liu Wei, Wang Xuanbo, Zhong Yiming, Ye Tingting, Guo Yanlong, Wang Shengrong, Wang Yiqing, Huang Lixin, Zhao Mengjun, Wang Yuling, Song Fuxing
Paper sample pictures	Photographed by: Huang Xiaofei
Paper sample	Sorted by: Tang Shukun, Zhu Yun, Ni Yingying, Zheng Bin, Fu Chengyun, Cai Tingting, Liu Wei, He Ai, Wang Shengrong, Ye Zhenzhen, Chen Yan, Wang Yiqing
Paper pictures showing the paper fiber	Produced by: Zhu Yun, Chen Yan, He Ai, Liu Wei, Wang Yiqing, Liao Yingwen

3. Preface, Introduction to the Writing Norms, Appendices and Epilogue

Preface

Writer	Tang Shukun

Introduction to the Writing Norms

Writers	Tang Shukun, Zhu Yun

Appendices

Terminology	Zhu Yun, Chen Denghang, Fu Chengyun, Qin Qing
List of figures	Zhu Yun, Wang Yiqing, Wang Shengrong, Ye Zhenzhen, Liao Yingwen
List of tables	Zhu Yun, Wang Yiqing, Wang Shengrong, Ye Zhenzhen, Liao Yingwen

Epilogue

Writer	Tang Shukun

4. Modification and Translation

Director of modification and verification	Tang Shukun
Modification planner	Zhu Yun, Zhu Zhenghai
Chief translator and director of Translation	Fang Yuanyuan
Other translators	Liu Li, Wang Xiaojing, Gao Qian, Gao Dingyi, Hu Xin, Liu Huimin

在历时3年半的多轮修订、增补与统稿工作中，汤书昆、黄飞松、朱赟、朱正海、方媛媛、陈敬宇、郭延龙等作为主持人或重要内容模块的负责人，对文稿内容、图片与示意图的修订增补，代表性纸样的测试分析，英文翻译，文献注释考订，表述格式的规范化，数据与表述的准确性核实等方面做了大量扎实而辛苦的工作。而责任编辑团队、北京敬人工作室设计团队、北京雅昌艺术印刷有限公司印制团队精益求精、力求完美的反复打磨，都是《安徽卷》书稿从最初的田野记录式提炼整理，到以今天的面貌和质量展现不容忽视的工作。

在《安徽卷》的田野调查过程中，先后得到中国宣纸股份有限公司胡文军先生、黄山市地方志办公室陈政先生、岳西县文化馆汪淳先生、金寨县政府王玉华先生等多位手工造纸传统技艺和非物质文化遗产研究与保护专家的帮助，在《中国手工纸文库·安徽卷》正式出版之际，我谨代表田野调查和文稿撰写团队，向记名与未曾记名的支持者表达真诚的谢意！

汤书昆

2019年12月于中国科学技术大学

Tang Shukun, Huang Feisong, Zhu Yun, Zhu Zhenghai, Fang Yuanyuan, Chen Jingyu, Guo Yanlong et al., who were in charge of the writing, modification and other related works, all contributed their efforts to the completion of this book in the past three and half years. Their meticulous efforts in writing, drawing or photographing, mapping, technical analysis, translating, format modifying, noting and proofreading should be recognized and eulogized in the achievement of the high-quality work. The editors of the book, Beijing Jingren Book Design Studio, Bejing Artron Art Printing Co., Ltd. have been dedicated to the polishing and publication of the book, whose efforts enable a field investigation-based research to be presented in a stylish and quality way.

Many experts from the field of handmade paper production and intangible cultural heritage research and protection have helped in our field investigations: Hu Wenjun from China Xuan Paper Co., Ltd., Chen Zheng from the Office of Chronicles in Huangshan City, Wang Chun from Cultural Center in Yuexi County, Wang Yuhua from local government in Jinzhai County, et al. On the verge of publication, sincere gratitude should go to all those who have supported and recognized our efforts!

Tang Shukun
University of Science and Technology of China
December 2019